# Motor Control Fundamentals

Steve Senty

Australia • Brazil • Japan • Korea • Mexico • Singapore • Spain • United Kingdom • United States

**Motor Control Fundamentals**
**Steve Senty**

Vice President, Editorial: Dave Garza
Director of Learning Solutions: Sandy Clark
Acquisitions Editor: Stacy Masucci
Managing Editor: Larry Main
Senior Product Manager: John Fisher
Editorial Assistant: Andrea Timpano
Vice President, Marketing: Jennifer Baker
Marketing Director: Deborah Yarnell
Marketing Manager: Erin Brennan
Marketing Coordinator: Jillian Borden
Production Director: Wendy Troeger
Production Manager: Mark Bernard
Content Project Manager: Barbara LeFleur
Production Technology Assistant: Emily Gross
Art Director: David Arsenault
Technology Project Manager: Joe Pliss



Library of Congress Control Number: 2011927125

ISBN-13: 978-0-8400-2462-6

ISBN-10: 0-8400-2462-2

**Delmar**
5 Maxwell Drive
Clifton Park, NY 12065-2919
USA

Cengage Learning is a leading provider of customized learning solutions with office locations around the globe, including Singapore, the United Kingdom, Australia, Mexico, Brazil, and Japan. Locate your local office at: **international.cengage.com/region**

Cengage Learning products are represented in Canada by Nelson Education, Ltd.

To learn more about Delmar, visit **www.cengage.com/delmar**

Purchase any of our products at your local college store or at our preferred online store **www.cengagebrain.com**



Printed in the United States of America
1 2 3 4 5 23 22 21 20 19

# Table of Contents

# Preface

The word "motors," as stated in the title, represents a diverse and complicated topic of study. There are many different types of motors, and several different methods of converting electrical energy into rotating mechanical energy via electromagnetism. The focus of this book is to understand basic single-phase and three-phase induction motor theory and operation, common motor control circuit schemes, reading, interpreting, and documenting motor control circuit diagrams, and providing practice circuits for connecting the actual motor control circuit components from ladder diagrams. The focus will be limited to single- and three-phase induction motors because they constitute the vast majority of motors the construction electrician will encounter in the field.

In the article "Minimizing AC Induction Motor Slip" in the trade magazine *Electrical Construction and Maintenance* (EC & M), dated April 1, 2004, Mauri Peltola of ABB Drives and Motors estimates that more than 90% of all motors used in worldwide industry are AC induction motors. If that number is adjusted to compensate for the specialized motor types used in specific industries, it becomes apparent that construction electricians will encounter an even higher percentage of induction motors in their everyday work. Very few construction electricians will ever encounter a wound-rotor motor, and because this type of motor is so rare, their employers will understand if they require help dealing with it. Every employer, however, is going to expect employee competence in the induction motors that constitute such a large part of their work.

The focus of this book also is confined to only the most basic content necessary for the beginning learner to understand induction motors and motor control, and build a strong knowledge base to build on in the field. There is a seemingly endless amount of content available on motors, and it is impossible for a single textbook to cover all of it. Trying to cover too much content in a short period of time can overwhelm learners, and leave them confused and unable to assimilate even the most important basic content. Like the math learner who has never understood the purpose of the parentheses in an algebra problem, not understanding the most basic concepts of any subject will lead to problems when more advanced material is encountered. As instructors we sometimes forget that some of the content we want to cover is very advanced, and we try to teach someone new to the electrical industry in a semester that which has taken us years to master.

## COMMENTS

### Electric Motor Standards

The electric motor standards that will be referenced in this study are the standards most applicable to the construction electrician: the National Electrical Manufacturers Association (NEMA), and the International Electrotechnical Commission (IEC). In general, the NEMA standard was adopted by North America, and most of the rest of the world adopted the IEC standard. NEMA standard MG-1 (Motors and Generators 1) sets the standard for the construction and manufacture of AC and DC motors and generators, and will be the main standard referenced. IEC standard 60034-1 pertains to all rotating electrical machines and will be referenced for comparison with the NEMA standard. Both NEMA and IEC motors and motor control components are common in the electrical industry today, so it is important to have knowledge of both standards, and to understand their interchangeability.

## SUPPLEMENTS

The *Motor Control Fundamentals Lab Manual* contains 16 lab exercises on various topics related to motor control, with some of the labs correlating to specific chapters. The labs state the purpose of the exercise, list equipment needed, provide step-by-step instructions, and contain review questions for reinforcement at the end of the labs. ISBN: 0-8400-2463-0

An *Instructor Companion Website* is available for this text. An Instructor Guide contains answers to the review questions at the end of each chapter, as well as answers to the Lab Manual review questions. Chapter presentations in PowerPoint are provided for each chapter, as well as additional test banks for each chapter. An image library containing low resolution figure files is posted as well. ISBN: 0840024649

## ABOUT THE AUTHOR

Steve Senty worked as a commercial and industrial electrician in IBEW L.U. #110 for 18 years before leaving the trade to teach full-time. His experience in the trade included many areas of the industry, from production floors to new construction. Specializing in motors and motor controls, he taught in the Construction Electrician Programs at Anoka Technical College and St. Paul College for 15 years. Before that he also taught Electrical Apprenticeship classes, and Journeyman Extension Classes for the IBEW L.U. #110 Joint Apprenticeship and Training Committee. Steve Senty maintains his membership in the IBEW, holds a 2 year Technical College Electrical Diploma, a B.A. in business, and a Class "A" Master electrical license.

## ACKNOWLEDGMENTS

The author and publisher would like to express thanks to those reviewers who provided insightful feedback throughout the development of this text:

Mike Melaney, Moraine Park Technical College, West Bend, WI

Marvin Moak, Hinds Community College, Raymond, MS

Don Pelster, Nashville State Community College, Nashville, TN

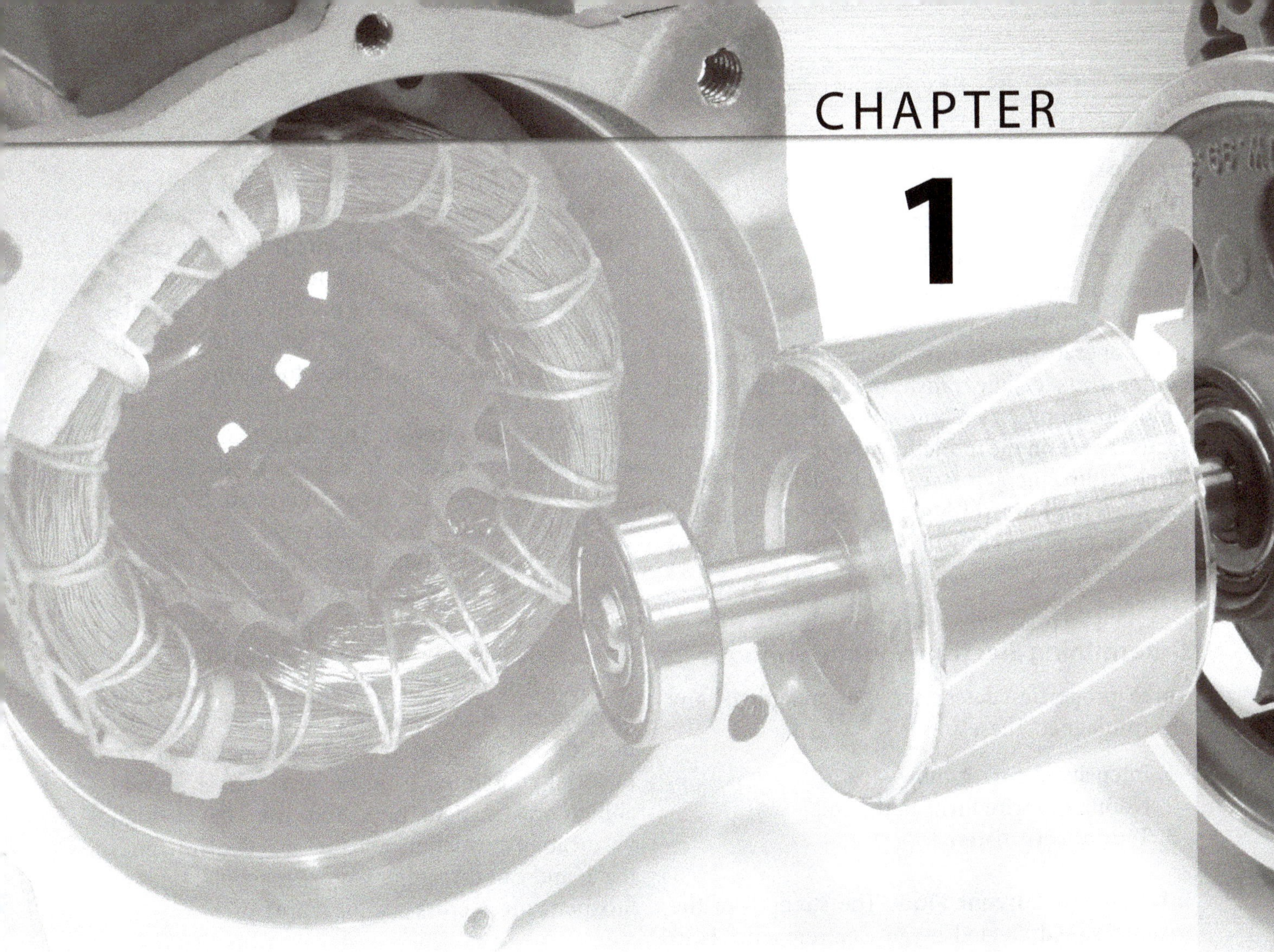


# Electromagnetic Induction Theory

## PURPOSE

To familiarize the learner with electromagnetic induction theory, electromagnetism theory, and how the two phenomena function in alternating current (AC) circuits.

## OBJECTIVES

After studying this chapter on electromagnetic induction and electromagnetism theory, the learner will be able to:

- Explain electromagnetism
- Explain electromagnetic induction
- Explain the volts-per-turn principle
- Explain mutual induction
- Explain self-induction
- Explain a counter-electromotive force (CEMF)
- Explain how inductors oppose a change in current, and cause a phase shift between voltage and current

## THE PRINCIPLE OF ELECTROMAGNETISM

Electromagnetism and electromagnetic induction are important electrical principles for electricians to understand, because many electrical components operate on these principles. These electrical components include transformers, generators, and induction motors, among other devices; but the focus here is on the induction motor. The principle of electromagnetism states that when an electrical current flows through an electrical conductor, a magnetic field will be created around that conductor.

### Determining Electromagnetic Strength

Three properties determine the strength of any electromagnet. These are:

1. Intensity of current flow
2. Number of wire turns in the coil
3. Type of core material.

**Intensity of Current Flow.** The strength of the magnetic field created by an electromagnet is in part determined by the amount of current flow in the inductor coil. With the other two factors held constant, a small current will produce a small magnetic field with few magnetic lines of flux, and a large current will produce a large magnetic field with many magnetic lines of flux. The magnetic polarity of the electromagnetic field is determined by the direction of current flow through the coil; if the electrical current through the conductor is reversed, the polarity of the magnetic field surrounding the conductor also will reverse. As the current intensity and the direction of current flow through the inductor coil change, as is true in an alternating current (AC) circuit, the magnetic field created by the current flow will change correspondingly.

**Number of Wire Turns in the Coil.** All of the small magnetic fields created around each turn of wire in the coil will add up to become a larger magnetic field for the whole coil. When the electrical current flows through the inductor coil, the current will create a magnetic field around each individual turn of wire in the coil that is proportional to the amount of current flow. Because the current flows through each wire turn of the coil in the same direction, the magnetic polarity of each individual wire turn will be the same, and all of the magnetic fields from the individual turns of wire will add together to equal their total sum.

**Type of Core Material.** This factor requires some additional explanation. In this instance, the property in which we are most interested is permeability. Permeability is the measure of how easily a material may become magnetized, and it determines how many lines of magnetic flux the material can carry per a given cross-sectional area, which is flux density. Air has a permeability of one (1), which is poor; this means that it is not easily magnetized and it cannot carry many magnetic lines of flux per a given cross-sectional area. Any material containing iron, however, has a permeability greater than one. Soft iron has a high permeability, which means that it is easily magnetized, and can carry many more magnetic lines of flux per a given cross-sectional area.

### Ampere-Turns Principle

Normally when talking about an electromagnet, "ampere-turns" is a term commonly used to explain that the strength of the electromagnetic field is determined by both the number of wire turns in the coil and the intensity of current flow through the coil, as indicated previously. A more practical method for the technician to use when thinking about the strength of the electromagnetic field of an electromagnet is the intensity of the current flow through the coil, because it is the only factor that can be controlled in the field. The type of core material used in the inductors electricians will encounter will always be iron, because of its superior permeability; and the number of turns of wire in an inductor coil is determined at the time of manufacture and cannot be changed in the field. Of the three factors that can affect the strength of the electromagnet in the field, the only value that can be controlled is the intensity of current flow, which is determined by the voltage applied to the coil.

## THE PRINCIPLE OF ELECTROMAGNETIC INDUCTION

The principle of electromagnetic induction states that if an electrical conductor is exposed to a changing magnetic field (magnetic lines of flux), a voltage will be induced into that conductor.

### Factors Needed for Electromagnetic Induction

Three factors are required to be present for electromagnetic induction to occur:

1. An electrical conductor
2. Magnetic lines of flux
3. Relative motion between the two.

**Relative Motion.** The first two factors for electromagnetic induction are self-explanatory, but the third factor, relative motion between the electrical conductor and the magnetic lines of flux, requires some clarification. The term "relative motion" means that it may be the electrical conductor, the magnetic lines of flux, or both that move to create the relative motion. A stationary magnetic field may have a moving electrical conductor pass through it to induce a voltage into the conductor, or a stationary electrical conductor may have a moving magnetic field pass through it. Of course, if both are moving it is still considered to be relative motion, unless they are both moving in the same direction and rate in relation to each other.

### Magnitude of Induction

The magnitude of the induced voltage from electromagnetic induction can be controlled, and is proportional to the following three factors:

1. Number of turns of wire
2. Magnetic flux density
3. Speed of cutting action.

**Number of Turns of Wire.** If the electrical conductor is wound into a coil, each wire turn of the coil will have a voltage induced into it, and the induced voltages from all of the turns of wire will be of the same polarity. The effect of all the wire turns of the coil will be like a series circuit of many batteries connected together in the same polarity, and each will add together to equal their total sum. The more turns of wire in a coil, the greater the induced voltage will be.

**Magnetic Flux Density.** The strength of the magnetic field is measured by the number of magnetic lines of flux there are in a given cross-sectional area; the more lines of magnetic flux, the stronger the magnetic field. The greater the magnetic flux density, the more lines of magnetic flux are available to cut through the electrical conductor(s) for a given speed of cutting action, and the greater the induced voltage will be.

**Speed of Cutting Action.** Relative motion between the electrical conductor and the magnetic lines of flux is the speed of cutting action; the faster the magnetic lines of flux cut through the electrical conductor, the greater the induced voltage will be.

## RELATIONSHIP BETWEEN ELECTROMAGNETIC INDUCTION AND ELECTROMAGNETISM

It is helpful to study electromagnetism and electromagnetic induction together, because in electric induction motors, both electromagnetism and electromagnetic induction work together to produce a turning torque. There is a special relationship between magnetism and electricity (specifically a current flow), because either can be used to produce the other. If voltage is applied to an inductor coil, the resulting current flow will create an electromagnetic field; and likewise, if an electrical conductor is subjected to a changing magnetic field, it will induce a voltage that will cause a current to flow if there is a complete electrical path for the current. Using electrical energy to create an electromagnetic field is the principle of electromagnetism, and using a magnetic field to create electrical energy is the principle of electromagnetic induction.

### Volts-per-Turn Principle

To help understand electromagnetic induction, another electrical principle must be introduced: the volts-per-turn principle. Simply stated, the volts-per-turn principle says that when an inductor coil is energized with a source voltage, that voltage will distribute itself evenly across all the wire turns of the energized coil and produce a ratio. The volts-per-turn ratio is determined by the winding energized with the source voltage, and once established the volts per turn will carry over to the winding where the voltage is being induced. The current flow caused by the power source voltage will create an electromagnetic field around each of the wire turns in the coil, and any voltage that is induced into another conductor or wire turn of a coil from that created magnetic field will be induced at the same potential as the wire turn from which the magnetic field originated.

## INDUCTION MAY BE EITHER MUTUAL INDUCTION, OR SELF-INDUCTION

### Mutual Induction

A very simple illustration of mutual induction is when there are two or more coils on a common core, with one coil being energized with a source voltage, and one or more other coils having a voltage induced into them. An illustration of mutual induction using two coils on a common core, as shown in Figure 1-1, sometimes makes the energized coil and induced coil concept easier to understand. The energized coil has 200 turns of wire, and the induced coil has only 50 turns of wire. If the energized coil is energized with 100 volts, those 100 volts will distribute evenly among all of the 200 turns, which establishes a volts-per-turn ratio of 0.5 V/t. Once the volts-per-turn ratio is established in the energized winding, any voltage that is induced by the magnetic field from this coil will induce 0.5 volts for every electrical conductor turn it cuts through. The induced coil has only 50 turns, so when the magnetic flux from the energized winding cuts through the 50 turns at 0.5 V/t, 25 volts will be induced.

To help clarify the volts-per-turn principle, Figure 1-2 shows an iron core where both the energized coil and the induced coil have 100 turns each. The energized winding has 100 turns, and is energized with 100 volts, which establishes a one volt-per-turn ratio. Once the volts-per-turn ratio is established in the energized winding, it carries through to the induced winding, inducing one volt for each of the 100 turns of the induced winding; or 100 volts. But, the output of the induced winding is actually slightly less than 100 volts because of the inductor losses.

In Figure 1-3, the displayed values are labeled "L," for lines of magnetic flux. Using Figure 1-3 as an example, assume that 100 volts on the energized winding caused enough current to flow in

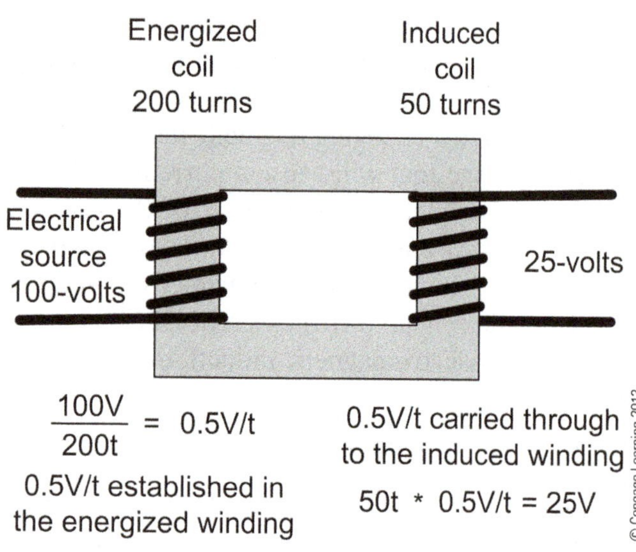



**FIGURE 1-1** Transformer core with 200 turns and 50 turns

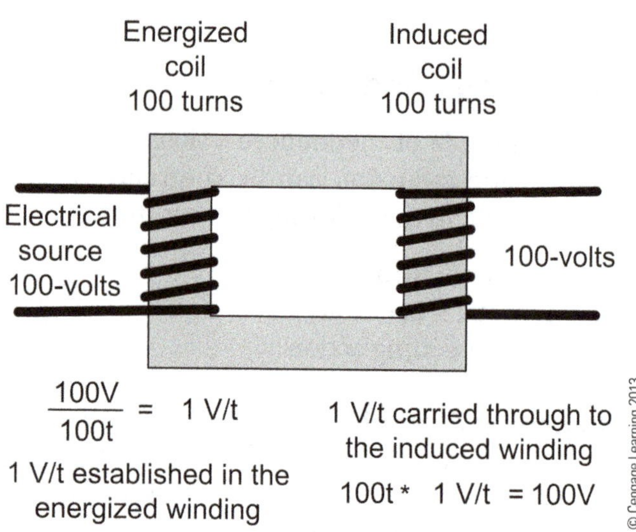



**FIGURE 1-2** Transformer core with 100 turns on each side

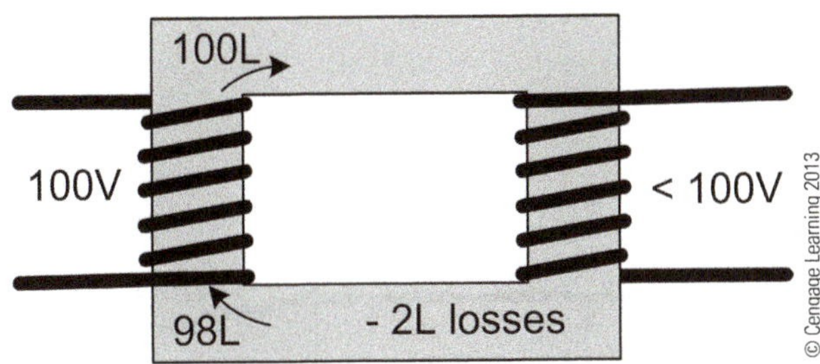


FIGURE 1-3 Transformer core with magnetic flux lines labeled

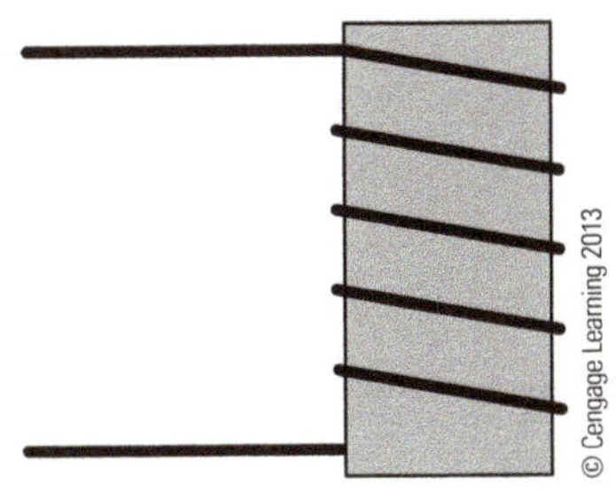


FIGURE 1-4 Inductor core with 5 turns of wire

the inductor to create 100 lines of magnetic flux. Those 100 lines of magnetic flux will circulate through the iron core, but because there are iron losses in the inductor, we will arbitrarily say that two of those lines of magnetic flux will be converted over to another form of energy, and are no longer available to induce voltage into the induced coil wire turns. In fact, in this example of completely fictitious numbers, only 98% of the magnetic flux would be available to induce a voltage, so only 98% of the source voltage that created the 100 lines of flux can be induced as a CEMF, which is 98 volts for this example.

## Self-induction

The term "self-induction" means that there is only one wire coil wound on a core that both the principle of electromagnetism and the principle of electromagnetic induction act on; it is both the energized and the induced winding. Two points that need to be clarified for self-induction are:

1. Inductors oppose a change in current
2. Two properties oppose current flow in an inductor: resistance and reactance.

Actually, the explanation for the first item in this list (inductors oppose a change in current) will explain the other as well, so this explanation will center on inductors opposing a change in current. Figure 1-4 is a diagram of an inductor coil with an iron core that has five turns of wire. For the start of the discussion we will assume that the inductor is perfect; that is, the wire has no resistance and the iron core has no losses. An inductor is nothing more than a coil of wire wrapped around an iron core; how can it oppose a change in current?

First of all, what does it mean to "oppose a change in current"? To oppose a change in current does not mean that inductors will prevent current from flowing in the inductor. Rather, it means that inductors will stabilize the current flow through the inductor. If the current through the coil is zero amperes, the inductor will try to keep the current in the inductor at zero amperes as the coil experiences changes in the applied voltage. If the current through the coil is five amperes, the inductor will try to keep the current in the inductor at five amperes as the coil experiences changes in the applied voltage. The following discussion will explain how the principles of electromagnetism and magnetic induction work together to cause this current-stabilizing behavior of an inductor coil.

**Inductors Oppose an Increase in Current.** Applying, or increasing, voltage to the inductor coil will cause an increased current to flow in the coil, but to keep the numbers relatively simple, 100 volts will be used for the potential in this example. A coil of wire is a series circuit, meaning that the current flow would be the same throughout the entire coil, and so would the magnetic behaviors associated with current flow. Rather than thinking in terms of the entire coil, however, to simplify this explanation the focus will be on a single electron, and the magnetic behaviors of that single electron at one point in one turn of the coil.

***Expanding magnetic field.*** For the sake of this discussion, in Figure 1-5, the 100-V supply voltage

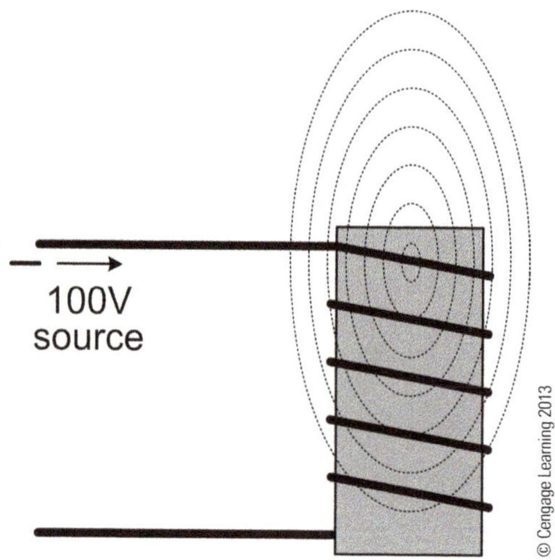


FIGURE 1-5 Inductor with 100-V source voltage

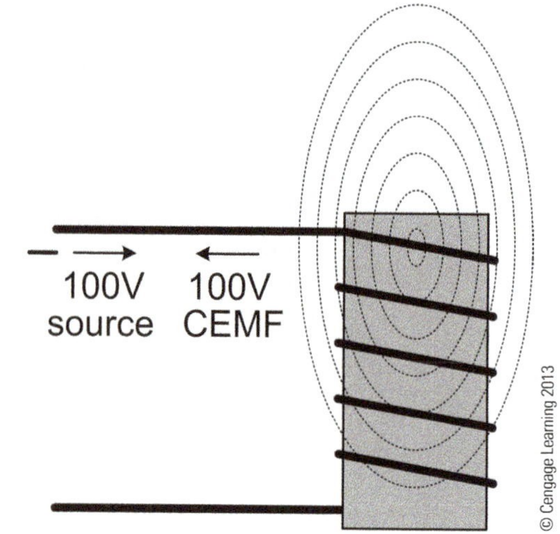


FIGURE 1-6 Inductor with 100-V source and 100-V CEMF

is going to push a single electron through the inductor coil. The instant that electron moves through the coil, it becomes current flow, which creates a magnetic field that will be built up around the conductor (shown as dotted circles). The instant before the 100-V was applied to the coil there was no current through, or magnetic field around, the wire turns of the coil. The current in the coil went from nothing to something, and the magnetic field around the wire went from nothing to something; this is a change, or motion, to the magnetic field. As the magnetic field expanded, it cut through all of the other turns of wire in the coil and induced a voltage into each.

***Counter electromotive force (CEMF).*** Electromotive force is another name for voltage, and for that reason, CEMF is sometimes called a counter voltage, but the term "CEMF" will be used here. The principle of electromagnetic induction states that when an electrical conductor is exposed to a changing magnetic field, it will induce a voltage into the conductor. The magnitude, or strength, of induction depends on three values identified previously: the number of wire turns in the coil, the magnetic flux density, and the speed of cutting action. Lenz's law (Heinrich Lenz was a 19th-century physicist) states that an induced voltage will always be the opposite polarity of the electrical source voltage that created it, which in self-induction is called a CEMF. The voltage that is induced into the wire turns of the coil is in the opposite polarity of the source voltage pushing the electron in, and it will push the electron right back out, as shown in Figure 1-6. The current in the inductor coil was zero before the supply voltage was applied, and the CEMF will try to stabilize the inductor coil current at zero.

**Resistance and Reactance.** Two properties in an inductor oppose current flow: resistance and reactance.

***Resistance.*** Resistance, which restricts or opposes the flow of electrical current, is a physical property of every electrical conductor and (for the purposes of this statement) does not change with time. As verified with Ohm's law, voltage and current are directly proportional in a resistive circuit; when the supply voltage is increased, the circuit current increases proportionally. Resistance is easy to identify, because it may be read with an ohmmeter, and when current flows through resistance, heat is dissipated; a power loss, $P = I^2R$. Resistance does not cause a phase shift between the voltage and current, as shown in the oscilloscope waveforms in Figure 1-7. The actual wire resistance of an inductor coil is very low, possibly only a couple of ohms. If the wire conductor were straightened out (not wound into a coil) and connected to a power source, it would draw close to short-circuit current and open the circuit overcurrent protective device,

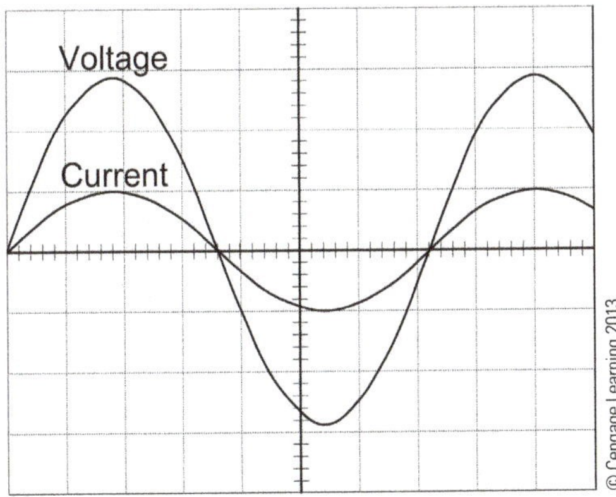


FIGURE 1-7 Graticule with in-phase waveforms

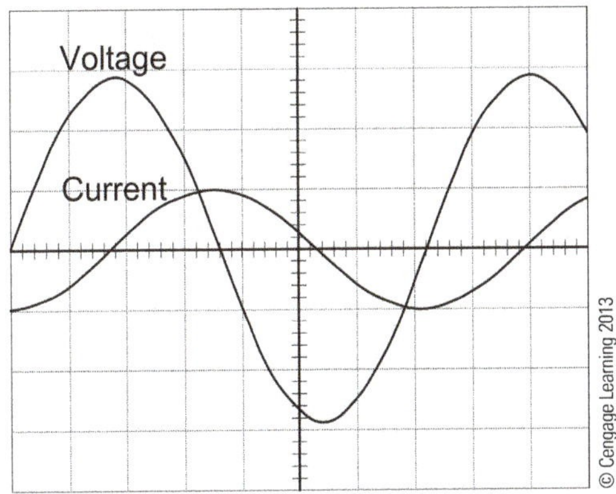


FIGURE 1-8 Graticule with out-of-phase waveforms

because the conductor resistance would be the only property limiting the circuit current.

***Reactance.*** Reactance, on the other hand, is an opposition to current flow in an electrical circuit that prevents the electrical current from flowing in the first place. When an electrical conductor is wound into an inductor coil, the reactive opposition to current flow, the CEMF, will be many times higher than the actual conductor resistance, and it will significantly limit current flow when the supply voltage is applied. Reactance is quantified in ohms, like resistance, but it cannot be measured with an ohmmeter; it is calculated. If the CEMF could equal the applied supply voltage, it would be like connecting two batteries of the opposite polarity in series. The net voltage across the two would be zero, and no current would flow in the circuit. Because reactance prevents a current flow from happening in the circuit, there are no $I^2R$ losses, or heat, associated with reactance.

***Reactance Causes a Phase Shift of Voltage and Current.*** Initially, when the source voltage is first applied to an inductor coil, the CEMF will be very high and strongly delay the flow of current. The CEMF, however, will diminish with time and current will start to flow in the inductor coil, as shown by the waveforms in Figure 1-8, demonstrating that the inductor delays the flow of current. The difference in time between when the voltage was applied to the inductor coil and current started to flow through the coil is called a phase shift. In a perfect inductor, current would lag voltage by 90°, but the actual phase shift will always be something less than 90°. When there is a phase shift between voltage and current in an electrical circuit, voltage and current are not proportional to each other, as is true in a resistive circuit (refer back to Figure 1-7).

**Self-induction Losses.** Return to the self-induction concept discussed earlier in this chapter, and relate the principles from the fictitious mutual induction example, as shown in Figure 1-9. In self-induction it is the same coil that is both energized and induced. Say, as in the previous examples, the 100-volt source voltage across the 100 turns of the coil produces a 1V/t ratio, which causes sufficient current flow to create 100 lines of magnetic flux. Because of inductor losses some of the magnetic lines of flux in the core will be converted into other

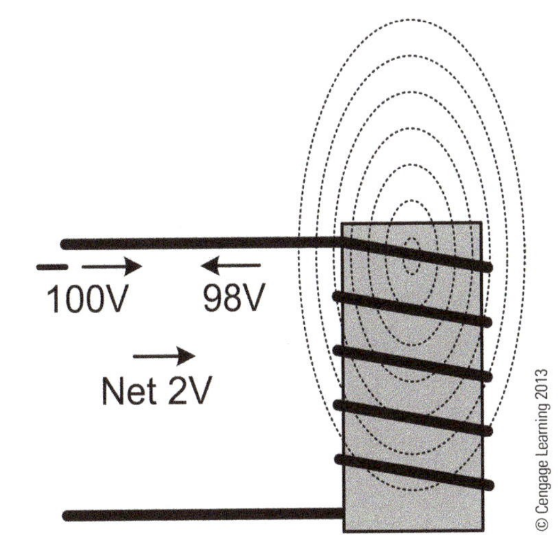


FIGURE 1-9 Inductor core with 2 volts of net voltage

forms of energy, and they will not be available to induce a counter EMF. The CEMF associated with self-induction always will be less than the supply voltage.

Once again, take only a single electron and a single turn of wire in the coil to make this concept easier to see. The magnetic field created by the current flow will expand from the center of the conductor and cut through all of the other turns in the coil, inducing a voltage of 1 V/t from the example, which will oppose the applied voltage to the inductor coil. In the example there is a 2% loss, which means the CEMF can only be 98% of the applied voltage, and that leaves a 2% net voltage in favor of the applied voltage.

## Summary of Inductors Opposing an Increase in Current

Inductors, then, do oppose or delay a change in current, but do not prevent the current flow from eventually happening. When observing inductors in electrical power applications, it appears that the entire process of allowing current to flow is instantaneous, but it is not. The amount of time inductors oppose a change in circuit current is calculated from circuit component values, and is used to control circuit operation. This time is expressed in time constants, and one time constant is calculated by multiplying the resistance of the circuit by the inductance of the circuit (1 TC = R* L). Without getting into time constants too much, it is generally accepted that the inductor will not oppose a change in current beyond the time of five time constants. As an example: an inductive circuit with 0.1 ohms of resistance and 10 mH of inductance will have a time constant of 1 ms (0.1 ohm *0.01 mH = 0.001 seconds). Five time constants, then, would equal 5 ms. This particular inductor example would have little effect on a 60-Hz current and tend to pass it unimpeded, because one cycle at 60 Hz takes so much longer to complete (T = 1/60 Hz = 16.67 ms). A transient spike, which takes place in microseconds, however, would be severely impeded (or attenuated), because its duration is very short compared to the circuit RL time constant.

## Inductors Oppose a Decrease in Current

Refer back to Figure 1-8, which shows the voltage waveform increasing from zero volts, but the current waveform is lagging behind. The current flow is not proportional to the voltage. When the voltage started increasing from zero volts, the inductor opposed a change in current from zero amperes, and the inductor did delay the current flow for some time. Now, look at the voltage and current waveforms where the voltage waveform reaches peak voltage and starts to go down. The affect of the inductor at this point is just as important as when the voltage waveform was at zero volts. As the current in the inductor continued to increase, it supported a larger and larger magnetic field in the inductor coils. Now when the voltage starts to decrease, it no longer can produce a large enough current flow to keep the magnetic field at its full size.

As the current in the circuit starts to decrease, the inductor will try to stabilize the current at its peak value. When the current drops below the level necessary to maintain the magnetic field at its maximum, the magnetic field will begin to collapse. As the magnetic field collapses, this, again, is motion of the magnetic field. The collapsing magnetic field will cut through the inductor coil windings and induce a voltage into each of them. This induced voltage is opposite the polarity of the collapsing magnetic field, which was opposite the supply voltage; a double negative. So, the voltage induced from the collapsing magnetic field is the same polarity as the supply voltage, and it will try to keep the inductor current from decreasing.

Anytime the voltage across an inductor changes, causing a corresponding change in current, the inductor will oppose the change. If the current in an inductor starts to change in either direction, up or down, the magnetic field will experience change that will provide the necessary relative motion between the electrical conductors and the magnetic field in order for magnetic induction to induce a voltage into the wire turns of the inductor. And, as Lenz's law states, the induced voltage will oppose the force that made it, and the change in current through the inductor will be opposed.

## CHAPTER SUMMARY

- The principle of electromagnetism states that when an electrical current flows through an electrical conductor, a magnetic field will be created around that conductor.
- The three properties that determine the strength of any electromagnet are: intensity of current flow, number of wire turns in the coil, and type of core material.
- The term "ampere-turns" indicates that the strength of the electromagnetic field is determined by the number of wire turns in the coil, and the intensity of current flow through the coil.
- The principle of electromagnetic induction states that if an electrical conductor is exposed to a changing magnetic field (magnetic lines of flux), a voltage will be induced into that conductor.
- The three factors that are required to be present for electromagnetic induction to occur are: an electrical conductor, magnetic lines of flux, and relative motion between the two.
- The magnitude of the induced voltage from electromagnetic induction can be controlled, and is proportional to the following three factors: number of turns of wire, magnetic flux density, and speed of cutting action.
- There is a special relationship between magnetism and electricity (specifically a current flow), because either can be used to produce the other.
- The volts-per-turn ratio says that when an inductor coil is energized with a source voltage, that voltage will distribute itself evenly across all the wire turns of the energized coil, and once established the volts-per-turn ratio will carry over to the winding where the voltage is being induced.
- Mutual induction is when more than one inductor coil is mounted on a common iron core.
- The term "self-induction" means that there is only one wire coil on which both the principle of electromagnetism and the principle of electromagnetic induction act.
- To oppose a change in current does not mean that inductors will prevent current from flowing in the inductor; it means that inductors will stabilize the current flow through the inductor.
- Lenz's law states that an induced voltage will be the opposite polarity as the electrical source voltage that created it, which in self-induction is called a CEMF.
- An induced CEMF can never equal the source voltage, because the inductor has losses.
- Two properties in an inductor oppose current flow: resistance and reactance.
- The difference in time between when the voltage was applied to the inductor coil and current started to flow through the coil is called a phase shift.

## REVIEW QUESTIONS

1. What is the principle of electromagnetism?
2. What three properties determine the magnetic strength of an electromagnet?

   1. ______________________________

   2. ______________________________

   3. ______________________________

3. What is meant by the term "ampere-turns" in relation to electromagnetism?
4. What is the principle of electromagnetic induction?
5. What three factors are necessary for magnetic induction to occur?

   1. ______________________________

   2. ______________________________

   3. ______________________________

6. The magnitude of the induced voltage can be controlled, and is proportional to what three factors?

   1. ______________________________

   2. ______________________________

   3. ______________________________

7. What is it called when there is only one wire coil on which both the principle of electromagnetism and the principle of magnetic induction act?
8. What two points are important to clarify about self-induction?

   1. ______________________________

   2. ______________________________

9. What is meant by the statement that inductors oppose a change in current?
10. What is another name sometimes used for CEMF?
11. According to Lenz's law, an induced voltage will have the opposite polarity of what other voltage?
12. How can the resistance of an inductor coil be determined?
13. What happens whenever an electrical current flows through resistance?
14. Reactance is the CEMF that does what to the current flow in an inductor?
15. The volts-per-turn principle is established in which coil?
16. If an inductor coil with 200 turns is energized with a source voltage of 600 volts, what is the volts-per-turn ratio that will be established?
17. Why can the induced voltage never be as great as the source voltage?

CHAPTER

2

# Induction Motor Overview

**PURPOSE**

To familiarize the learner with the terminology associated with induction motors converting electrical energy into mechanical energy, and induction motor operating characteristics.

**OBJECTIVES**

After studying this chapter on the terminology and operating characteristics of induction motors, the learner will be able to:

- Identify the parts of an induction motor
- Explain each part of an induction motor
- Explain electromagnetic induction to the rotor of an induction motor
- Explain the difference between conducting and inducing electrical energy to the rotating member of an electric motor
- Explain what power factor is
- Define the electrical/mechanical terminology associated with power factor
- Explain how induction motors cause a power factor problem
- Explain the current draw and torque-speed curve operating characteristics of induction motors

## THE PARTS OF AN INDUCTION MOTOR

Figure 2-1 shows a cutaway view of an induction motor with all of the parts labeled, and shows how they are assembled in relation to each other. A short description to explain the function of each part is provided here.

### Conduit Box

The conduit box, sometimes called a terminal box, is labeled on the top right of Figure 2-1, and is where the leads of the stator (stationary member of the motor) coil wires are brought out to be connected to the power supply. Normally the leads are just individual conductors brought out to the conduit box, and labeled with the designator "T" (T1, T2, etc.) for "Terminal."

### End Bells and Through Bolts

The end bells and through bolts are labeled on the top left and bottom right of Figure 2-1. The end bells are manufactured to seat on the motor housing for mounting, and receive the outer race of the rotor shaft bearing to hold the rotor centered in the stator core. For smaller motors, such as the one pictured in this figure, to hold the motor together through bolts are inserted from one end bell, through the motor, to the other end bell. For larger motors the end bells will normally bolt on to their respective ends of the motor housing individually.

### Cooling Fan

The cooling fan, labeled on the top left of Figure 2-1, is mounted directly on the rotor shaft, and circulates cooling air when the rotor is turning. Depending on

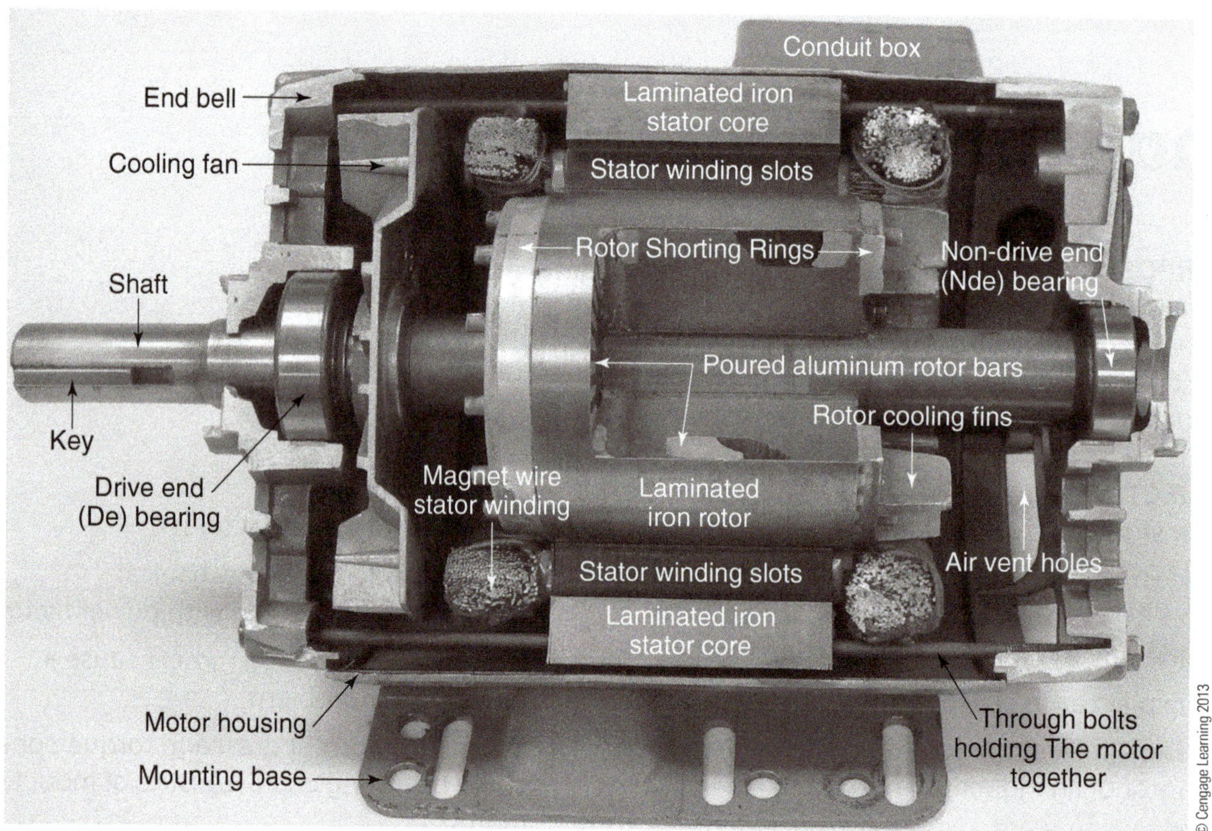


FIGURE 2-1 Induction motor cutaway view with labels

the design of the motor, the cooling fan may be on either end of the motor, and even on the exterior of the motor with a cowling to direct the airflow over the motor's cooling fins.

### Air Vent Holes

The air vent holes of the motor pictured in Figure 2-1 are labeled on the bottom right of the picture. Not all motors have air vent holes; some are totally enclosed for use in dirty areas. When a motor does have air vent holes, they should be checked and cleaned periodically to keep them open for unobstructed air flow.

### Shaft and Shaft Key

The shaft and shaft key are labeled on the left center of Figure 2-1. The shaft travels the full length of the motor, and any part of the motor that rotates is mounted to the shaft. The purpose of the shaft key is to prevent the mechanical connection between the shaft and the load from slipping under very high torque conditions, because the key is half on the shaft and half on the load connection. Not all motor shafts have key slots. Many smaller motors use a flat area on the shaft where a set screw can be seated.

### Motor Housing

The motor housing, labeled on the lower left of Figure 2-1, holds all the parts of the motor together. All of the stationary parts of the motor, the stator core, conduit box, and mounting base are all mounted to the motor housing. The end bells, which hold the rotor and all moving parts of the motor centered in the stator core, are also mounted to the motor housing.

### Mounting Base

The motor base for the motor pictured in Figure 2-1 is labeled on the lower left. There are many different methods of fastening motors in place, so not all motors have mounting bases. For the motors that do have mounting bases, this is how the motor is fastened in place. The bolt-hole dimensions and patterns are defined in the NEMA standard with the frame information, which will be more fully explained in the chapter on motor nameplate information.

### Drive End and Opposite Drive End Bearings

The bearings are the mechanical connection between the stationary and moving parts of the motor that allows the rotor to turn inside the stator core. The bearings are also the only wear parts of three-phase induction motors. The drive end bearing for the motor pictured in Figure 2-1 is labeled on the center left, and the opposite drive end bearing is labeled on the center right. The inner race of motor bearings, as pictured in Figure 2-2, is pressed onto the rotor shaft, and the outer race of the bearing is friction fit into the motor's end bell. The drive end bearing is larger because it is going to shoulder additional stresses associated with driving the load, such as belt tensioning if it is being used in a belt driven application.

### Laminated Iron Stator Core

Figure 2-3 shows the end view of a complete motor stator core that is removed from the motor housing. The stator iron consists of many iron laminations built up to form a larger cross-sectional area, which will carry the magnetic lines of flux from the stator coil windings to create an electromagnet.

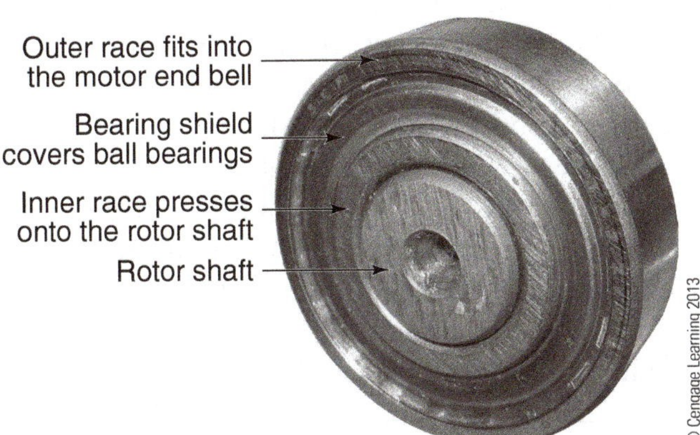
Outer race fits into the motor end bell
Bearing shield covers ball bearings
Inner race presses onto the rotor shaft
Rotor shaft

**FIGURE 2-2** Bearing showing inner and outer races

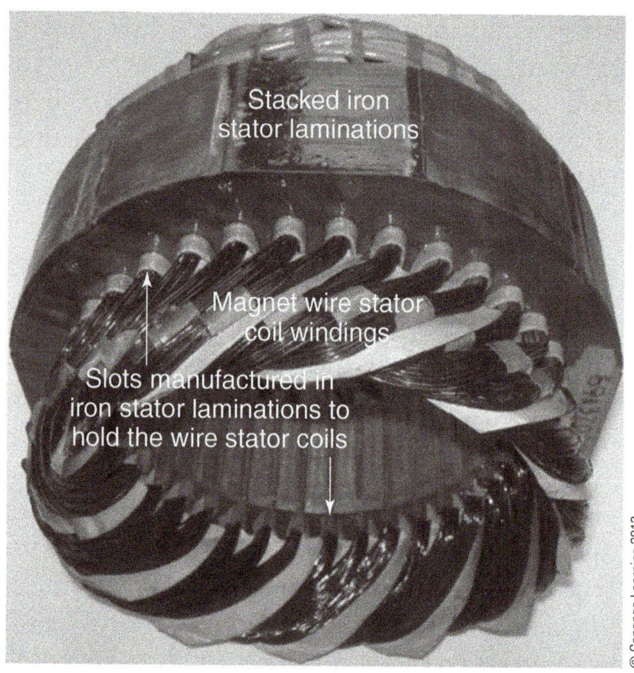




FIGURE 2-3 The motor stator

## Stator Winding Slots

Compare the stator winding slots in Figure 2-1 with the slots in Figure 2-3 to see how these slots are recessed into the iron stator core, and the magnet wire stator winding coils are placed inside them. This assures that the stator coils are completely surrounded by iron, which has superior permeability to carry magnetic lines of flux.

## Magnet Wire Stator Windings

Induction motor stator coils are wound with many turns of magnet wire and placed in special slots recessed into the iron stator core, which is clearly seen in Figure 2-3. Magnet wire is a special type of wire that is insulated with only a thin layer of varnish, rather than a thicker type of insulation construction electricians normally encounter on wire. The thin insulation is necessary for two reasons: first, it makes it possible to fit many more turns of wire into the limited space slots in the iron stator core; second, inductive (magnetic) coupling requires that the conductors be as close together as possible.

FIGURE 2-4 Poured rotor bars in iron laminations

## Laminated Iron Rotor

The laminated iron rotor with poured aluminum rotor bars is labeled in the center of Figure 2-1. The rotor consists of many iron laminations built up to form a solid piece of iron. When the rotor iron laminations are assembled, cavities will be formed where molten aluminum will be poured to form the rotor bars. Figure 2-4 is a magnified end view of the rotor shown in Figure 2-1, which has part of the iron laminations machined out to see the poured aluminum rotor bars. Each of the iron laminations have a tight friction fit to the motor shaft to hold them in place.

## Rotor Shorting Rings, Poured Aluminum Rotor Bars, and Rotor Cooling Fins

Figure 2-5 is a more inclusive end view of the same motor rotor as in Figure 2-1, and shows a different perspective of how the entire rotor is assembled. After the rotor iron laminations are assembled, molten aluminum is poured through the laminations into the cavities designed to form the rotor bars (which are the electrical conductors of the rotor). When the bar cavities are filled, additional molten aluminum is poured on each end of the rotor to form shorting rings and electrically short all of the aluminum bars together to form current paths between the rotor bars. Sometimes aluminum cooling fins are poured as part of a shorting ring to circulate cooling air inside the motor housing. Figure 2-6 is a side view of a different rotor, and identifies the same rotor components without any part of the rotor being machined away.

## Squirrel Cage Rotor

The rotors of induction motors are commonly referred to as squirrel cage rotors, because the rotor bars assembly and shorting rings, without the iron laminations, resemble a pet squirrel's running wheel, as shown in Figure 2-7. The reason the conductor bars are shorted together on each end of the rotor is so that when a voltage is induced into the conductor bars, there will be a complete electrical path for current to flow around in the rotor, as shown in Figure 2-8, to produce a magnetic field in the rotor to interact with the rotating magnetic field of the stator.

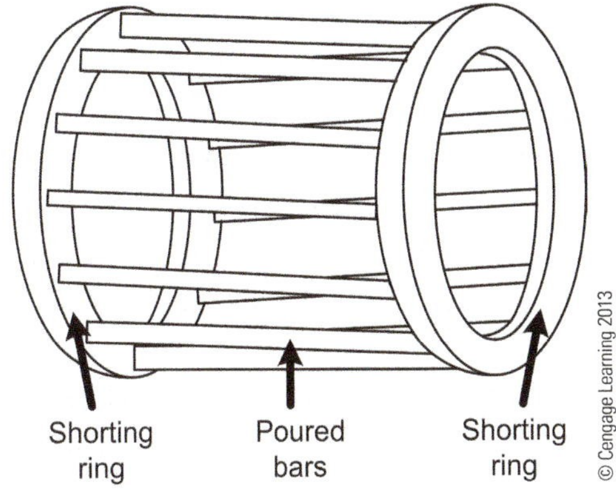


FIGURE 2-7 Squirrel cage rotor without iron laminations

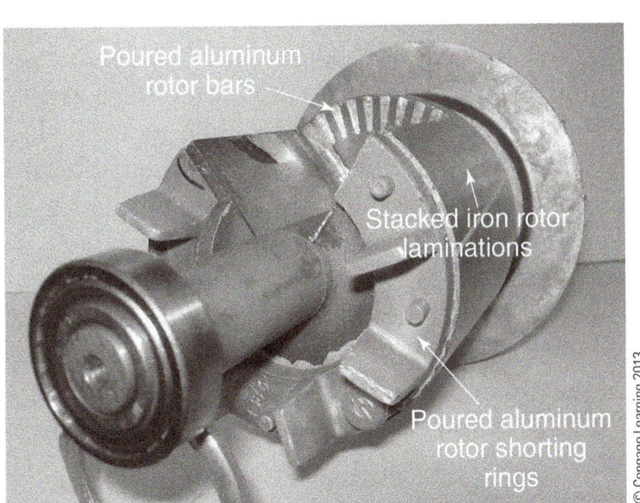


FIGURE 2-5 Poured rotor shorting rings

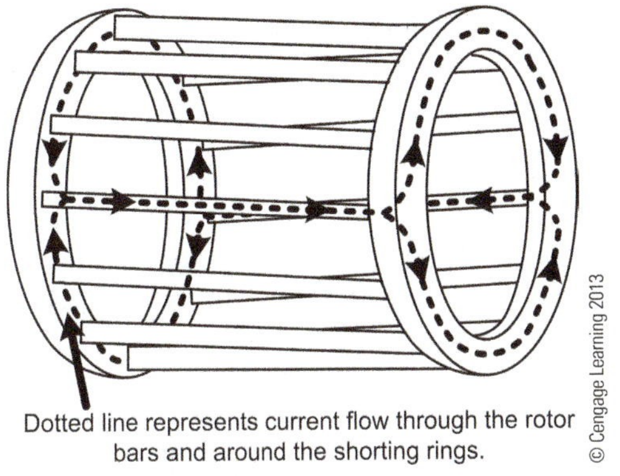

Dotted line represents current flow through the rotor bars and around the shorting rings.
© Cengage Learning 2013


FIGURE 2-8 Current paths in rotor

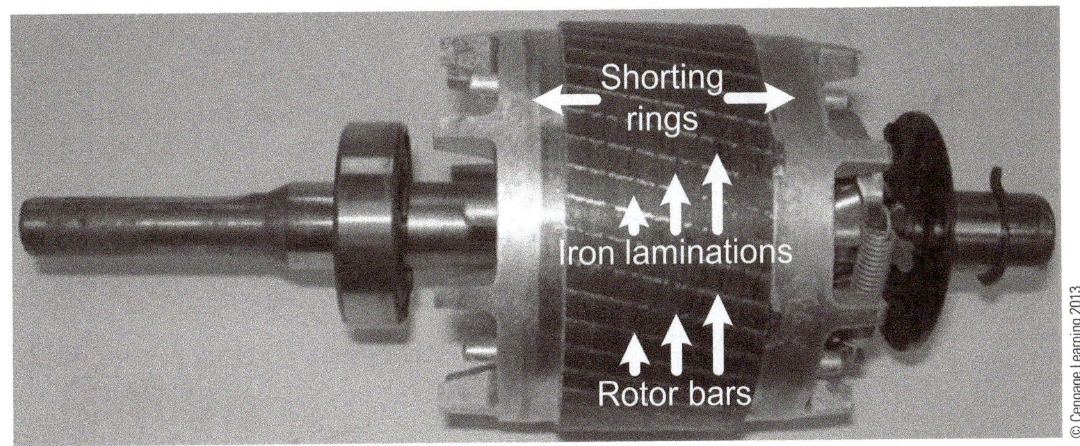


FIGURE 2-6 Rotor showing bars laminations and shorting rings

## WHAT IS AN INDUCTION MOTOR?

With the exception of some very specialized motors, electric motors can be classified as either conduction or induction, and the classification pertains to the method that electrical energy is delivered to the rotating member of the motor. Conduction motors require electrically conductive brushes, held in an assembly mounted on the stationary member of the motor, which press against the commutator or slip rings mounted on the rotating member of the motor. This provides a direct electrical connection between the stationary and rotating members of the motor, which conducts electrical energy to the rotating member. The brushes, brush assembly, and commutator or slip rings present wear issues that that require maintenance on these types of motors.

Induction motors, on the other hand, do not have a direct electrical connection between the stationary member and the rotating member. They use magnetic induction to induce a voltage in the rotating member. The absence of brushes and a commutator or slip rings makes the induction motor more reliable and maintenance free. In fact, the only wear parts of three-phase induction motors are the bearings, and with today's sealed bearings, they are virtually maintenance free. The simplicity and dependability of induction motors has earned them the reputation of being the workhorses of the electrical industry.

### Electromagnetic Induction Theory and Induction Motors

When electrical energy is induced into the rotor of the induction motor, it is induced through electromagnetic induction from the changing magnetic field of the stationary stator coils. When voltage is induced into the rotating member, the magnitude will be determined by the volts-per-turn principle. The stator windings have many turns of wire, and as the windings are energized, this creates a low volts-per-turn ratio. The rotor's conductor bars are all shorted on the ends by shorting rings (refer back to Figure 2-7) and they act like a single winding for the induced coil, which means that the induced voltage in the rotor will be very low. The volt-amperes in the rotor are going to equal the volt-amperes in the stator, so the lower voltage in the rotor means that the current is going to be very high. This high current requires that the conductors on the rotating member must be very large, and in fact they are poured aluminum bars. Rotors have very low voltage, high current, and depend on current intensity to build a strong magnetic field.

As an example to put the high rotor current in perspective, if the stator coil windings have 100 turns and the rotor bar circuit is the equivalent of one turn, the transformer volts-per-turn ratio would establish the current through the rotor bars as 100 times higher than the stator current. If a very small induction motor drew only one ampere from the power source, it could mean as much as 100 amps through the rotor bars. This very high current in the rotor bars explains why the rotor bars have to be so thick.

The solid poured aluminum bar conductors found in the induction motor rotor are designed with some resistance to reduce the motor locked rotor starting current operating characteristics. The lower the resistance of the rotor bars, the higher efficiency and torque the motor will have, but also the higher the locked rotor starting current of the motor will be. If the locked rotor starting current of a motor is too high, it may cause nuisance tripping of the motor circuit overcurrent device on startup.

Also, rotor bars are designed with a slight skewing. (Refer back to Figure 2-5 and notice that the rotor bars are angled rather than being straight across the rotor from one shorting ring to another.) This slight skewing helps reduce "cogging" torque. Cogging torque is a term used to describe when the rotor turns in jerks or increments rather than a smooth turning motion. Cogging torque may be caused when a motor has too few coil windings, too few rotor bars, or if the rotor bars are manufactured straight across the rotor from one shorting ring to the other. Any of these conditions would leave torque dead-spots as the rotor turns.

### Induction Motors and Power Factor

The main disadvantage of induction motors is that they cause a lagging power factor problem on the premises electrical distribution system. The power factor is a ratio that measures how effectively the electrical energy taken from the electrical power source is used to perform work, compared to the portion of the electrical energy that is stored and returned to the power supply unused. A poor power factor means that the productive current-carrying capacity of the premises electrical distribution system is reduced, because a portion of the capacity is used carrying unproductive electrical currents. To better understand what the power factor is, and how induction motors cause a power factor problem, it is necessary to define a few terms.

**Energy.** Even though there are more creative ways to say it, energy is nothing more than the capacity, or potential, to perform work. There are different types of energy, such as electrical and thermal (heat), and different mediums of storage, such as batteries and fossil fuels; but this discussion will focus on electrical energy. Electrical energy is the total volt-amperes (VA) available from a power supply, and determines the potential or capacity of work that may be accomplished. For example, the typical convenience outlet in the wall is 120 volts, and for this calculation it will be a 20-ampere branch circuit. This would constitute a relatively low-energy circuit of 2,400 volt-amperes (120 volts times 20 amperes). If the circuit voltage is increased to 480 volts, the same 20-ampere branch circuit will have four times the energy: 9,600 volt-amperes (480 volts times 20 amperes). Increasing the circuit voltage did not accomplish anything except to increase the potential, or capacity, to perform work.

**Work.** When electrical energy is applied to a device that can convert it to another form of energy, such as an electric motor converting it to mechanical energy, it performs work. Work cannot be performed by electrical energy unless the electrical energy is converted over to another form. Examples of work might include an electric motor moving an elevator car up and down, moving a box on a conveyor belt, or moving air with a fan blade. Using an electric heating element to heat air or water is another example of converting electrical energy into another form and performing work. One measure of mechanical work is a unit called "foot-pounds." A foot-pound is the work done when a one-pound weight is lifted vertically the distance of one foot. The product of any weight and distance will equal foot-pounds. So, if a 550-pound weight is lifted vertically one foot, regardless of how much time it may take, 550 foot-pounds of work is accomplished. The simple definition of work that will be used here is: work is performed when something is moved or heated.

**Torque.** Torque is a turning force, and when applied to electric motors, torque determines what mechanical work may be performed. As an example, if a belt conveyor requires more torque to overcome the belt friction and the weight of the material being moved than the electric drive motor is capable of producing, the motor will not be able to turn and no mechanical work will be performed. For mechanical work to be performed by an electric motor, the motor must be capable of producing enough torque to overcome the turning force requirements of the mechanical load. The most common torque unit of measure for NEMA-rated electric motors is foot-pounds, and for IEC-rated motors it is newton-meters. The torque foot-pound measurement is different from the work foot-pound measurement by the way it is calculated. A torque foot-pound is defined as the force of one pound acting on a radius of one foot. The product of any force in pounds and radius in feet will equal foot-pounds of turning torque. For example, if a motor drives a pulley with a two-foot radius, and the belt has eight pounds pulling force on it from the motor, the turning torque of this application is 16 foot-pounds (2 feet times 8 pounds).

**Power.** Power is the measurement of how much work is accomplished in a specific amount of time, with time being the differentiating focus. It is possible for a small propeller-engine plane to transport a person across the country, but a jet plane can transport that same person across the country much faster, because generally a jet plane is more powerful. The more powerful jet plane did not get more work done. It got the same work done faster. The most common power unit of measure for

NEMA-rated electric motors is horsepower, and for IEC-rated motors is watts. One horsepower is defined as performing 550 foot-pounds of work in one second, and the horsepower-to-watts conversion factor is 1 horsepower = 746 watts.

**Review of Terms.** For the purposes of this book, the following simple definitions of energy, work, torque, and power will be sufficient. Energy is the potential to perform work. Work is performed when something is moved or heated. Torque is the measure that determines what mechanical work may be performed by an electric motor. And power determines how fast the work may be accomplished.

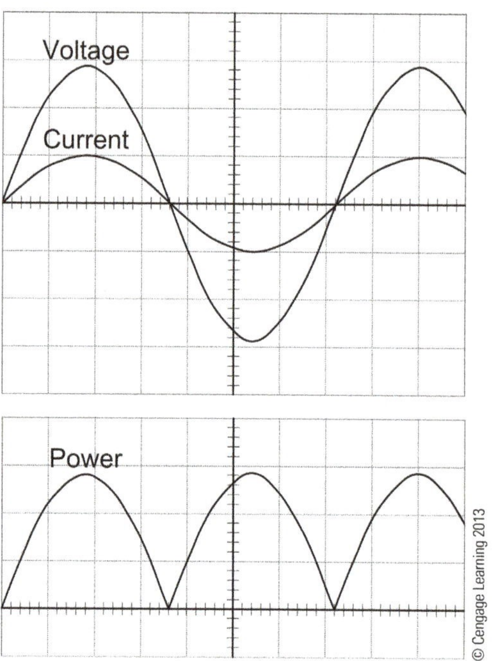


FIGURE 2-9 Graticule with in-phase waveforms showing power

## Waveform Analysis of Power Factor

In Chapter 1 waveform graphics were used to demonstrate the phase-shift between voltage and current caused by inductors. The same type of waveform graphic will be used here to demonstrate the effect that phase-shift has on power, and ultimately power factor.

**Resistive Load.** Figure 2-9 shows the voltage and current waveforms for a purely resistive load. Resistive loads are said to be in-phase, because voltage and current are always the same polarity, and are being used in the same proportion for all time points along the horizontal axis of the waveforms. Notice that the power waveform in this graphic shows that power is dissipated anytime voltage and current are present, because in a purely resistive circuit watts and volt-amperes are equal. This is an important point to make in this figure because power is always positive. Mathematically this concept of power always being positive holds true, as the product of a positive voltage times a positive current is a positive power, and likewise, the product of a negative voltage times a negative current is also a positive power. Purely resistive electrical loads do not cause a power factor problem, because there is no phase shift between voltage and current.

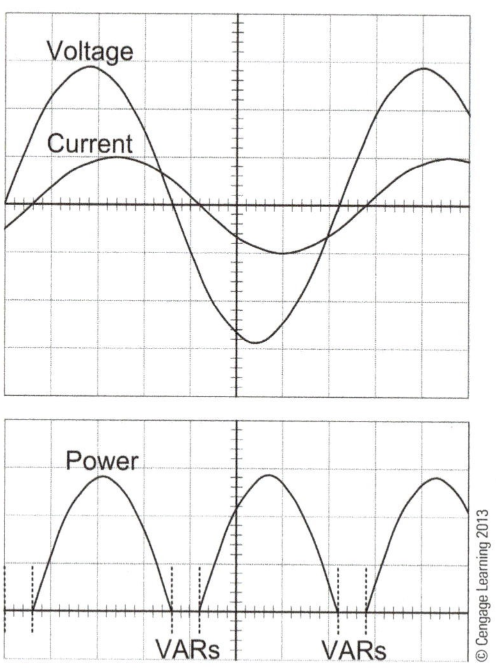


FIGURE 2-10 Graticule with out-of-phase waveforms showing power

**Inductive Loads.** Figure 2-10 shows the voltage and current waveforms out of phase for an inductive load. As demonstrated in Chapter 1, inductors, which include induction motors, cause a phase shift between voltage and current that causes current to lag voltage, and it is this phase shift that causes the power factor. This phase shift causes periods of time where the voltage and current values are of different polarities. Notice that the power waveform in this graphic shows that power is dissipated only during the periods of time that the voltage and current are the same polarity. During the periods of time that the voltage and current are of opposite polarities, the power waveform drops to zero and remains flat. The reason for this is that power is always positive, and the product of the voltage times amperes will be negative if only one of the values is negative. There are still volt-amperes of electrical energy, but they are volt-amperes reactive (VARs). VARs are the portion of the total electrical energy taken from the power source that are stored in the magnetic field of the inductor, and returned to the power source unused during the periods of time that the voltage and current are of different polarities.

## Power Triangle Analysis of Power Factor

Another method of graphically demonstrating the power factor relationship between energy and power is with the power triangle. Figure 2-11 shows the same power triangle three times, represented with three different sets of labels, all of which are just different ways to explain each respective side. The hypotenuse of the triangle is the total volt-amperes, the product of the voltage times the current taken from the electrical supply; in other words, the total electrical energy drawn from the power supply by the motor; the potential to perform work. Some of that total electrical energy that enters the motor is converted to other forms, such as mechanical energy or heat, and it produces work. That work, with time, is power, measured in watts, and becomes the in-phase side of the power triangle (side adjacent $\angle\theta$). And, finally, some of that total electrical energy that enters the motor is stored in the stator's magnetic field, and is returned to the power source during the part of the sine wave cycle where voltage and current are of opposite polarities (side opposite $\angle\theta$). These volt-amperes are not converted to another form of energy, and are called volt-amperes reactive (VARs); they do not produce work and no power is associated with them.

Remember that power factor measures the ratio of the volt-amperes of an electrical load that are converted to another form and produce work, to the total volt-amperes the load drew from the power supply. On the power triangle, the power factor is calculated by the cosine of $\angle\theta$, which is the side adjacent divided by the hypotenuse; the volt-amperes that produced work, divided by the total volt-amperes taken from the power supply. This calculation will produce a decimal answer, which is then multiplied by 100 to express the answer as a percentage. As an example, 100 watts divided by 120 volt-amperes = 0.833; multiplied by 100 = 83.3% power factor.

**Power Factor and Current-Carrying Capacity.** The power triangle is also an excellent method to demonstrate graphically how a poor power factor decreases the current-carrying capacity of an electrical distribution system. In a purely resistive circuit, watts equal volt-amperes and there is no

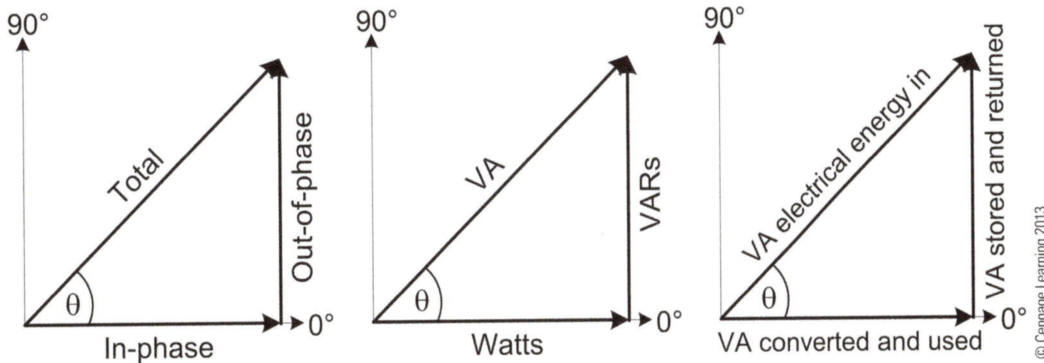



**FIGURE 2-11** The three power triangles

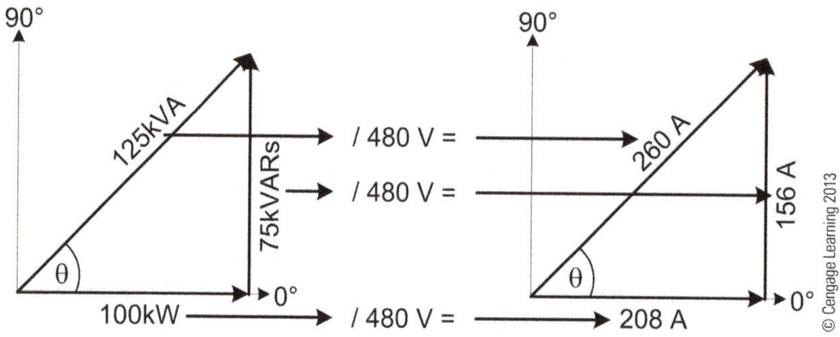


FIGURE 2-12 Power triangles labeled with currents

power triangle because the power factor is 100%; $\angle\theta = 0°$. A power factor of 100%, also called a unity power factor, indicates that 100% of the total volt-amperes taken from the power supply are converted to another form of energy and perform work. Whenever the power factor is anything less than 100%, there will be a triangle and the hypotenuse is always the longest side. The fact that the hypotenuse, or total volt-amperes taken from the power supply, is larger than the side adjacent $\angle\theta$, the volt-amperes that are converted to another form of energy and perform work, it means that some system capacity is wasted.

As an example of this point, Figure 2-12 shows a power triangle that is labeled with the values for a hypothetical induction motor with an 80% power factor, operating on a 480-volt system. Since each side of the power triangle is volt-amperes, the current associated with each side of the triangle may be found by dividing each volt-ampere value by 480 volts. The second triangle in Figure 2-12 demonstrates that 260 amperes will be drawn from the line, but only 208 amperes of that will contribute to work. Fifty-two amperes of the total 260 amperes (260 amperes minus 208 amperes = 52 amperes) are unnecessarily taking up current-carrying capacity on the electrical distribution system, moving back and forth between the power supply and the load, without producing work.

## Induction Motor Operating Characteristics

**Induction Motor Torque and Current.** The operating characteristics such as torque and current draw are determined by the design of the motor. NEMA classifies electric motors by design letters that specify standard ratings for each different design letter. Two of the most common operating characteristics specified are the locked rotor starting current requirements and torque-speed curves. NEMA has many motor letter classification designs by these characteristics (A,B,C, D, etc.), but the NEMA design B motor is the general purpose design, and is by far the most common. The design B motor provides a good balance between locked rotor torque, locked rotor current, and full-speed torque that make it suitable for a wide range of electric motor applications.

Figure 2-13 shows the full-voltage current curve for NEMA Design B motors. Notice from the graphic that the highest current draw is at locked rotor (zero RPM) speed, and does not start to drop off significantly until the rotor reaches 80–85% full-load speed.

The torque-speed characteristics of induction motors are a little more involved, and require a

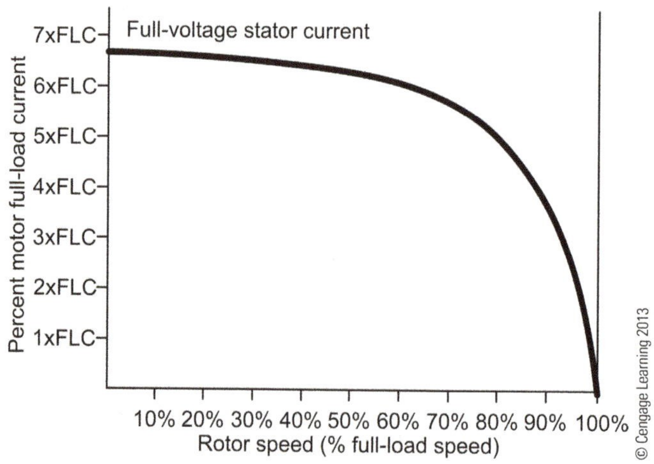


FIGURE 2-13 Full-voltage stator current graph

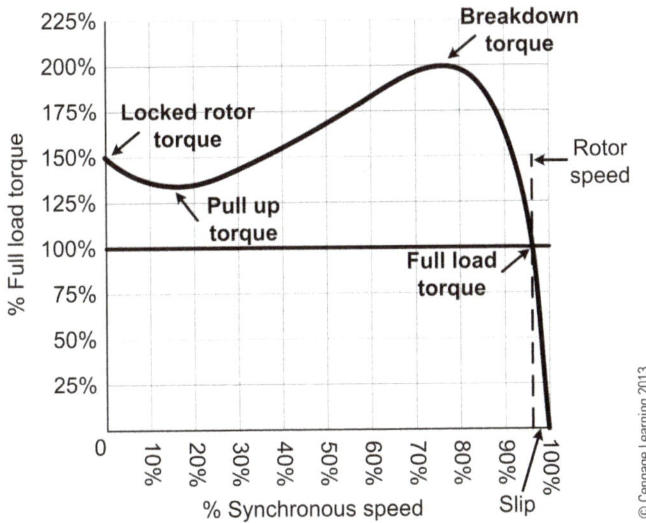


FIGURE 2-14 Design B induction motor torque curve

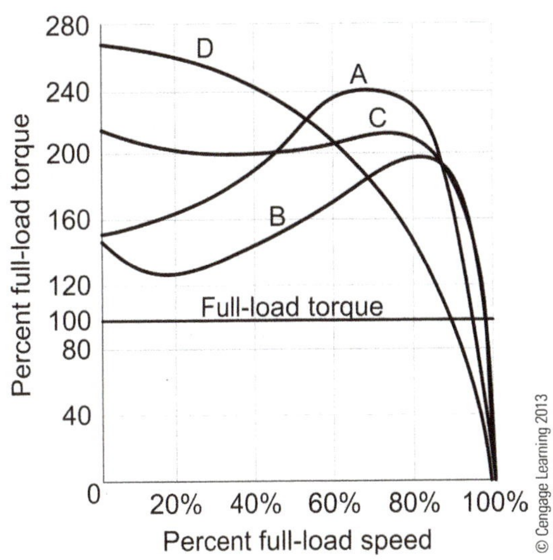


FIGURE 2-15 Design A, B, C, and D torque curves

more detailed explanation. The four torque measurements identified on the torque-speed graph in Figure 2-14 include:

- Locked rotor torque—the torque that the motor develops at zero speed (rotor not turning) when the rated voltage and frequency are applied.
- Pull-up torque—the minimum torque developed from locked rotor to breakdown torque; represents the maximum torque load that can be started by the motor.
- Breakdown torque—the maximum torque the motor can develop after having reached full operating speed at the rated voltage and frequency, before stalling (experiencing an abrupt drop in speed).
- Full-load torque—the torque produced at full-load speed that gives the rated horsepower output of the motor.

NEMA design types describe standard torque-speed design characteristics of induction motors. Figure 2-15 shows the torque-speed characteristics of the first four NEMA designs: A, B, C, and D.

The following list also identifies some of the most common NEMA motor design operating characteristics for the same four design letters.

A. NEMA design A motor characteristics
   - Starting current 6 to 10 times full-load current.
   - Starting torque is about 150% of full load torque.
   - Maximum torque is over 200% but less than 250% of full-load torque.
   - Slip is held to 5% or less.

B. Design B motor (often referred to as a general purpose motor) characteristics
   - Higher reactance than the Design A motor, obtained by means of deep, narrow rotor bars.
   - Starting current is held to about 5 times the full-load current.
   - Allows full-voltage starting.
   - Slip is held to 5% or less.

C. Design C motor characteristics
   - Has a "double layer" or double squirrel cage winding.
   - Starting current about 5 times full-load current.
   - Starting torque is high (200%–250%).
   - Slip is held to 5% or less.

D. Design D motor characteristics
   - Produces a very high starting torque-approximately 275% of full-load torque.
   - Has low starting current.
   - Low efficiency.
   - High slip (7–16%).

## CHAPTER SUMMARY

- The conduit box is where the leads of the stator coils are brought out to be connected to the power supply.
- The end bells seat on the motor housing, and receive the outer race of the rotor shaft bearing to hold the rotor centered in the stator core.
- The cooling fan is mounted directly on the rotor shaft, and circulates cooling air when the rotor is turning.
- The shaft travels the full length of the motor, and any part of the motor that rotates is mounted to it.
- The motor housing holds all of the stationary parts of the motor: the stator core, conduit box, end bells, and mounting base.
- The bearings are the mechanical connection between the stationary and moving parts of the motor that allow the rotor to turn inside the stator core.
- The stator core consists of many iron laminations stacked together to form a large cross-sectional area. The wire coils of the stator windings are insulated with only a thin layer of varnish, and are placed in slots that are recessed into the stator iron core to surround them completely with iron.
- The rotor core also consists of many iron laminations stacked together to form a large cross-sectional area. The design of the core leaves cavities where molten aluminum will be poured to form the rotor bars, which are the electrical conductors of the rotor (for electromagnetic induction). The shorting rings on each end of the rotor are poured at the same time to short all of the rotor bars together, creating the equivalent of one turn (for the volts-per-turn principle) in the rotor.
- The rotors of induction motors are commonly referred to as squirrel cage rotors, because the rotor bars assembly and shorting rings, without the iron laminations, resembles a pet squirrel's running wheel.
- The voltage induced into the rotor bars is very low, so the current intensity in the rotor must be very high in order for the volt-amperes of the stator coils and rotor bars to be equal.
- The main disadvantage of induction motors is that they cause a lagging power factor problem on the premises electrical distribution system.
- Power factor measures the ratio of how effectively the electrical energy taken from the electrical power source is used to perform work, compared to the portion of the electrical energy that is stored and returned to the power supply unused.
- Energy is the potential to perform work. Work is performed when something is moved or heated. Torque is the measure that determines what mechanical work may be performed by an electric motor. Power determines how fast the work may be accomplished.
- The most important fact to remember about purely resistive electrical loads is that they do not cause a power factor problem, because there is no phase shift between voltage and current.
- Inductors, which include induction motors, cause a phase shift between voltage and current that causes current to lag voltage, and it is this phase shift that causes the power factor.
- Power factor reduces the productive current carrying capacity of electrical distribution systems, because it causes current to move back and forth between the power supply and the load unnecessarily, without producing work.
- The current draw of an induction motor does not drop off significantly from the initial locked-rotor current until 80–85% of full-load rotor speed.

- Locked rotor torque is the torque that the motor develops at zero speed when the rated voltage and frequency are applied.
- Pull-up torque represents the maximum torque load that can be started by the motor.
- Breakdown torque is the maximum torque the motor can develop, after having reached full operating speed at the rated voltage and frequency, before stalling.
- Full-load torque is the torque produced at full-load speed.
- NEMA defines the operating characteristics for general purpose induction motors using a system of design letter designations.
- The NEMA design B motor provides a good balance between locked rotor torque, locked rotor current, and full-speed torque that make it suitable for a wide range of electric motor applications.

## REVIEW QUESTIONS

1. How are the motor connection leads in the motor conduit box labeled?
2. What is the purpose of the motor end bells?
3. Why is the drive end bearing larger than the opposite drive end bearing?
4. What is the reason for recessing the stator wire coils into the iron stator core?
5. What are two reasons the stator coils are wound with magnet wire?
6. How are the rotor bars formed inside the iron rotor core?
7. What is the purpose of the shorting rings on each end of the rotor?
8. What is a name commonly given to the rotors of induction motors?
9. What determines whether a motor is classified as a conduction or induction motor?
10. How is electrical energy delivered to the rotating member of induction motors?
11. When all of the rotor bars are shorted together, what does the turns equivalent become?
12. Would the volts-per-turn principle determine that the voltage induced in the rotor would be very high, or very low?
13. If the volt-amperes in the rotor are equal to the volt-amperes in the stator, does that make the rotor current very high, or very low?
14. What is it about inductive loads that causes power factor?
15. What is a simple definition of energy?
16. What is a simple definition of work?
17. What is torque?
18. What is a simple definition of power?
19. What is necessary between the polarities of voltage and current for power to be dissipated?
20. When a power triangle is made for an inductive load, how is the power factor calculated from the triangle?
21. When the voltage and current are of different polarities, what is that electrical energy called?
22. The locked rotor starting current of an induction motor does not start to drop off significantly until the rotor has reached what percentage of full-load speed?
23. What is the most common NEMA design letter induction motor, and why?
24. What are four torque measurements identified on a torque-speed curve?

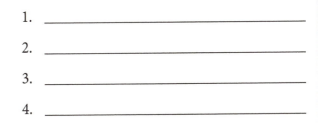

    1. ______________________________
    2. ______________________________
    3. ______________________________
    4. ______________________________

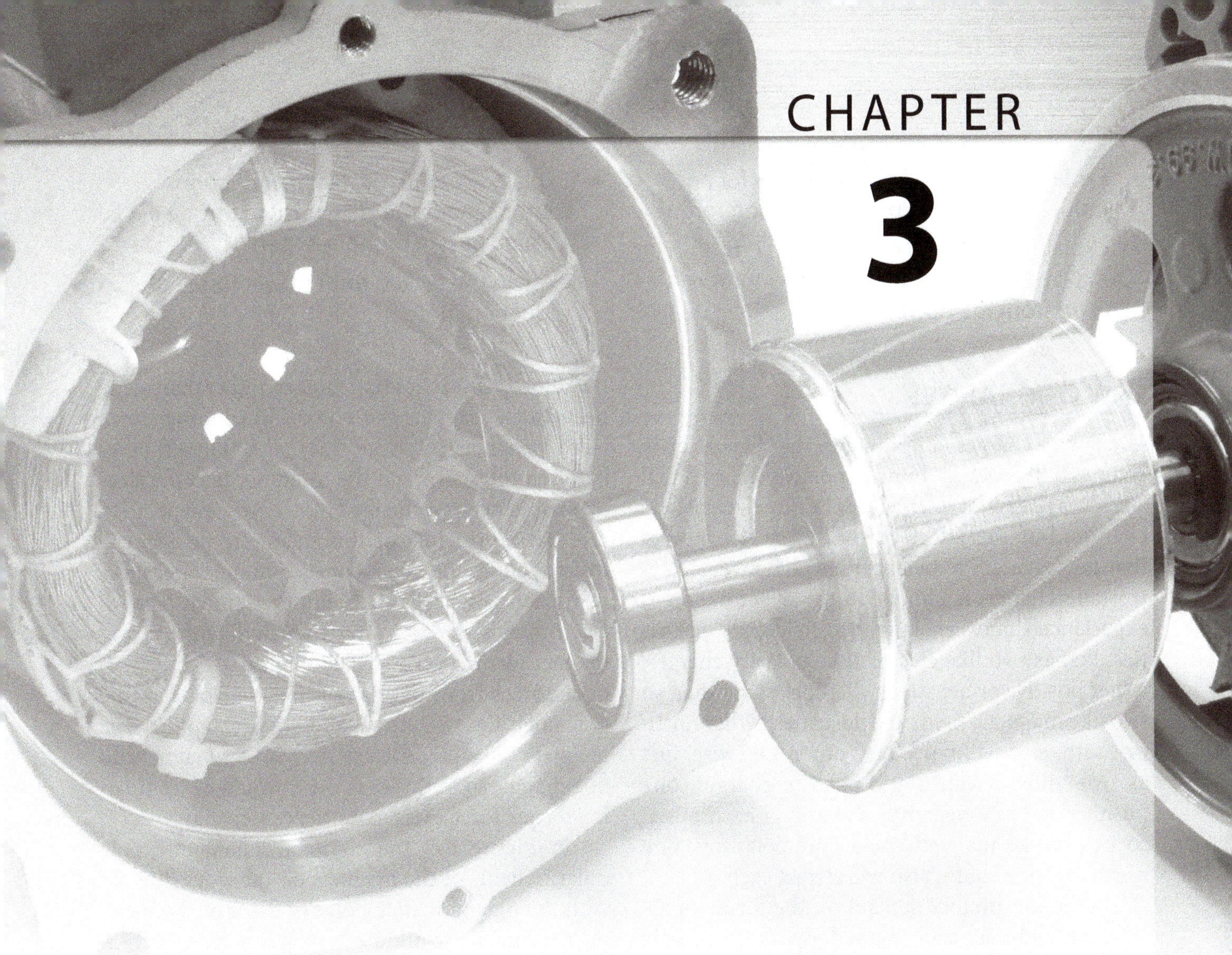

CHAPTER

# 3

# Three-Phase Motor Theory

**PURPOSE**

To familiarize the learner with the electromagnetic theory of the rotating magnetic field in the motor stator, and operation of three-phase electric induction motors.

**OBJECTIVES**

After studying this chapter on three-phase motor theory, the learner will be able to:

- Explain the law of charges
- Relate the forces of electrostatic charges to the forces of magnetic fields
- Explain two advantages of using electromagnets in induction motors
- Explain the physical relationships of three-phase induction motor stator coils
- Explain how the stator coils of a three-phase induction motor, with three-phase AC power, creates a rotating magnetic field in the motor stator
- Explain synchronous speed and how to calculate it
- Explain rotor slip and how to calculate it expressed as a percentage

- Explain why rotor slip is necessary for an induction motor to produce turning torque
- Explain why locked rotor current is the maximum current an induction motor can draw from the power supply
- Explain why the rotor of an induction motor cannot produce turning torque if it is turning at synchronous speed
- Explain the locked rotor power losses of induction motors

## LAW OF CHARGES

The law of charges simply states that like polarity charges will repel, and unlike polarity charges will attract. The law of charges applies to electrostatic charges, but the affect of the physical property that causes forces to be experienced between electrostatic charges is similar to the effect of forces experienced between magnetic field pole polarities. Just as two like electrostatic polarity charges, either positive or negative, will repel each other, so two like magnetic field pole polarities, either north or south, also will repel each other. Likewise, just as two unlike electrostatic polarity charges, one positive and one negative, will attract each other, so two unlike magnetic field pole polarities, one north and one south, also will attract each other.

Induction motors depend on the forces of attraction and repulsion caused by the magnetic field pole polarities interacting between the stator and the rotor to create a turning torque force on the rotor. Inside induction motors, rather than using permanent magnets the stator coil windings form electromagnets to create the magnetic field poles between the stator and the rotor, because the electromagnets can be changed in strength and polarity. The basic principle of this magnetic pole polarity attraction and repulsion is demonstrated here using permanent magnets, but the same is true for electromagnets. When free to turn, like the center magnet turning on a shaft in Figure 3-1, the magnetic forces from permanent magnets of opposite pole polarities will attract and hold each other. The magnetic pole polarities of permanent magnets do not change, so in this case there would be no turning torque force exerted on the center shaft.

In Figure 3-2 the stationary magnetic pole polarity fields of the magnets on the sides interact with the magnetic pole polarity fields of the rotating magnet in the middle to produce the attraction forces associated with opposite magnetic poles. With the rotating magnet in the middle moved out

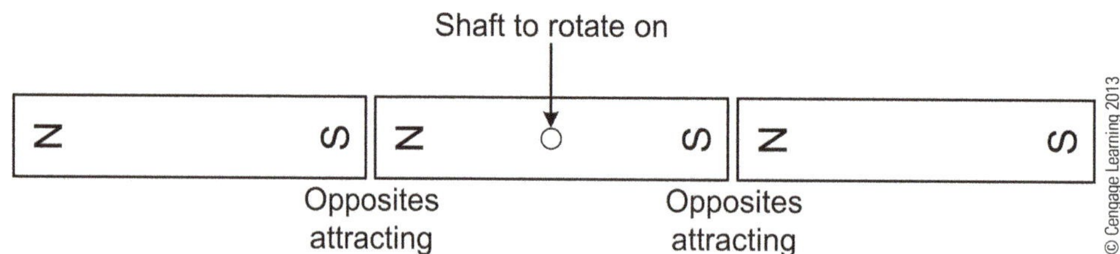


**FIGURE 3-1** Opposite magnetic poles attracting with no turning torque

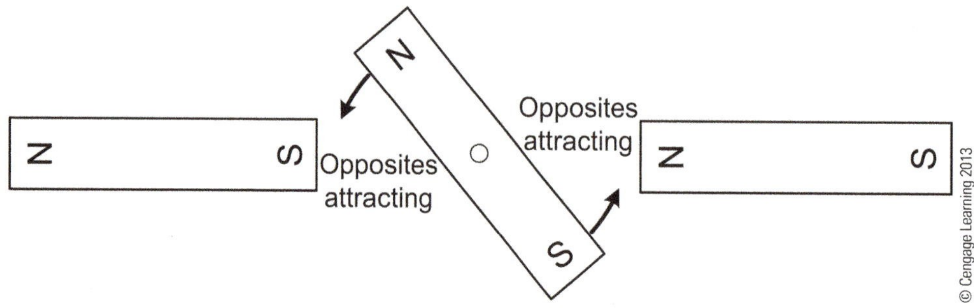


**FIGURE 3-2** Opposite magnetic poles attracting with turning torque

of alignment, the magnetic field forces will attract each other on each end to exert a rotating torque to turn the middle magnet until the magnetic field polarities are all lined up by attracting opposite polarity poles.

In Figure 3-3 the stationary magnetic pole polarity fields of the magnets on the sides interact with the magnetic pole polarity fields of the rotating magnet in the middle to produce the repelling forces associated with like magnetic poles. With the rotating magnet in the middle held in alignment with like magnetic pole polarities facing one another, the like magnetic pole polarity fields exert forces that produce a rotating torque to turn the middle magnet in either direction of rotation, and force them apart.

Figure 3-4 demonstrates further what happens in the stator of an induction motor, with both the magnetic forces of attraction and repulsion working together to cause a turning torque. Notice that when the rotating magnet in the middle is positioned between two stationary magnetic poles, each magnetic pole on the ends of the rotating magnet will experience magnetic forces of both attraction and repulsion to create a turning force. The stationary pole of opposite magnetic polarity will attract the rotating magnet, but the stationary pole of the same magnetic polarity will repel it. Inside an electric motor, when a voltage is induced into the rotor conductor bars and a resulting current starts to circulate in the rotor causing a magnetic field to build, that magnetic field will be both attracted and repelled on each side of the rotor bar at the same time, causing the torque that turns the rotor, similar to this example in Figure 3-4. The only difference between the permanent magnets in this example and the electromagnets of an induction motor is that the electromagnets in the induction motor are constantly changing magnetic pole polarity, which will cause the rotating member to continue turning.

## ELECTROMAGNETS IN INDUCTION MOTORS

The advantage of using electromagnets to create a magnetic field are twofold: first, the strength of the magnetic field is controlled by the intensity of the current flow through the wire coil; second, the magnetic pole polarity is reversed when the direction of current flow through the wire coil is reversed. Refer to the alternating current sine wave waveform in Figure 2-9 in Chapter 2. Notice that for one-half of the cycle waveform the current intensity goes from zero amperes to peak amperes and back to zero amperes in one polarity, which

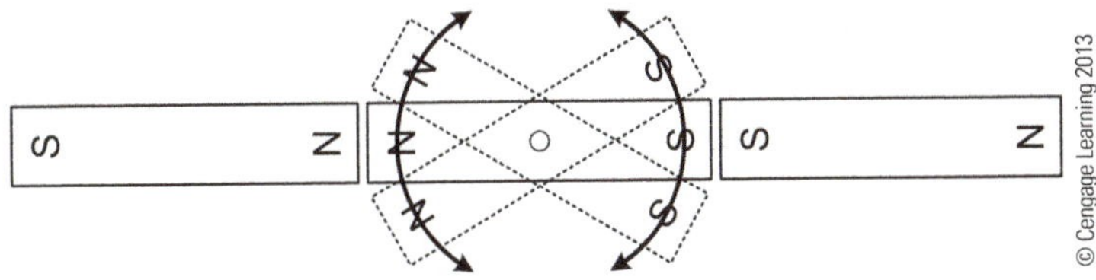


**FIGURE 3-3** Like magnetic poles repelling with turning torque either direction

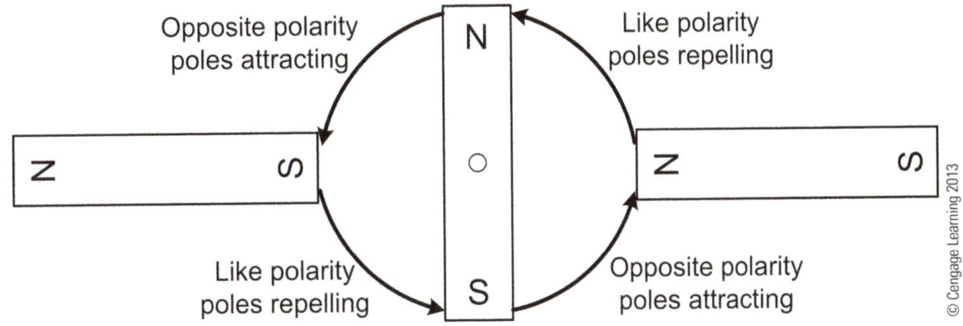


**FIGURE 3-4** Like and opposite magnetic poles causing both repulsion and attraction for turning torque

would cause the magnetic pole field of the inductor coil to do the same thing, creating a magnetic pole of one polarity. During the second half of the same sine wave cycle waveform, the sequence is repeated with the opposite current polarity, which would, again, cause the magnetic pole field of the inductor coil to do the same thing with the opposite magnetic pole polarity. This concept is important because it, with the stator inductor coil winding placement, is what produces the rotating magnetic field in the stator.

## Stator Poles

Refer to the induction motor stator core picture in Figure 2-2 in Chapter 2. Induction motor stators have coils of wire placed around the circumference of the stator core, which form electromagnets when energized and create magnetic pole fields. In most cases, you can look inside the motor and see the stator coils. They are distinct bundles of wire, evenly spaced around the stator core. Each wire coil has a corresponding coil on the opposite side of the stator core, which is connected in series with the first wire coil so it experiences the same current intensity and polarity, or direction of current flow, but is wound to produce the opposite magnetic pole field as its opposing stator coil. The result of having these opposing magnetic pole fields from corresponding wire coils on opposite sides of the stator is that the magnetic lines of flux that travel between them will travel through the rotor to complete the magnetic path. As the magnetic lines of flux travel through the rotor, they will cut through the aluminum rotor bars and induce the voltage that will cause a current flow, because all of the rotor bars are shorted together to provide a current path around the rotor.

## A Rotating Magnetic Field from a Stationary Stator Core

Producing a rotating, or moving, magnetic field from a stationary object can be a difficult concept to understand. The trick to understanding this concept is that the rotating magnetic field is not a physical or mechanical rotation, it is an electromagnetic rotation controlled by the direction and intensity of current flow through electromagnets in the motor stator. Similar to the stationary lights on a marquee sign that are turned on and off in a sequence that causes the light pattern to move across the face, the stationary electromagnets around the circumference of the motor stator are also turned on and off in a sequence that causes the magnetic pole polarity to rotate around the stator.

## Stator Core Examples to Explain the Current-Created Magnetic Field Concepts

**Interpretation of the Illustrations.** The illustration in Figure 3-5 demonstrates the concepts that will be used in this explanation of how alternating electrical current, in conjunction with electromagnets, creates magnetic fields that change both intensity and polarity and thus produce rotation in the stationary stator core. Suppose for the sake of this illustration, though not necessarily always true, that when a voltage sine wave waveform polarity is moving in the positive direction, it will cause current to flow through the coils in a direction that will create a (top to bottom) north to south pole polarity of the stator poles. And likewise, suppose that a voltage sine wave waveform moving in the negative direction will cause current to flow in a direction that will create a (top to bottom) south to north pole polarity.

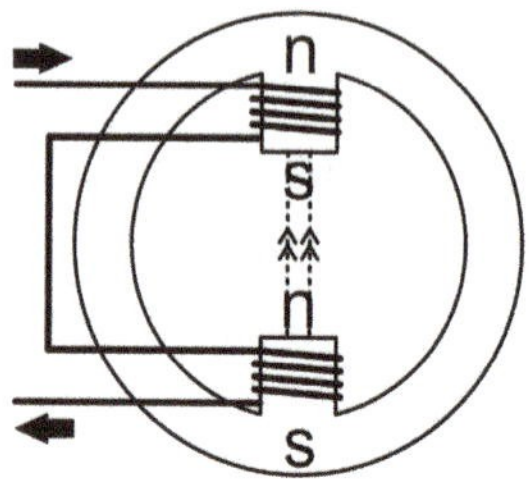

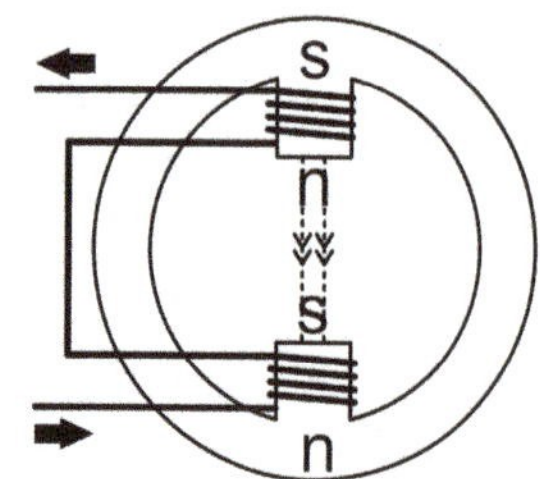

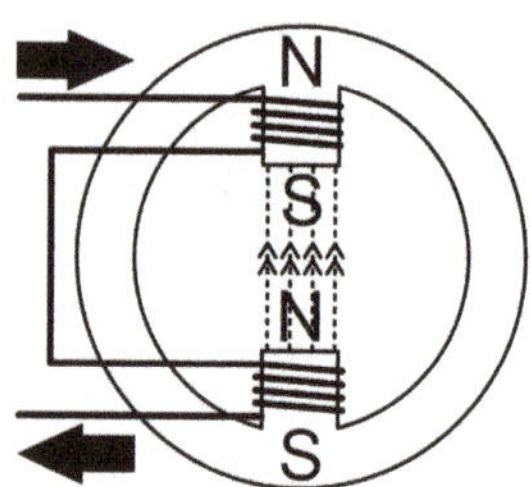

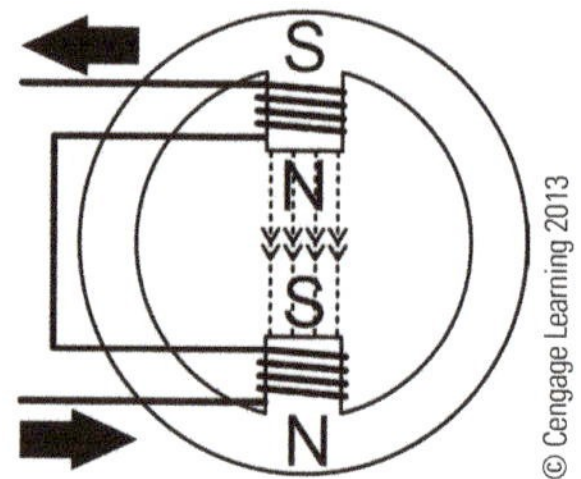




**FIGURE 3-5** Stator core examples to explain the current-created magnetic field concepts

In the illustrations, a small arrow signifies the direction of a small electrical current flow, and the corresponding weak magnetic field pole polarity associated with that small current will be represented with a lowercase letter of the magnetic field pole polarity (n = north, and s = south). A large arrow signifies the direction of a large electrical current flow, and the corresponding strong magnetic field pole polarity associate with that large current will be represented with an uppercase letter (N = north, and S = south). The wire coils on the stator poles are omitted from the illustrations to reduce clutter.

**360 Electrical Degree Explanation.** Figure 3-6 relates an example of the current intensity and polarity relationship to the magnetic poles created in the stator of an induction motor, in 45° increments along the 360° of sine wave waveform with a two-pole motor stator. At 0° there is no voltage on the coils, so no current would flow through the coils, which means there is no magnetic flux, and no magnetic field or pole polarity. At 45° a small positive voltage on the coils would cause a small positive current, which would create a small amount of magnetic flux, and a weak magnetic north polarity pole at each of the coils. At 90° the maximum positive voltage on the coils would cause the maximum positive current, which would create the maximum magnetic flux and the strongest magnetic north polarity pole at each of the coils. At 135° the voltage level has decreased to a lower level and the result will be less electrical current, which will create less magnetic flux and a weaker magnetic north polarity pole at each of the coils. At 180° everything is zero again. At 225° the voltage is starting to increase again, but in the opposite electric polarity. A small negative polarity voltage on the coils would cause a small negative current, which would create a small amount of magnetic flux and a weak magnetic south polarity pole at each of the coils. At 270° the maximum negative voltage on the coils would cause the maximum negative current, which would create the maximum magnetic flux and the strongest magnetic south polarity pole at each of the coils. At 315° the voltage level has decreased to a lower level and the result will be less electrical current, which will create less magnetic flux and a weaker magnetic south polarity pole at each of the coils. At 360° the cycle will start over.

## Three-Phase Stator Rotating Magnetic Field

The previous example using only two stator poles alternating back and forth between two magnetic polarities does not constitute a rotating magnetic field. If three sets of these poles were mounted

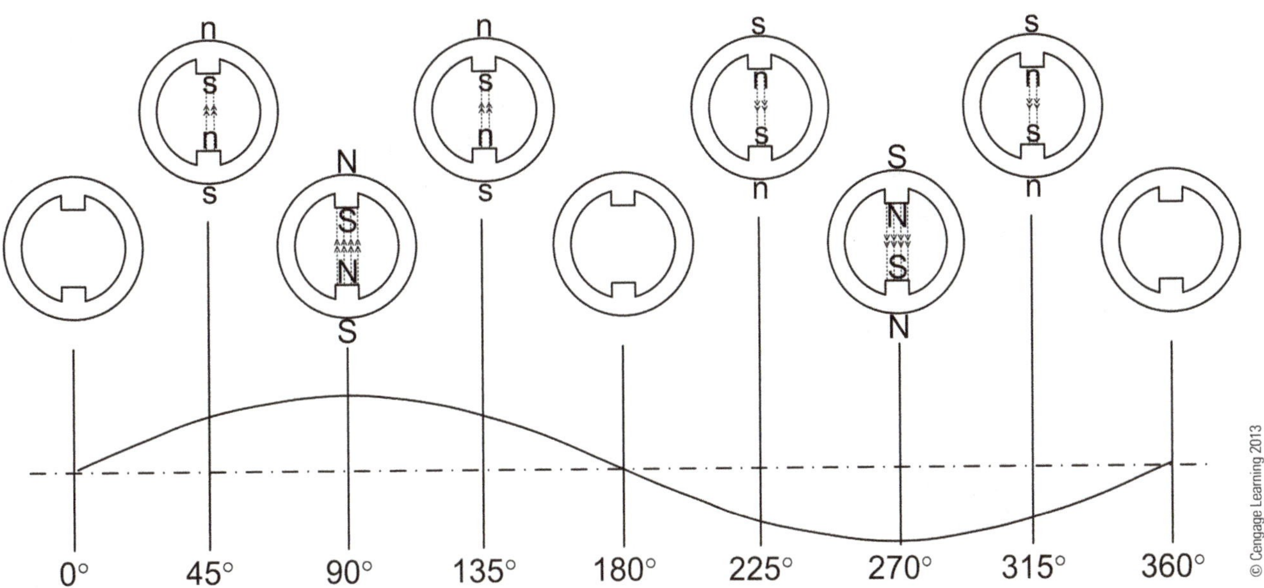


FIGURE 3-6 Rotating magnetic field of two-pole stator with sine wave reference

around a motor stator, as is the case in three-phase motors, a rotation would quickly become evident as each set of poles created their magnetic fields 120 degrees out of phase with each other. Each individual phase of a three-phase stator, consisting of two poles, will have the same effect in the three-phase motor stator as the two poles did in the previous example. The three-phase two-pole motor stator shown in Figure 3-7 shows that the two poles associated with each phase of a three-phase motor stator are 180° apart. Notice also that the phase coils with the same subnotations ($A_1$, $B_1$, and $C_1$) are 120° out of phase with each other, which is the actual separation for three-phase motors. If the subnotations are not observed, it would be easy to mistakenly think that the phase coils of three-phase motors are only 60° out of phase; the difference between phase coils $A_1$ and $B_2$ or $C_2$, for example.

**Explanation of the Subnotation.** The subnotations $C_1$ and $C_2$ are provided to help keep the magnetic polarities of the three-phase stator poles straight. If a positive voltage is applied to phase coil $A_1$ and the resulting current flow creates a magnetic north polarity pole in the stator, then applying a positive voltage to phase coils $B_1$ and $C_1$ also will create a magnetic north polarity pole in the stator for each of those coils. When the same polarity voltage is applied to each phase with the same subnotation, each will create the same magnetic polarity poles in the stator core.

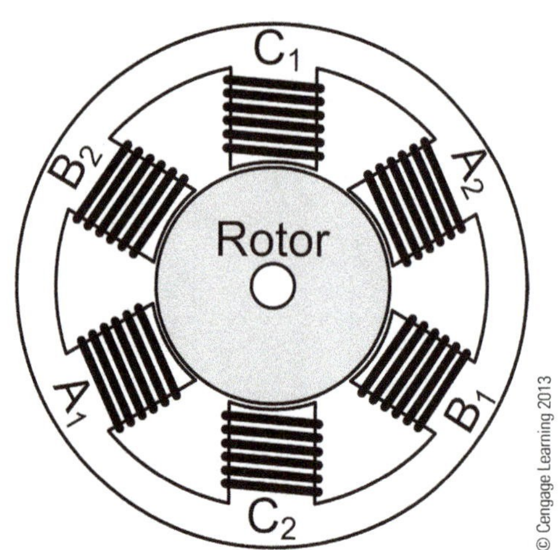


**FIGURE 3-7** Three-phase two-pole motor stator

**Three-Phase Alternating Current (AC) Power Supply.** Equally important to the concept that the stator coils in a three-phase induction motor are 120° out of phase with each other is that the three voltage sine waves of three-phase AC power are also 120° out of phase with each other. Figure 3-8 shows one complete cycle of 360° for the three voltage sine wave waveforms of a three-phase AC power supply, one for each of the three phases. Notice that the waveform for phase A peaks at 0°, phase B at 120°, and phase C at 240°, demonstrating that the waveforms are 120° out of phase with each other. It is this condition of both the physical location of the coils in the stator and the electrical phase relationship of the AC power supply energizing the coils out of phase by the same amount that produces the rotating magnetic field.

The following sequence of drawings demonstrates how a three-phase alternating voltage power creates a rotating magnetic field in the stator of induction motors. Figure 3-9 identifies each part of the drawings that will be used in the sequence: the stator, stator slots for wire coil windings, wire coil windings, rotor, rotor bars, magnetic flux lines, and air gap between the stator and rotor.

Figure 3-10 shows the magnetic pole fields in the stator of a 2-pole, three-phase motor at 0°, which is the same as 360° on the sine wave voltage waveforms. At this point on the voltage sine wave graphic, phase A is peak positive, creating the strongest north pole magnetic field on coil $A_1$, as depicted by the capital (N) and the four dotted lines representing magnetic flux lines. Phases B and C on the voltage sine wave graphic are less than peak negative voltage, creating only weak south pole magnetic fields on both coils $B_1$ and $C_1$, as depicted by the small (n) and only two dotted lines representing magnetic flux lines.

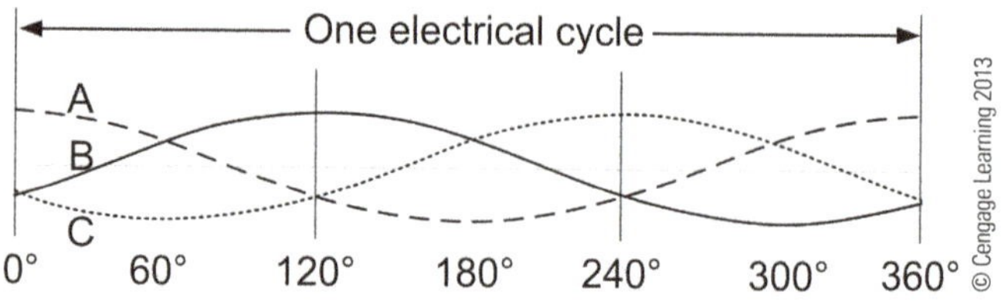


**FIGURE 3-8** Three-phase sine wave waveform diagram

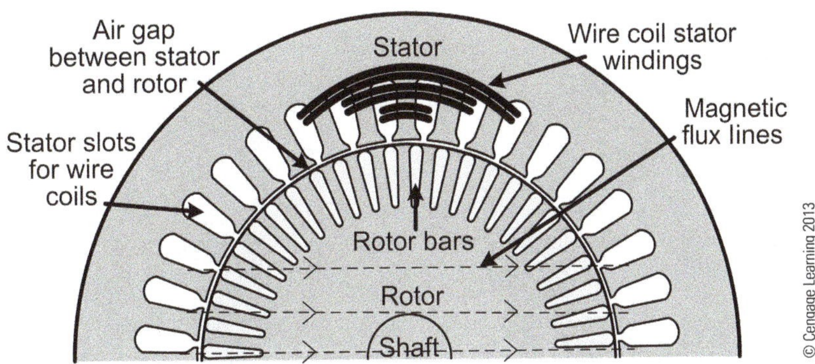


**FIGURE 3-9** One-half stator and rotor drawing with labels

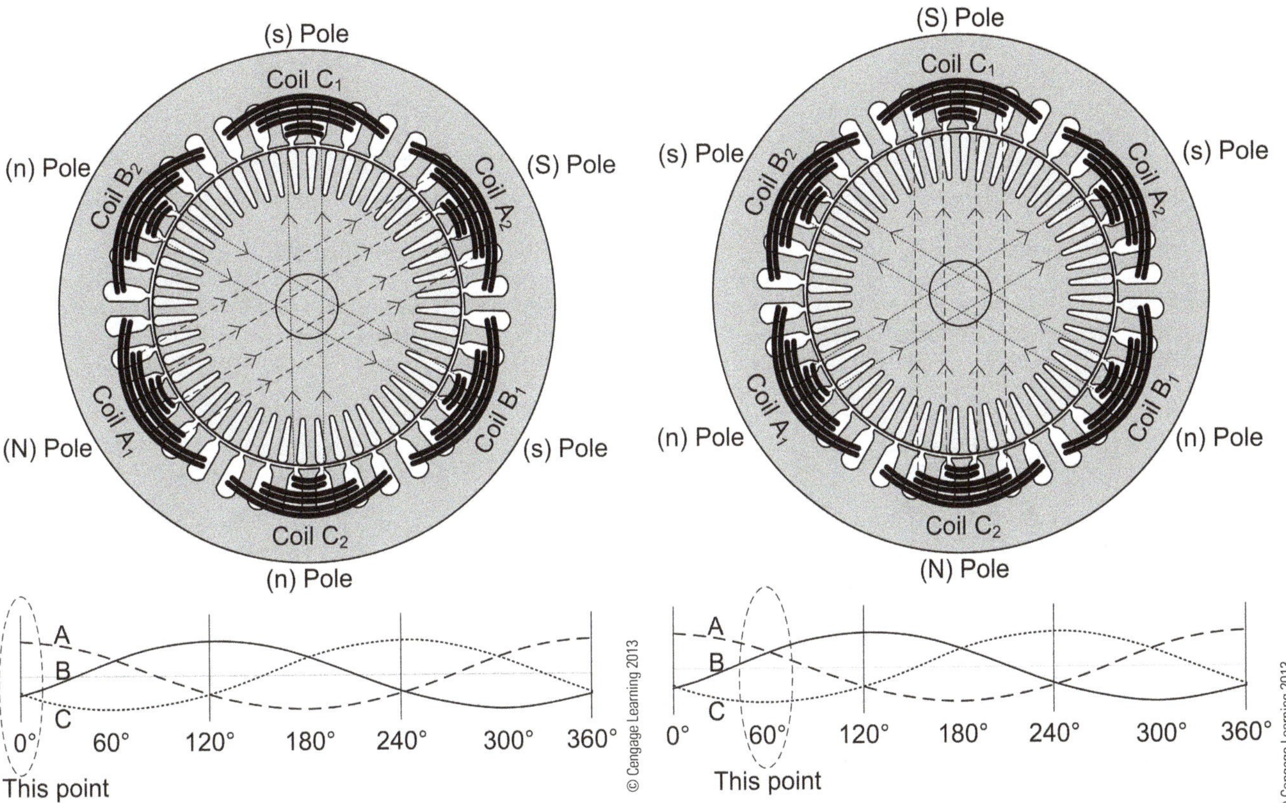


**FIGURE 3-10** Rotating stator field with sine wave reference at 0 and 360 degrees

**FIGURE 3-11** Rotating stator field with sine wave reference at 60 degrees

Figure 3-11 shows how the magnetic field pole polarities would change as the three-phase voltage sine wave waveforms reach 60°. At this point phase C is peak negative, creating the strongest south pole magnetic field on coil $C_1$, which causes the strongest north pole magnetic field on coil $C_2$, because the coil pairs are wound to create opposing pole polarity fields. Notice that the strongest north pole magnetic field has now rotated counterclockwise 60°.

Figure 3-12 shows how the magnetic field pole polarities would change as the three-phase voltage sine wave waveforms reach 120°. At this point on the voltage sine wave graphic, phase B is peak positive, creating the strongest north pole magnetic field on coil $B_1$. Phases A and C on the voltage sine wave graphic are less than peak negative voltage, creating only weak south pole magnetic fields on both coils $A_1$ and $C_1$.

Follow the remaining graphics of Figures 3-13, 3-14, and 3-15 to follow the rotation of the magnetic field poles around the stator as the three-phase voltage sine wave waveforms change in magnitude

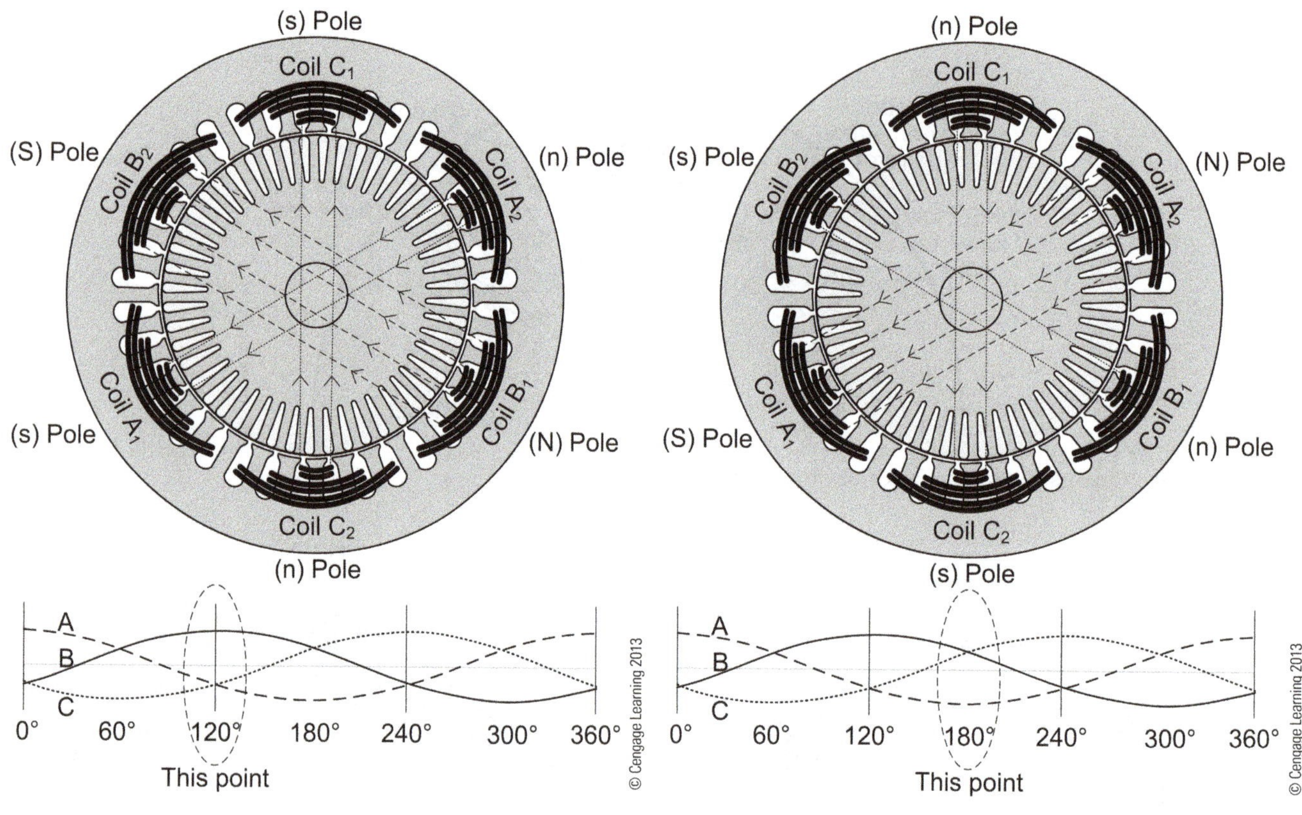


**FIGURE 3-12** Rotating stator field with sine wave reference at 120 degrees

**FIGURE 3-13** Rotating stator field with sine wave reference at 180 degrees

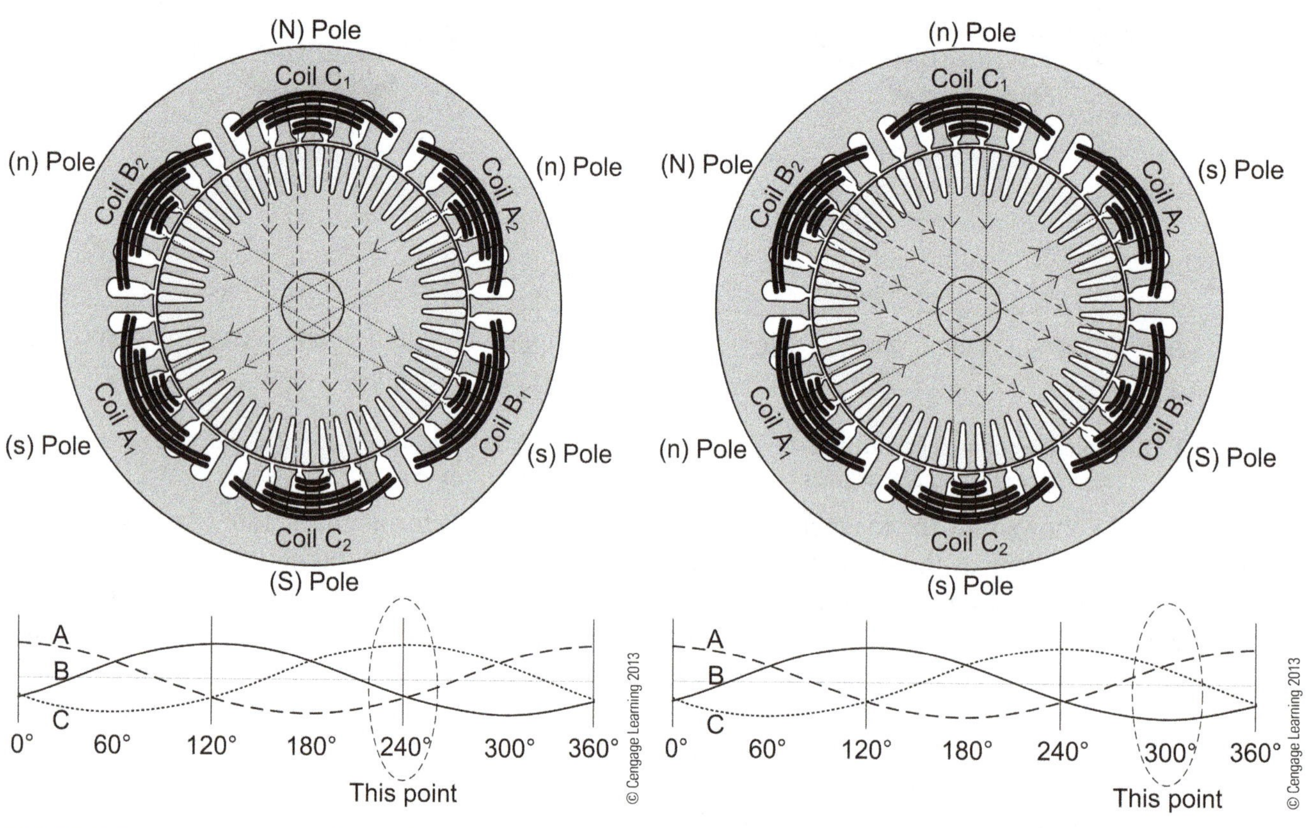


**FIGURE 3-14** Rotating stator field with sine wave reference at 240 degrees

**FIGURE 3-15** Rotating stator field with sine wave reference at 300 degrees

and polarity with time. The magnetic field poles in the stator are only documented in these graphics every 60°, but the magnetic fields actually rotate around the stator as smoothly and continuously as the sine wave waveforms themselves.

## SYNCHRONOUS SPEED

The synchronous speed of an electric induction motor is the speed at which the magnetic field rotates around the stator. In the example of Figures 3-10 through 3-15, the magnetic field rotated around the stator one full revolution for one full cycle of the sine wave. The number of poles, combined with the alternating current line frequency (HZ), determines the synchronous speed revolutions per minute (RPM) of induction motors. Synchronous speed is defined as the speed of the rotating magnetic field in the motor stator, but does not represent the actual motor rotor RPM under load. The mathematical formula for making a synchronous speed calculation is the number of cycles (HZ), times 60 (for seconds in a minute), divided by the number of pairs of poles.

There are two caveats to keep in mind when using this formula. The standard method of counting poles of an electric motor for identification and labeling purposes is to count all magnetic poles in the stator. A two-pole motor actually has two magnetic poles in the stator, which are 180 mechanical and electrical degrees out of phase with each other. Neither pole can work independently from the other. The magnetic poles on the stator of an induction motor work (and are wired) in pairs to always be the opposite magnetic polarity of each other. The formula for synchronous speed, however, requires pairs of poles to be counted, rather than poles. A two-pole motor, for the synchronous speed calculation, would actually be counted as one pair of poles. A four-pole motor would be counted as two pairs of poles, etc.

The second caveat to keep in mind is that when counting the poles in a three-phase motor, only the coils of one phase are counted. Look at the left-most drawing in Figure 3-16. Phase A has two poles. Hence for the synchronous speed calculation this is regarded as a two-pole motor, which is counted as one pair of poles for determining synchronous speed. The drawing in the middle has four phase A poles. Hence this is regarded as a four-pole motor, which is two pairs of poles for the synchronous speed calculation. How many poles is the third drawing?

**Varying Motor Applications Based on Different Synchronous Speed Calculations.** A two-pole, 60-Hz motor operates at 3,600 RPM (3,600 divided by one pair of poles) synchronous speed, and approximately 3,450 RPM at full-load. Two-pole motors often are found in pump applications, such as water recirculating equipment. One thing to keep in mind is that the higher the RPM, the "noisier" a motor may sound.

A four-pole, 60-Hz motor operates at 1,800 RPM (3,600 divided by two pairs of poles) synchronous

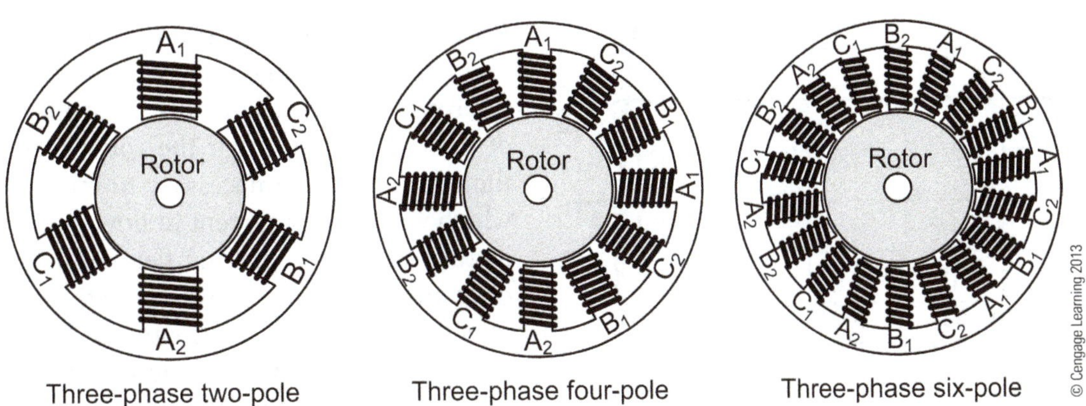


**FIGURE 3-16** Two-, four-, and six-pole stators

speed, and approximately 1,750 RPM at full-load. Four-pole motors are commonly found in belt-driven applications such as blowers, fans, and air handling equipment.

Six-pole, 60-Hz motors run at 1,200 RPM (3,600 divided by three pairs of poles) synchronous speed, and approximately 1,050 RPM at full-load. They are often used for direct-drive applications, such as furnace blowers, room air conditioners, and other equipment that requires the motors to be in close proximity to people, because the slower motor speed makes for quieter operation.

## Rotor Slip

Squirrel cage induction motors are sometimes referred to as constant speed motors, because their RPM speed changes only a small amount through the entire torque load range of the motor (see Figure 3-17). Regardless whether an AC motor is mechanically loaded or unloaded, the rotating magnetic field in the stator will rotate at synchronous speed, because it is determined by the power supply line frequency and the number of motor stator poles. What does change with mechanical load on the motor is the rotating speed of the rotor. As the torque requirement of a mechanical load increases, the motor rotor is restrained from turning at the same speed as the rotating magnetic field. Even if no mechanical load is connected to the motor, the rotor will not turn at synchronous speed, because the motor itself has friction losses in the bearings that will mechanically load the motor to a small degree.

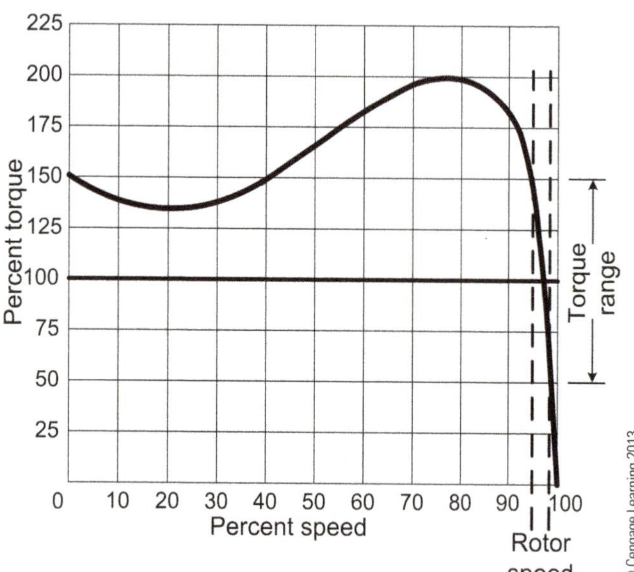


**FIGURE 3-17** Rotor speed-torque curve full-load range

The RPM speed difference between the stator rotating magnetic field, synchronous speed, and actual rotating speed of the rotor is called slip. As an example, the synchronous speed of a four-pole, 60-Hz AC motor is 1,800 RPM but the actual rotor speed at full-load is approximately 1,750 RPM. Rotor slip is calculated using the following formula:

$$\text{Percent slip} = \frac{(\text{synchronous speed} - \text{rotor speed})}{\text{synchronous speed} \times 100}$$

$$= \frac{(1{,}800\text{ RPM} - 1{,}750\text{ RPM})}{1{,}800\text{ RPM}} \times 100$$

$$= \frac{50}{1{,}800\text{ RPM}} \times 100$$

$$= 0.0278 \times 100$$

$$= 02.78\text{ percent}$$

The slip varies with the mechanical load on the motor, but as the speed-torque curve for a NEMA design B induction motor in Figure 3-17 shows, a small change in rotor speed (slip) can produce a large torque operating range for the motor. It is important to understand that the speed listed on the nameplate of an electric motor is the manufacturer's declaration of the motor's RPMs under full-load, not synchronous speed.

**Rotor Slip Is Necessary To Produce Torque.** The rotor must always turn sufficiently slower than the rotating magnetic field so that magnetic lines of flux will cut through the rotor bars and have the relative motion necessary to induce a voltage, which will cause a current to flow in the rotor and in turn create a magnetic field sufficient to produce enough torque to overcome the motor losses plus the motor load. As the motor load increases, the rotor will continue to slow down, inducing a greater voltage in the rotor conductor bars, causing a greater current to flow, producing a stronger magnetic field in the rotor, and increasing the produced torque as the larger magnetic fields interact to drive the greater load.

## Locked Rotor Current

When first energized with the full supply voltage, the motor will draw the highest current the motor can draw, which is called the locked rotor current (LRC). The motor draws this high current, because the rotor is not yet turning, and the stator's rotating magnetic field is instantly turning at synchronous speed, cutting through the rotor bars. The induced voltage in the rotor is proportional to the number of lines of magnetic flux that are cut, so the induced voltage will be at its maximum, because the maximum lines of flux are cut when the rotor is standing still. With the maximum voltage induced into the rotor bars, and the rotor current limited only by the resistance of the large poured aluminum rotor bars, the rotor current will be very high. The high current will create a strong magnetic field in the rotor that will interact with the rotating magnetic field of the stator (attracting and repelling), which will cause the rotor to produce a turning torque.

To help demonstrate the lines of magnetic flux rotating through the rotor conductor bars when the rotor is standing still, Figure 3-18 shows the rotor with only one conductor bar in the rotor. The magnetic lines of flux between the stator coils are drawn as dotted lines. At the instant the motor is energized, the magnetic field in the stator will start to rotate around the stator at synchronous speed. If the rotor is not yet turning, all of the lines of flux of the rotating magnetic field will rotate around, and cut through the rotor bar.

Once the rotor starts to turn, the conductor bars in the rotor will cut through fewer lines of magnetic flux of the rotating magnetic field in the stator, because the conductor bars will be turning closer to the speed of the rotating magnetic field. The closer rotor RPM is in relation to the synchronous speed of the rotating stator magnetic field, the less relative motion exists between the synchronous rotating magnetic field of the stator and the rotor conductor bars. Less relative motion means fewer magnetic lines of flux are cut, which means less induced voltage and lower rotor and stator winding currents. If the rotor does reach synchronous speed, the voltage induced into the rotor, and the rotor current, will drop to zero.

## Rotor at Synchronous Speed

It is not possible for the rotor of an induction motor to turn at synchronous speed and still produce turning torque. Look at the drawing in Figure 3-19, and notice that when the rotor turns at the same speed as the rotating magnetic field in the stator, no lines of magnetic flux are cut, and no voltage is induced into the rotor bars. If no voltage is induced into the rotor bars, there will be no resulting current flow through them, and therefore no magnetic field around them to interact with the rotating magnetic field of the stator. Without a magnetic field, the rotor would not experience any torque, and the rotor would start to slow down because of the bearing friction. To help demonstrate the rotor turning at the same speed as the rotating magnetic lines of flux in the stator (no magnetic lines of flux being cut), Figure 3-19 again shows only one rotor bar in the rotor.

As the rotor starts to slow down from turning at synchronous speed, the faster rotating magnetic field in the stator will again start to rotate through the slower turning rotor bars. As the magnetic lines

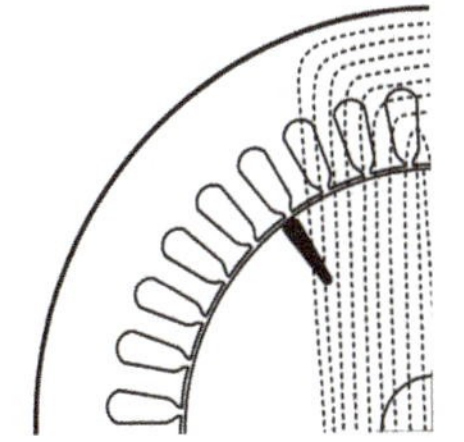
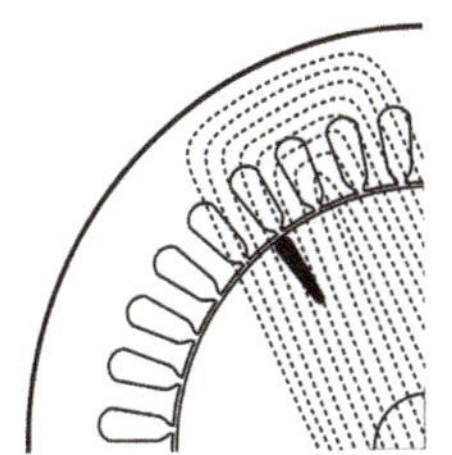
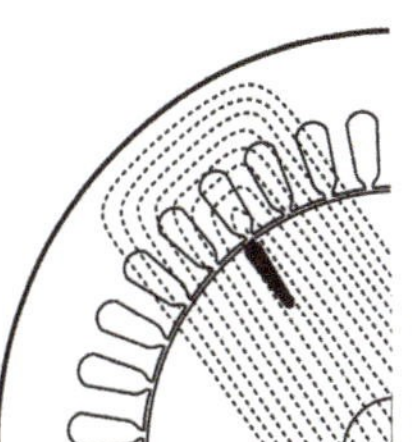
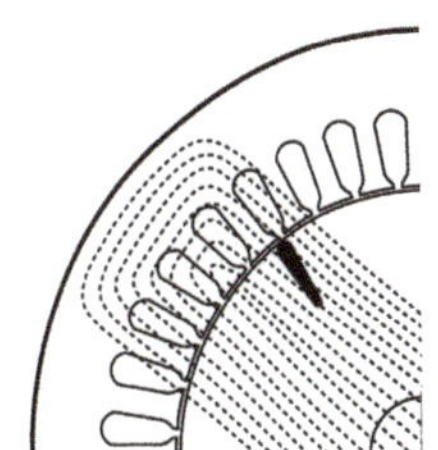
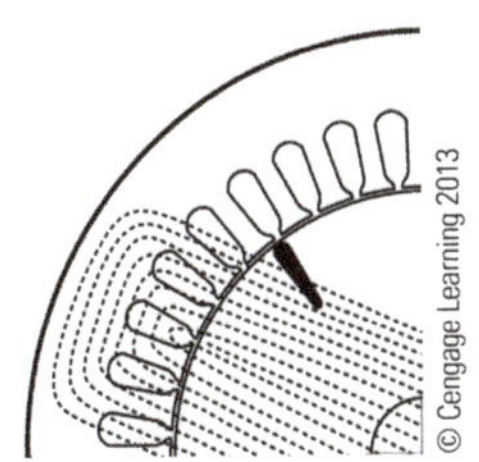


**FIGURE 3-18** Stator rotating magnetic field with locked rotor

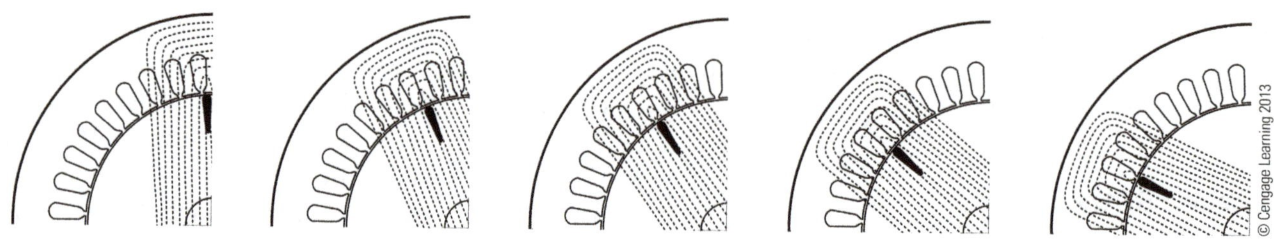


FIGURE 3-19 Stator rotating magnetic field with rotor turning synchronous speed

of flux start to cut through the rotor bars again, a voltage will be induced, which will cause a current to flow through the rotor bars, creating a magnetic field that will interact with the rotating magnetic field of the stator, and torque will begin to be produced again.

## Locked Rotor Power Losses

The power loss dissipated by a motor at the instant the motor is first powered (before the rotor starts to turn), is almost fifty times the power loss dissipated at full speed. A couple of simple calculations will help make the increased starting losses more tangible. Start with a motor that has a 2-ampere current rating, and 6 ohms of resistance. Using the power formula $P = I^2R$: $P = 2$ amperes$^2$ times 6 Ω; $P = 2^2$ times 6; $P = 24$ watts. NEMA design B induction motors normally have locked rotor currents of six to eight times their full load current; use seven times for the calculation. $P = 14$ amperes$^2$times 6 Ω; $P = 14^2$ times 6; $P =$ 1,176 watts; 49 times more power loss at locked rotor thanrunning at full load. The locked rotor power losses drop off significantly as soon as the rotor starts to turn, but coil windings of the motor still have to dissipate the power loss from the excess current as heat. This is the reason that induction motors are rated for the number of times they can be started per hour, because the coil windings need time to cool by dissipating their heat into the mass of the motor.

## CHAPTER SUMMARY

- The law of charges states that like polarity charges will repel, and unlike polarity charges will attract.
- Induction motors depend on the forces of attraction and repulsion caused by the magnetic field pole polarities interacting with each other, to create a turning torque force.
- The advantage of using electromagnets to create a magnetic field are twofold: first, the strength of the magnetic field is controlled by the intensity of the current flow through the wire coil; and second, the magnetic pole polarity is reversed when the direction of current flow through the wire coil is reversed.
- Induction motor stators have coils of wire placed around the circumference of the stator core, which form electromagnets when energized, and create magnetic pole fields.
- Each wire coil has a corresponding coil on the opposite side of the stator core, which is connected in series with the first wire coil so it experiences the same current intensity and polarity, or direction of current flow, but is wound to produce the opposite magnetic pole field as the first wire coil.
- The synchronous speed of an electric induction motor is the speed at which the magnetic field rotates around the stator.
- Synchronous speed = Hz times 60 divided by number of pairs of poles.
- Squirrel cage induction motors are sometimes referred to as constant speed motors, because their RPM speed only changes a small amount through the entire torque load range of the motor.
- Regardless whether an AC motor is mechanically loaded or unloaded, the rotating magnetic field in the stator will rotate at synchronous speed, because it is determined by the power supply line frequency and the number of motor stator poles.
- The RPM speed difference between the stator rotating magnetic field, synchronous speed, and the actual rotating speed of the rotor is called slip.
- Percent slip = ((synchronous speed–rotor speed) / synchronous speed) x 100
- The speed listed on the nameplate of an electric motor is the manufacturer's declaration of the motor's RPMs under full-load, not synchronous speed.
- The rotor must always turn sufficiently slower than the rotating magnetic field so that magnetic lines of flux will cut through the rotor bars and induce a voltage.
- When first energized with the full supply voltage, the motor will draw the highest current the motor can draw, which is called the locked rotor current.
- Once the rotor starts to turn, the conductor bars in the rotor will cut through fewer lines of magnetic flux of the rotating magnetic field in the stator, because the conductor bars will be turning closer to the speed of the rotating magnetic field.
- It is not possible for the rotor of an induction motor to turn at synchronous speed and still produce turning torque.
- The power loss dissipated by a motor at the instant the motor is first powered can be fifty times the power loss dissipated at full speed, which limits the number of times the motor can be started per hour and safely dissipate heat.

## REVIEW QUESTIONS

1. What does the Law of Charges state about how objects will react to each other?

   Like charges ____________, unlike charges ____________

2. What are two advantages of using electromagnets rather than permanent magnets to create the magnetic field in electric motors?

   1. ________________________________

   2. ________________________________

3. Why does each stator coil have an opposing magnetic pole coil on the opposite side of the stator?
4. What is the definition of synchronous speed?
5. What is the formula for calculating synchronous speed of an induction motor?
6. What would the synchronous speed of a four-pole motor be, when operated at 60 Hz?
7. When counting the poles in a three-phase motor, the coils of how many phases are counted?
8. Why are induction motors sometimes referred to as constant speed motors?
9. How is synchronous speed affected by the mechanical load on a motor?
10. What is the definition of slip?
11. What is the calculation for percent slip?
12. How is slip affected by the mechanical load on a motor?
13. Why is rotor slip necessary to produce torque?
14. At what rotor speed is the maximum voltage induced into the rotor bars of an induction motor?
15. As the rotor starts to turn, does the relative motion between the rotating magnetic field of the stator and the conductor bars in the rotor increase or decrease?
16. As related to the relative motion between the magnetic field and the electrical conductor necessary for magnetic induction, why is no voltage induced into the rotor bars if the rotor is turning at synchronous speed?
17. Nameplate RPM of an induction motor is the manufacturer's declaration of the motor's RPMs under what conditions?
18. Why are induction motors rated for a maximum number of starts per hour?

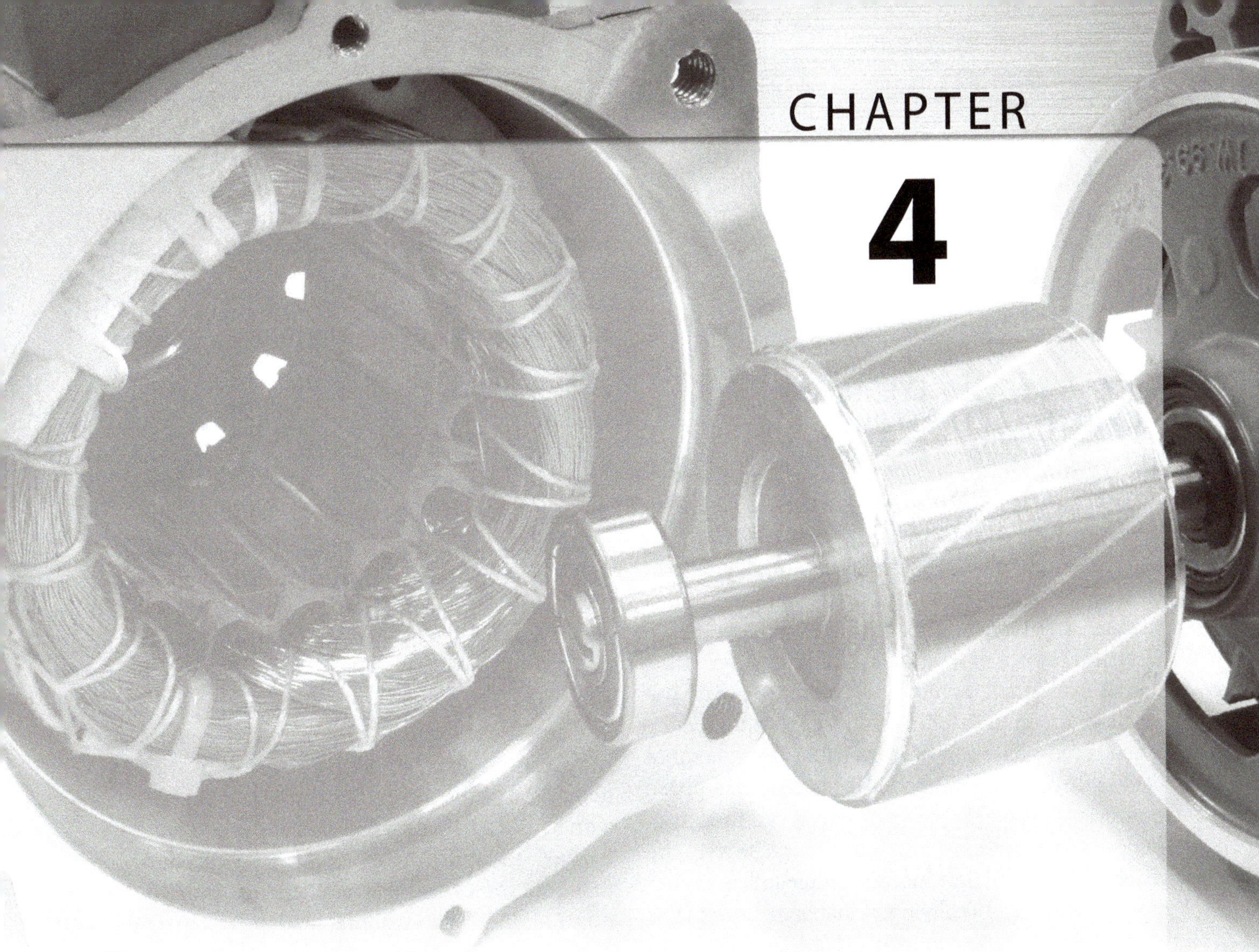


# Single-Phase Induction Motors

## PURPOSE

To familiarize the learner with the electromagnetic theory, advantages and disadvantages, and operation of single-phase electric induction motors.

## OBJECTIVES

After studying this chapter on single-phase induction motors, the learner will be able to:

- Identify the advantages and disadvantages of single-phase induction motors over three-phase induction motors
- Identify the parts of a single-phase induction motor
- Explain the function and operation of the auxiliary windings in single-phase motors
- Explain the function and operation of the centrifugal switch in single-phase induction motors
- Name, identify the wiring diagram for, and discuss the operating characteristics of the four most common split-phase induction motors
- Explain the differences between motor starting capacitors and motor running capacitors

- Explain a safe method for discharging capacitors in an electrical circuit
- Explain the difference between motor nameplate voltage ratings of motors and voltage of electrical supply distribution systems

## SINGLE-PHASE ALTERNATING CURRENT (AC) MOTORS

The main advantage of single-phase induction motors is that they can be used where three-phase power is not available, but that is where the advantages of single-phase motors over three-phase motors end. For residential use, all motor applications require single-phase motors, because three-phase electrical service to residential buildings is very rare. Sometimes small commercial buildings may use single-phase motors for light-duty applications, even when three-phase power is available, but it is more likely a three-phase motor will be used if three-phase power is available. Large commercial and industrial building applications will almost exclusively be three-phase induction motors. Single-phase induction motors are typically fractional horsepower (less than 1 HP), but sizes up to 10 HP are common, and a little research can locate sizes up to about 30 HP. It may seem that these low horsepower ratings would represent a huge application limitation for single-phase motors, but some estimates are that only 1% of all motors in use are rated over 20 horsepower.

### Disadvantages of Single-Phase Motors

When single-phase motors are used, they will be limited to relatively low horsepower and low torque load requirement applications. Single-phase induction motors are larger per horsepower, draw more current to generate the same horsepower, present an unbalanced load to the electrical distribution system, and are more complicated than three-phase motors. The most complicating factor of single-phase induction motors over three-phase induction motors is that they require additional auxiliary windings and additional mechanical wear parts to operate. These additional mechanical parts in single-phase induction motors simply present more things that can go wrong with the motor, which causes them to be less reliable than three-phase induction motors. Even with all of these disadvantages of single-phase motors, some estimates are that up to 80% of all motors in use today are single-phase, probably a testament to residential use.

### Main and Auxiliary Windings

Single-phase motors have two different windings in the stator, the main winding and the auxiliary winding. The main winding is often called the run winding, and both terms will be used in this text depending on the purpose of the winding in a particular motor design application. The auxiliary winding, which may be used for starting purposes only, or starting and running purposes is often called the start winding, and both terms will be used in this text depending on the purpose of the winding in a particular motor design application. The auxiliary winding is necessary in single-phase induction motors to create a rotation in the magnetic field of the stator, because the main winding alone will not create a rotating magnetic field. Refer back to Figure 3-6 in Chapter 3 to see how the magnetic poles in the stator change polarity, but no rotation of the magnetic field is established.

### Purpose of Auxiliary Windings

Without additional stator coil windings, the magnetic poles of the run windings will alternate between north and south at the synchronous speed of the power supply frequency, and not produce any rotation. This is the locked rotor condition where the motor rotor will stand still and growl, which is the name given to the noise the motor will make as the rotor vibrates in the stator in reaction to the strong changing magnetic fields. With just the run windings energized, the motor could be started in either direction, and it would start and run equally well in either direction. Single-phase induction motors require auxiliary start windings for the two following reasons: to establish a rotation to the stator's magnetic field, and to predetermine the motor's direction of rotation.

### Creating Rotation

The auxiliary start winding is usually placed over, and out of phase with, the run winding, as shown in Figure 4-1. The main run winding is generally a

**FIGURE 4-1** Main stator windings with auxiliary start windings

larger coil (more turns of wire), made with a larger gauge magnet wire, and the auxiliary start winding is generally a smaller coil (fewer turns of wire), made with a smaller gauge magnet wire. The result is that the run winding is a higher inductance value, lower resistance winding, and the auxiliary start winding is a lower-inductance, higher-resistance winding. This difference in inductance values causes the current in the auxiliary start winding to slightly lead the current in the run winding. The current reaching the auxiliary winding before the main winding will cause a magnetic pole to be created in the auxiliary slightly before the main winding, as shown in Figure 4-2, which shows a counterclockwise rotation.

Both the main run and auxiliary start windings are connected to the line when the motor is first energized, because both windings are needed to produce the rotation of the magnetic field necessary to cause the rotor to start turning. As the rotor reaches approximately 75% of synchronous speed the auxiliary start winding is no longer necessary, because the rotor will be able to follow the magnetic poles of the main run windings switching back and forth between magnetic poles. At this point the weights of the centrifugal weight assembly on the rotor will be flung outward, pulling the collar back, and opening the auxiliary start winding contact in the motor end bell, or another stationary part of the motor housing, to disconnect the auxiliary start winding. The motor starting currents through the smaller wire auxiliary start winding are relatively high, so it is important to remove the auxiliary start windings from the starting circuit as soon as possible to avoid damage.

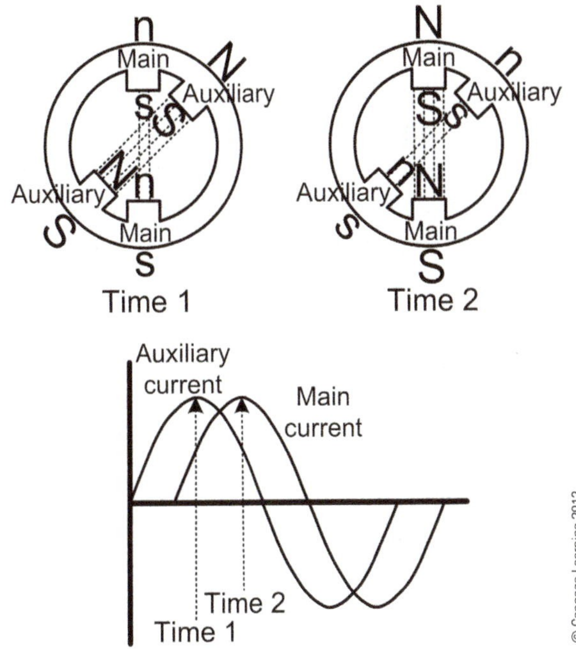


**FIGURE 4-2** Main and auxiliary magnetic field offset

## Direction of Rotation

The direction of rotation for single-phase motors is determined by the minute time differences between when the main run winding and the auxiliary start winding build a strong magnetic pole field, and the magnetic pole polarity differences between them. To change the direction of rotation for a single-phase motor, the magnetic pole polarity relationship between the auxiliary start windings and the run windings must be changed. Normally polarity is not a consideration with an alternating current power supply, but with motor coils and magnetic poles it is. Actually, it is an instantaneous magnetic pole polarity that determines which direction the motor will turn at the start. The graphic on the left of Figure 4-3 demonstrates that for the first instantaneous polarity relationship between the main and auxiliary windings, the auxiliary winding will get a strong magnetic pole of the same polarity slightly before the main winding, and cause one direction of rotation.

The graphic on the right of Figure 4-3 demonstrates that reversing the instantaneous magnetic polarity of the auxiliary winding in relation to the main winding will cause a different magnetic polarity relationship between them that will cause the direction of rotation to reverse. The wiring connection for either the main coil winding or the auxiliary coil winding may be reversed to reverse the single-phase motor. If both the main coil winding and the auxiliary coil winding are reversed, the magnetic pole polarity relationship between the two will remain unchanged, and the motor's direction of rotation will remain unchanged. Reversing a single-phase motor is usually just a matter of moving female quick-connect jumpers in the terminal box of the motor.

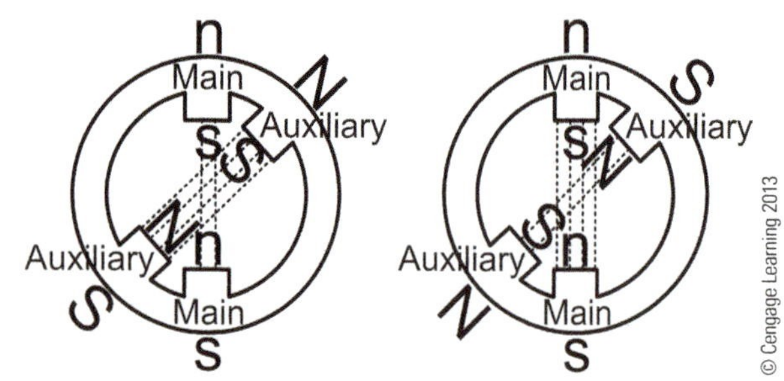


FIGURE 4-3 Reversing a single-phase motor graphic

## Centrifugal Switch Weight Assembly

The centrifugal switch assembly of the single-phase induction motor is designed to open the auxiliary start winding circuit once the rotor has reached a predetermined speed. The auxiliary start winding is only necessary to start the rotor turning from a standing start, because the magnetic poles of the main winding are too far apart for the rotor to follow. Once the rotor is turning close to operating speed, the rotor is able to follow the magnetic poles of the main winding, and the auxiliary start winding is no longer needed and must be disconnected. Single-phase induction motors measure the rotating speed of the rotor with a spring-loaded centrifugal weight assembly that is solidly mounted on the rotor, so that it turns at the same speed as the rotor. Figure 4-4 shows the centrifugal weight assembly mounted on the rotor, with the weights pulled in by the spring tension as if the rotor was not turning. Notice that the shoulder of the weight assembly is pushed out to approximately 1 5/8", which would press against the start switch contacts to hold them closed, and energize the auxiliary start windings when the motor is energized. Figure 4-5 shows the centrifugal weight assembly with the weights held in the outward position as if the rotor is turning at full speed. Notice that the shoulder of the weight

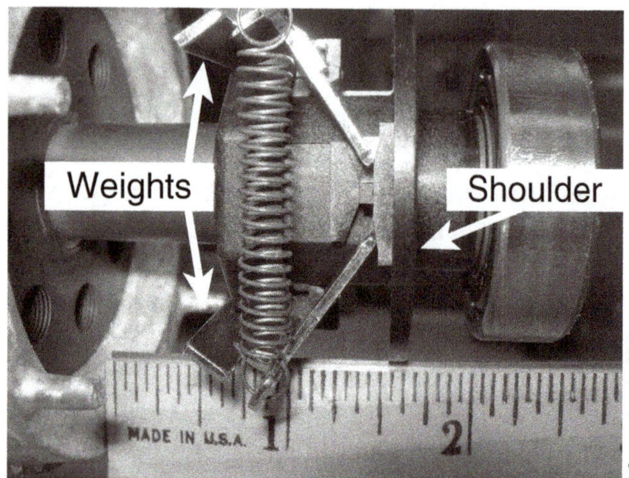


FIGURE 4-4 Centrifugal weight with shoulder pushed out

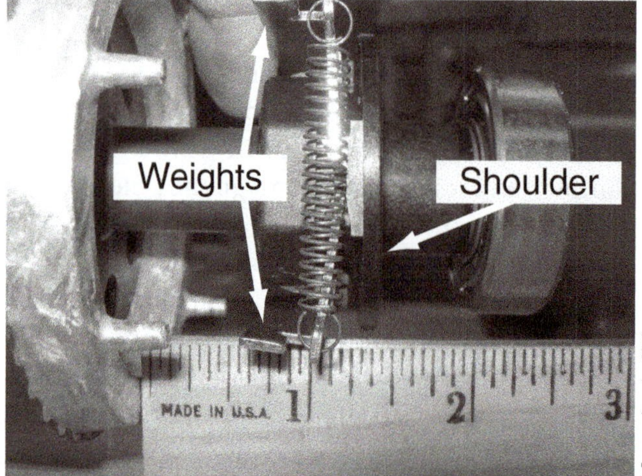


FIGURE 4-5 Centrifugal weight with shoulder pulled in

assembly has been pulled back by the weights to approximately 1 3/8", which would pull the weight assembly shoulder back from the start switch contact lever and open the contacts of the start switch contacts to de-energize the auxiliary start winding.

## Centrifugal Switch Contact Assembly

The centrifugal switch contact assembly shown in Figure 4-6 is not mounted on the rotor and does not turn; it is mounted on a stationary part of the motor housing or on the end bell, as shown in Figure 4-7. These electrical contacts are connected in series with the auxiliary start windings of single-phase induction motors, to close when the motor is stopped so that the winding will energize when the motor is started, and open when the motor approaches operating speed to de-energize the auxiliary windings. The mechanical connection to transition from the rotating rotor with the centrifugal weight assembly to the stationary switch contact assembly is accomplished by having the shoulder of the weight assembly rub against the switch contact lever to open and close the switch contacts as the weight assembly moves the shoulder in and out.

Figure 4-8 is provided to show how the rotating weight assembly shoulder mounted on the rotor is going to line up with the stationary switch contact lever of the switch assembly connected to the end bell when the bearing on the end of the rotor is seated into the end bell. Figure 4-9 shows

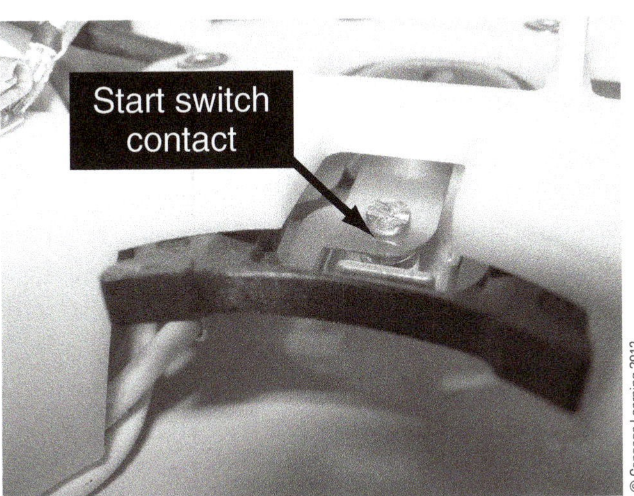


FIGURE 4-6 Start switch contact

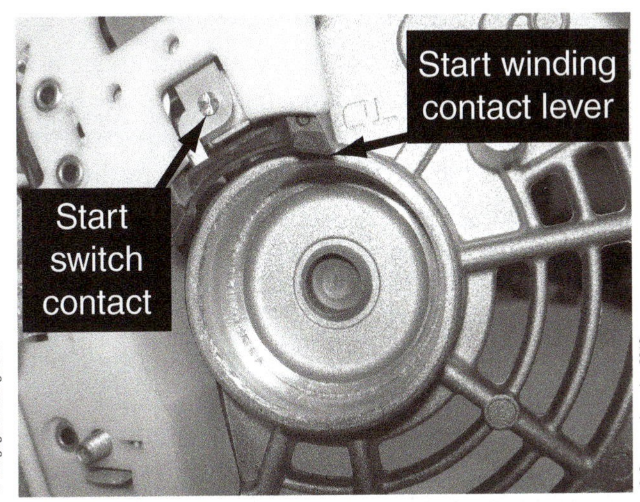


FIGURE 4-7 Start switch contact mounted in the end bell

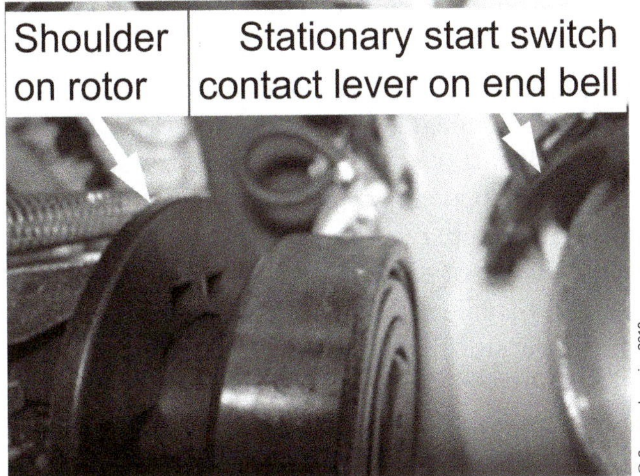


FIGURE 4-8 Shoulder mating with start switch contact lever

FIGURE 4-9 Weight assembly and start winding contact assembled

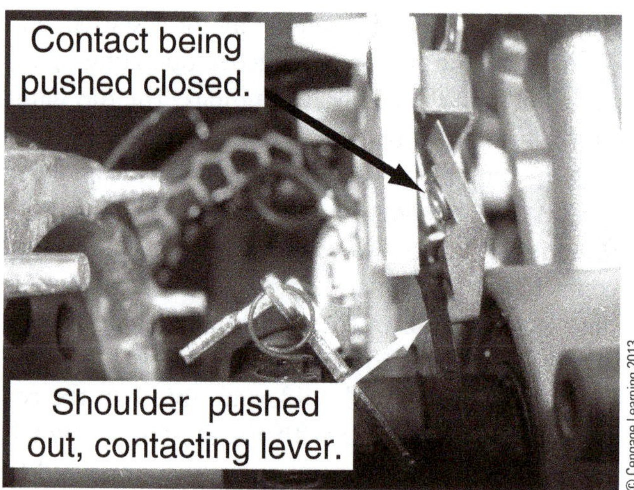


**FIGURE 4-10** Shoulder pushing starting switch closed

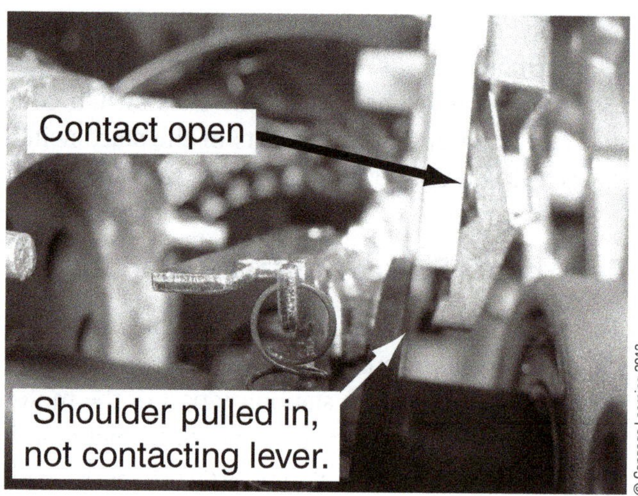


**FIGURE 4-11** Shoulder pulled away starting switch open

the entire assembly of the rotating shoulder and the stationary contact lever from the perspective of inside the motor. Notice from this perspective that as the weight assembly shoulder is pushed out it will push against the contact lever and close the switch contacts. When the shoulder is pulled in it will pull away from the contact lever and allow the switch contacts to open.

Figures 4-10 and 4-11 show yet another perspective of the rotating shoulder on the rotor, making and breaking the electrical contacts of the switch assembly on the end bell by pressing on the contact lever. Figure 4-10 shows that when the rotor is not turning, the weights of the weight assembly are pulled in by the spring tension, causing the shoulder to push out. When the weight assembly shoulder is pushed out, it makes contact with the lever of the switch assembly and closes the start winding contact to connect the auxiliary start windings for the next motor start. Figure 4-11 shows the position of the weight assembly shoulder when the rotor is turning at near operating speed, so the shoulder is pulled back from the contact lever, allowing the contacts of the switch assembly to open and de-energize the auxiliary start winding.

Figure 4-12 shows the actual contacts of the switch contact assembly, and the short travel distance necessary to make and break the contacts to close and open the auxiliary start windings.

### Summary of the Centrifugal Switch Operation

The auxiliary start winding is in the circuit when the motor is at rest, because the weight assembly weights are pulled in by spring tension, which pushes the shoulder out and holds the switch assembly contacts closed. When the motor is started, and the rotor spins close to the operating speed, the weights of the weight assembly are thrown out against the spring tension holding them in. When the weights are thrown out, the shoulder of the weight assembly is drawn in away from the switch contact assembly lever, allowing the auxiliary start winding contacts to open and de-energize the auxiliary start winding. The auxiliary start winding is not in the circuit when the motor is at operating speed, and there is no physical contact between the rotating member and the stationary member of the motor. When the motor is de-energized and coasting to a stop, there will be a "click" and then a rubbing noise until the rotor comes to a complete stop. The "click" that is heard is the spring-loaded weight assembly pulling back in when the rotor speed is no longer sufficient for the centrifugal force to keep the weights out against the spring tension holding them in. The rubbing noise is the shoulder of the weight assembly, extending out and rubbing against the contact assembly lever, re-closing the auxiliary start winding contacts for the next start.

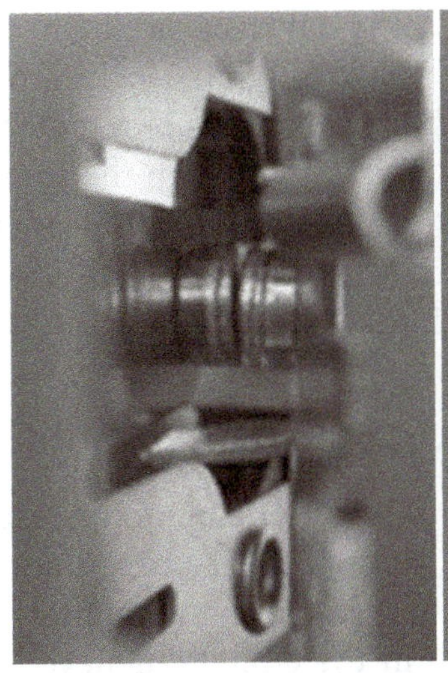
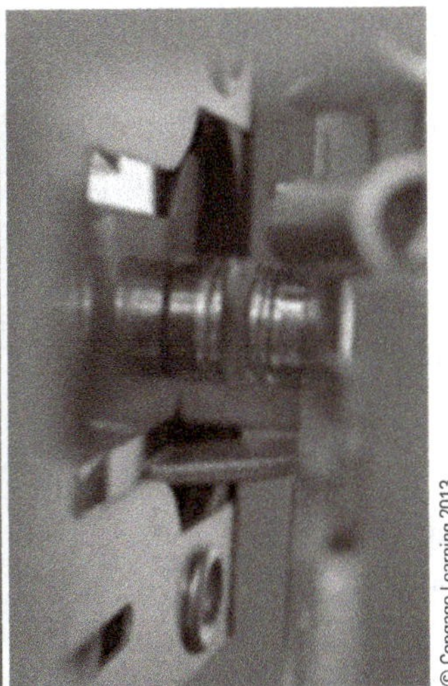


FIGURE 4-12 Start switch contacts open and closed

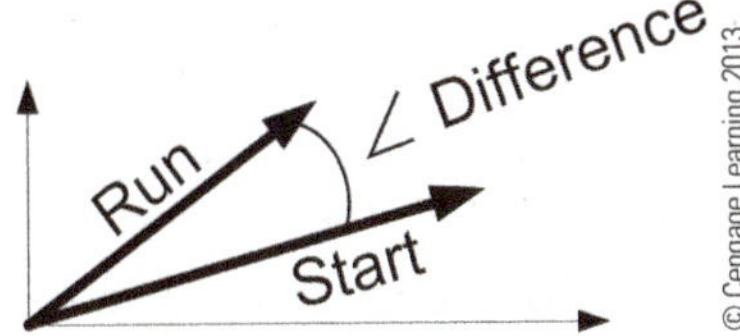


FIGURE 4-13 Split-phase phase shift

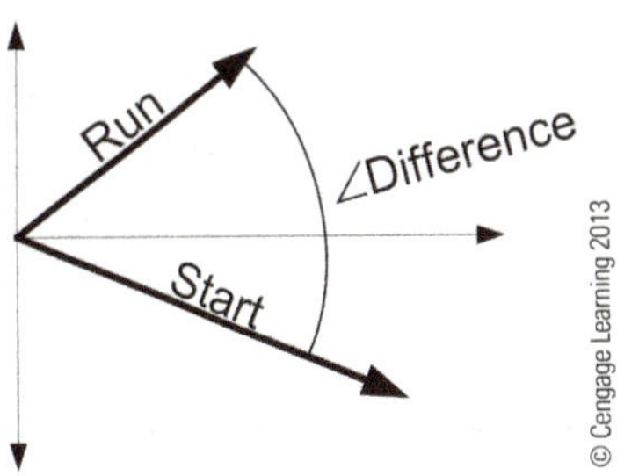


FIGURE 4-14 Split-phase capacitor start phase shift

## SINGLE-PHASE INDUCTION MOTOR CLASSIFICATIONS

There are many different classifications of single-phase motors. Listed below are four common types of single-phase induction motors that electricians are likely to encounter:

- split-phase
- capacitor start
- capacitor run
- capacitor start and run

### Split-Phase Induction Motors

As the drawing in Figure 4-13 shows, the phase shift created by the inductance difference between the main and auxiliary windings is relatively small. This small phase shift will produce a limited amount of turning torque, which makes the split-phase motor suitable only for loads with low locked rotor torque, such as fans and circulation pumps. The main advantages of split-phase motors over other single-phase induction motor types is that they are less expensive, have fewer parts, and are more dependable when applied correctly. The two main limitations of split-phase motors are that they are limited to fractional horsepower (less than 1-HP) sizes, and they have limited locked rotor starting torque.

### Capacitor Motors

Capacitor motors are split-phase motors with a capacitor added to the auxiliary start winding, main run winding, or both to improve the operating torque characteristics of the motor by causing a greater phase shift in the current between the main and auxiliary start windings, as shown in Figure 4-14. Remember from Chapter 1 that current lags voltage in inductors. Capacitors have the exact opposite effect in electrical circuits, and current leads voltage in capacitors. When large μF capacitors are connected in series with the inductor coils, which cause a lagging current, the leading current effect of the capacitor can completely mitigate the lagging current effect of the inductor coil and significantly increase the phase shift between the coil windings.

## Capacitor Start Motors

Capacitor start motors use very high μF capacitance rating AC electrolytic type capacitors in series with the auxiliary start winding to produce locked rotor starting torques of 200% to 400% of full-load torque, as shown in Figure 4-15. The AC electrolytic capacitor can be energized only for a short period of time, so a centrifugal switch must be connected in series with the capacitor and auxiliary start winding to de-energize them once the rotor approaches full speed. The operating characteristics of capacitor start motors make them suitable for applications like conveyors and compressors, because of their high locked rotor starting torque to get a load moving.

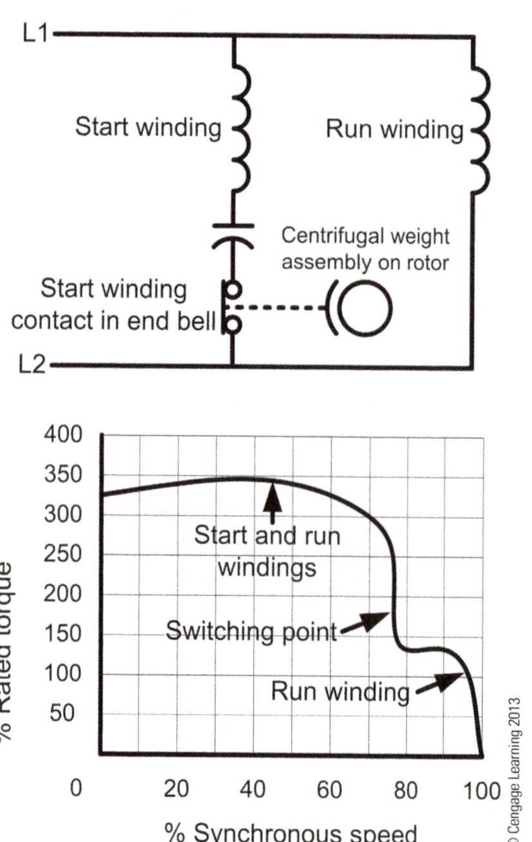


**FIGURE 4-15** Capacitor start motor diagram and torque curve

The start capacitor in the auxiliary start winding improves both the locked rotor torque and the pull-up torque ratings of a motor. Once the start capacitor is removed from the auxiliary start winding by the centrifugal switch at operating speed, the motor's operating characteristics are no different than the split-phase motor.

## Capacitor Run Motors

Capacitor run motors use very low μF capacitance rating, paper and foil-type capacitors in series with the auxiliary winding of the motor, which stays in the circuit whenever the motor is running, as shown in Figure 4-16. The capacitor run motor has no centrifugal switch assembly, auxiliary start winding contact, or start capacitor. Instead, it has a run-type capacitor permanently connected in series with the auxiliary winding, providing an increased phase shift between the run and auxiliary windings. This increased phase shift increases locked rotor starting torque (though not as much as a starting capacitor does) and breakdown torque to a limited extent over the split-phase motor. The run capacitor must be rated to be connected to an

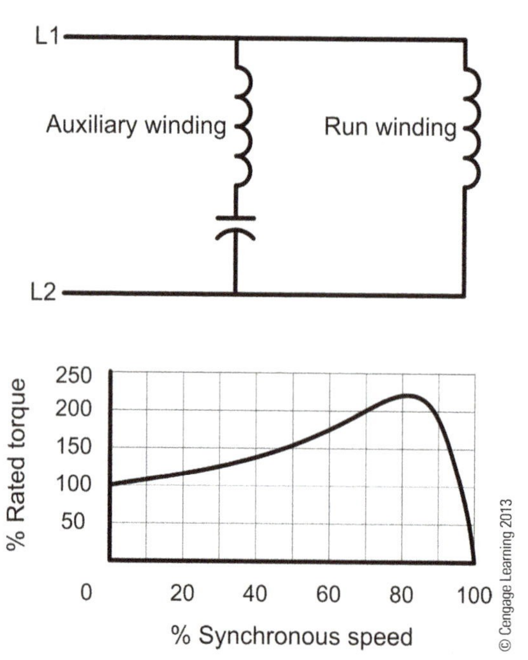


**FIGURE 4-16** Capacitor run motor diagram and torque curve

AC source indefinitely, as it is energized anytime the motor is operating.

## Capacitor Start and Run Motors

The capacitor start and run motor combines the best operating performance enhancements of the capacitor start motor and the capacitor run motor. It has a start capacitor in series with the auxiliary winding like the capacitor start motor for improved high locked rotor starting torque, as shown in Figure 4-17, which must be removed from the circuit with a centrifugal switch as the motor reaches full speed. And, like a capacitor run motor, it also has a run capacitor that is in series with the auxiliary winding all the time, which improves the locked rotor starting torque and the breakdown torque. Locked rotor, pull-up, and breakdown torque characteristics are all optimized with the capacitor start and run motor design to produce a single-phase motor on steroids.

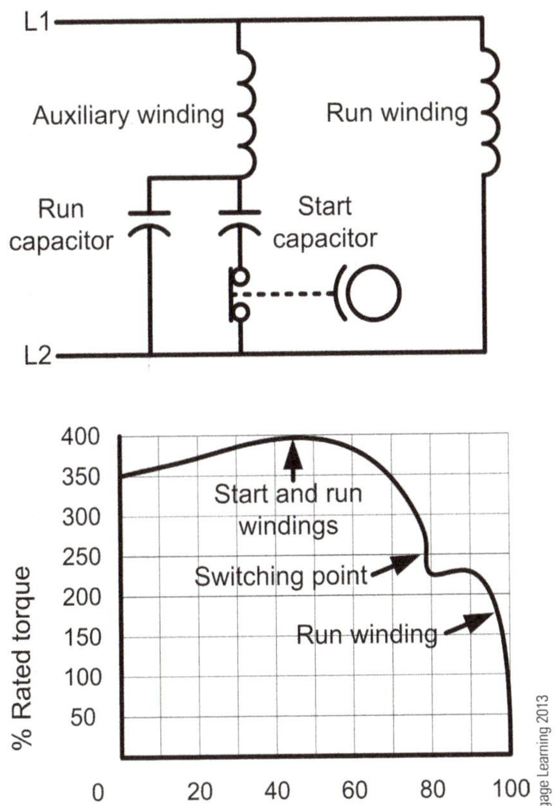


**FIGURE 4-17** Capacitor start and run motor diagram and torque curve

## Motor Capacitors

**Start and Run Capacitors.** The two capacitors encountered with single-phase AC induction motors, the start and run capacitors, are different from one another, and are not interchangeable. Start capacitors are usually large microfarad (100 μF to 1,000 μF) AC electrolytic, and are designed to be in the energized circuit for only a short period of time. Run capacitors are usually small microfarad (5 μF to 50 μF) paper and foil type, and are designed to be in the energized circuit for extended periods of time.

**Start Capacitors.** The start capacitor requires a high μF capacitance (necessary for a large phase shift) in a small enough package to fit on the motor. The best way to get high μF capacitance in a small package is to use an electrolytic capacitor. Normally electrolytic capacitors are polarity sensitive, and connecting them to AC would cause them to explode. For starting AC motors, however, a special AC-rated electrolytic capacitor is made by placing two electrolytic capacitors back-to-back in the one container. A common type of start capacitor is the AC electrolytic capacitor in a Bakelite container, and a "vent hole", or pressure safety valve on the top. Look at the picture in Figure 4-18, the round hole on the top with paper

**FIGURE 4-18** AC electrolytic motor starting capacitor

across it is the vent hole. If something goes wrong and the starting capacitor is energized for too long, hot gases may build up pressure inside the Bakelite container. The purpose of the vent hole is to release the pressure on the inside of the capacitor before it explodes. Anytime the vent hole paper is broken on this type of capacitor, it should be discarded.

Electricians really do not need to understand the chemistry of the AC electrolytic capacitor to use it, but it is important to understand that the AC electrolytic capacitor requires a chemical reaction to work, and leaving it across the AC line for too long can cause it to explode.That is the purpose of the centrifugal switch on the motor. Within a second or so of the motor starting, the rotor will reach a high enough speed that the centrifugal switch will open the auxiliary start winding contacts to disconnect the start capacitor from the AC supply.

**Run Capacitors.** Motor run capacitors are nonpolarized, and are much larger physically for a much smaller μF capacitance (pictured in Figure 4-19). Run capacitors do not depend on a chemical reaction to achieve their capacitance, so they can be across the AC power supply line indefinitely. The run capacitor is normally a simple paper and foil capacitor, which is two strips of aluminum foil with a paper dielectric between them, and rolled together to occupy a smaller area as shown in Figure 4-20. Run capacitors are pretty much bullet proof, and seldom fail.

**Discharge Resistor.** Some capacitors have a bleeder resistor connected across the terminals to safely discharge the capacitor after it has been de-energized. Refer back to Figure 4-18, and notice the resistor soldered across the terminals of the capacitor. The resistor is sized to have a high enough resistance so that it does not interfere with circuit operation when the capacitor is energized, but will still be low enough resistance to discharge the capacitor to a safe voltage level within one minute after being de-energized.

Some electricians have a dangerous habit of discharging capacitors by bridging the contacts

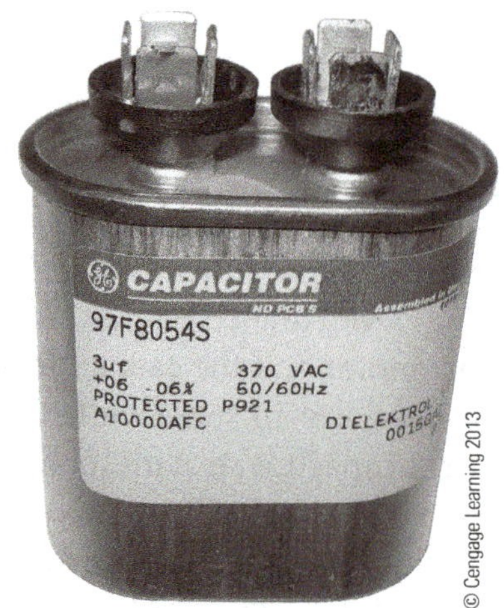




**FIGURE 4-19** AC run capacitor

**FIGURE 4-20** Paper and foil capacitor drawing

with the tip of a screwdriver, or shorting the terminals with a piece of wire. Such a practice is dangerous because, first of all, there are some capacitors that can be damaged by discharging them so rapidly; and second, if, inadvertently, the circuit is not de-energized, the electrician could be injured seriously. If the capacitor does not have a bleeder resistor across its terminals, a safer method of discharging the capacitor is to connect a lower input impedance voltmeter across the terminals. A low input impedance voltmeter will normally discharge the capacitor in a reasonable amount of time (and verify it with a reading), and if, inadvertently, the circuit is not de-energized, the impedance of the voltmeter will not cause a dangerous short circuit situation.

## Electrical Supply Distribution System

The nameplate voltage found on motors is a terminal voltage, meaning that this is the voltage that should exist at the motor terminals after experiencing the feeder and branch circuit voltage drops of the electrical system. There are no standards for nameplate voltages, but the following list indicates the diversity between motors and manufacturers. In the case of electric motors, the operating voltages are usually multiples of 115 volts; for example, 115, 230, 460, and 575 volts. These operating voltages have been picked deliberately to be slightly lower than the utility delivery voltages, because in an industrial plant or large commercial building there may be several hundred feet between the incoming service point and the equipment being powered. The distances involved always will lead to some voltage drop through the wiring system. On short runs this might be very small, even less than a volt, but on long, heavily loaded runs it might be as much as 3% or 4% of the operating voltage. So designing motor terminal voltages to be less than the utility service voltage assures the likelyhood that the motor will be operated closest to its rated voltage.

## CHAPTER SUMMARY

- The main advantage of single-phase induction motors is that they can be used where three-phase power is not available.
- Single-phase induction motors are typically fractional horsepower; less than 1 HP.
- Single-phase motors are limited to relatively low horsepower and low torque load requirement applications.
- Single-phase induction motors are larger per horsepower, draw more current to generate the same horsepower, present an unbalanced load to the electrical distribution system, and are more complicated than three-phase motors.
- Single-phase motors have two different windings in the stator: the main winding, sometimes called the run winding, and the auxiliary winding, which may be used for starting purposes only, or starting and running purposes depending on the motor design.
- The auxiliary winding is necessary in single-phase induction motors to create a rotation in the magnetic field of the stator, because the main winding alone will not create a rotating magnetic field.
- The run winding is generally a larger coil (more turns of wire), made with a larger gauge magnet wire, and the auxiliary start winding is generally a smaller coil (fewer turns of wire), made with a smaller gauge magnet wire.
- Both the main and auxiliary windings are connected to the line when the motor is first energized, but as the rotor reaches approximately 75% of synchronous speed the auxiliary start winding is removed from the electrical circuit.
- The standard direction of rotation for single-phase motors is counterclockwise from the front of the motor, the end opposite the drive.
- The centrifugal switch assembly of the single-phase induction motor is designed to open the auxiliary start winding circuit once the rotor has reached a predetermined speed.
- The auxiliary start winding is only necessary to start the rotor turning from a standing start, because the magnetic poles of the main winding are too far apart for the rotor to follow. Once the rotor is turning close to operating speed, the rotor is able to follow the magnetic poles of the main winding, and the auxiliary start winding is no longer needed.
- The centrifugal weight assembly is mounted on the rotor, and turns the same speed as the rotor.
- The centrifugal switch contact assembly is mounted on a stationary part of the motor housing.
- The shoulder of the weight assembly is moved in and out by the weights of the assembly, and rubs against the switch contact lever to open and close the switch contacts.
- Four of the most common types of single-phase induction motors are split-phase, capacitor start, capacitor run, and capacitor start and run.
- Capacitor motors are split-phase motors, with a capacitor added to the auxiliary start winding, main run winding, or both to improve the operating torque characteristics of the motor by causing a greater phase shift in the current between the main and auxiliary start windings.
- Capacitor start motors use very high μF capacitance rating, AC electrolytic type capacitors in series with the auxiliary start winding to increase the locked rotor starting torque.
- Capacitor run motors use very low μF capacitance rating paper and foil type capacitors in series with the auxiliary winding of the motor, which stays in the circuit whenever the motor is running to increase the locked rotor starting torque and breakdown torque over the split-phase motor.

- The capacitor start and run motor combines the best operating performance enhancements of the capacitor start motor, and the capacitor run motor.
- Start capacitors are usually large microfarad AC electrolytic, and are designed to be in the energized circuit for only a short period of time.
- Run capacitors are usually small microfarad paper and foil type, and are designed to be in the energized circuit for extended periods of time.
- Some capacitors have a "bleeder" resistor connected across the terminals to safely discharge the capacitor after it has been de-energized.
- The nameplate voltage found on motors is a "terminal voltage," meaning that this is the voltage that should exist at the motor terminals after experiencing the feeder and branch circuit voltage drops of the electrical system.

## REVIEW QUESTIONS

1. What is the main advantage of single-phase induction motors?
2. What horsepower ratings are typical for single-phase induction motors?
3. What are the names of the two split-phase stator windings?

   1. ______________________________

   2. ______________________________

4. What are two purposes of auxiliary start windings in single-phase electric motors?

   1. ______________________________

   2. ______________________________

5. What is the name given to the sound the motor will make as the rotor vibrates inside the stator, when a motor is energized but cannot turn?
6. Which winding, main or auxiliary, is the highest resistance?
7. Which winding, main or auxiliary, has the most inductance?
8. What is the point of making the main and auxiliary windings different inductance values?
9. What is the purpose of adding a starting capacitor in series with the auxiliary winding?
10. What is the standard direction of rotation for single-phase induction motors?
11. How is the direction of rotation reversed for single-phase induction motors?
12. What would happen if both the main and auxiliary windings were reversed?
13. What is the purpose of the centrifugal switch on single-phase induction motors?
14. What are four common types of single-phase induction motors electricians are likely to encounter?

    1. ______________________________

    2. ______________________________

    3. ______________________________

    4. ______________________________

15. What are the two main limitations of single-phase, split-phase motors?

    1. ______________________________

    2. ______________________________

16. Which two operating torque ratings of a split-phase induction motor are improved when a starting capacitor is added to the auxiliary start winding?

    1. ______________________________

    2. ______________________________

17. Which two operating torque ratings of a split-phase induction motor are improved when a run capacitor is added to the auxiliary winding?

    1. ______________________________

    2. ______________________________

18. What types of capacitors are used for motor starting applications to achieve the necessary high μF ratings?

19. What safety precaution must be observed with these special starting capacitors?

20. What practice should never be used to discharge a capacitor in an electrical circuit?

21. What type of capacitors are used for motor running applications that allow the capacitor to be connected in an energized electrical circuit indefinitely?

22. What is the purpose of having a bleeder resistor connected across the terminals of a capacitor?

23. What is the name for the operating voltage found on motor nameplates, which indicates that the voltage drops of the electrical distribution system have been taken into consideration?

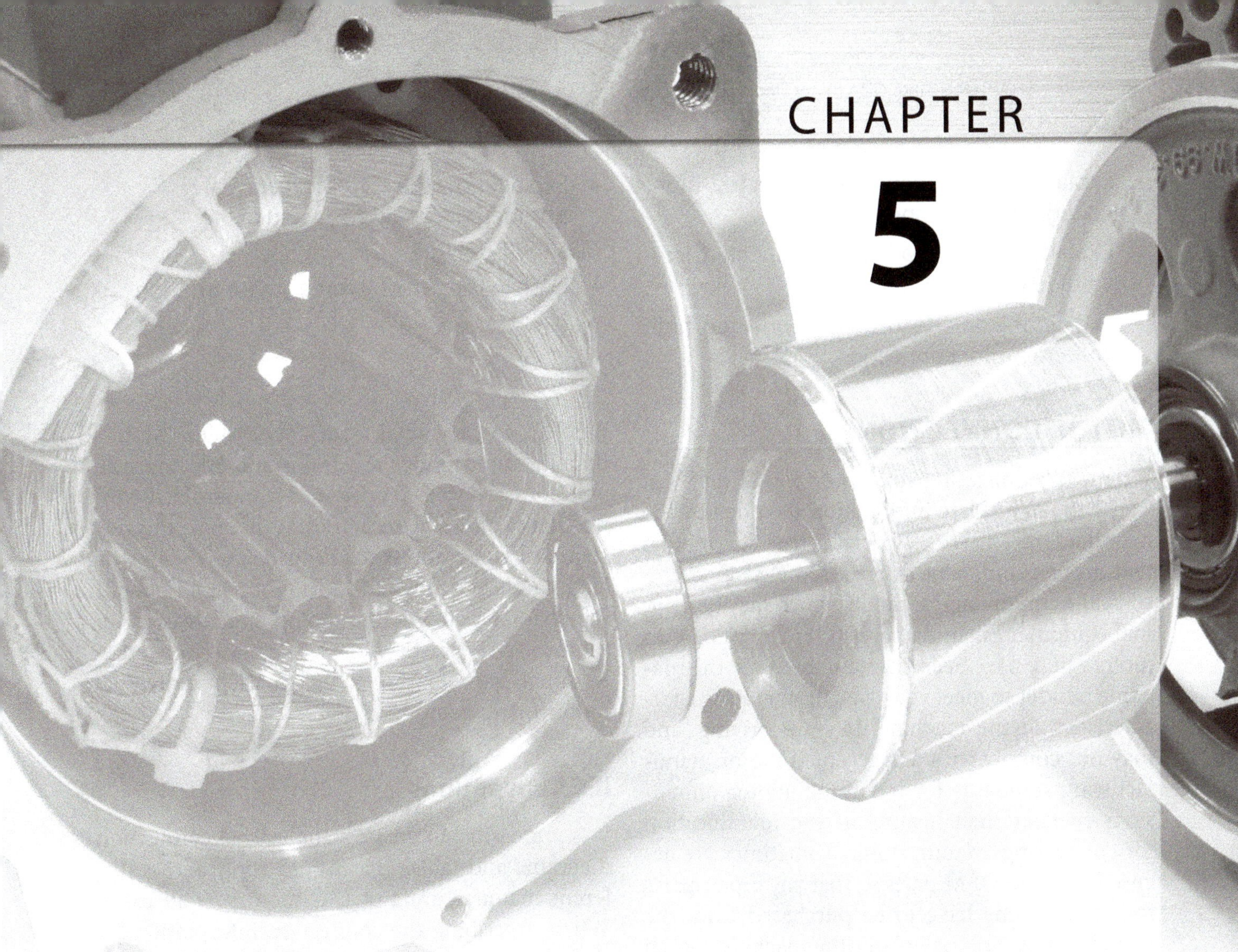

CHAPTER

5

# Induction Motor Electrical Connections

**PURPOSE**

To familiarize the learner with the electrical connections of single-phase and three-phase induction motors for both high and low voltages of dual voltage motors.

**OBJECTIVES**

After studying this chapter on induction motor electrical connections, the learner will be able to:

- Explain how to reverse a single-phase induction motor
- Explain a single-phase, single voltage, reversible motor connection
- Explain a single-phase, dual voltage, reversible motor connection
- Explain how to reverse a three-phase induction motor
- Explain the nine-lead, wye-connected, low voltage motor connection

- Explain the nine-lead, wye-connected, high voltage motor connection
- Explain the nine-lead, delta-connected, low voltage motor connection
- Explain the nine-lead, delta-connected, high voltage motor connection
- Explain how the stator coil ends are numbered for both wye and delta configurations
- Explain the use of a three-phase rotation tester

## MOTOR CONNECTIONS

Whenever connecting any motor, it is always best to use the nameplate information provided by the manufacturer. This is especially true for motors that are a component of an assembled machine, such as a compressor or pump, which may be original equipment manufacturer (OEM) parts. OEM motors may have been manufactured for the machine builder to meet very specific operating characteristics of the machine they are driving, and may not comply with any accepted motor manufacturing standards for mounting, motor dimensions such as shaft height, or lead identification for connecting. Manufacturers sometimes are accused of using OEM parts so that any repair parts for the machine have to be purchased from the company that originally built the machine, which will ensure future revenue for the manufacturer. The more likely reason for using OEM parts is that the machine builder knows that a standard general purpose motor will not drive the load properly, so they build the machine so that only an OEM replacement motor will fit in the provided space and mounting provisions of the machine.

### NEMA Conventions

Most of the motors that construction electricians will encounter in the field will not be OEM. They will be general purpose induction motors that have been manufactured to accepted motor manufacturing standards. Motors that are manufactured under the NEMA and IEC standards are basically the same motor functionally, but the wire color codes, labels, terminal markings, and voltage ratings are different. Motors manufactured under the NEMA standard will have well-defined and standardized connection terminal wire markings: the "T," for terminal, marking system. Motors manufactured under the IEC standard, however, will not have well-defined and standardized connection terminal wire markings, because the IEC standard does not specifically define them. The IEC standard defines some mounting and operating characteristics of the motor, and leaves other motor characteristic details such as connection terminal wire markings up to each individual motor manufacturer.

### IEC Conventions

It is impossible to create standardized connection diagrams and tables that would apply across different manufacturers of IEC motors in all cases. When describing the connection information for IEC motors throughout this chapter, the conventions of a single motor manufacturer, which is not named, will be used for demonstration purposes only, to explain each IEC motor connection. The best rule to follow when connecting IEC motors in the field is to consult the manufacturer's connection diagram for the correct terminal markings, because differences from one motor manufacturer to another are common.

### Manufactured for Both NEMA and IEC Markets

With all of the generalizations that can be made about the differences between motors manufactured under the NEMA and IEC standards, the fact is that most motor manufacturers have started producing motors that are rated to be used in either market, and the trend seems to favor that direction. Other motor manufacturers have dual-rated only certain motor characteristics, such as rated voltages, mounting dimensions, or shaft height, to make them more interchangeable. And, of course, other motor manufacturers have made no attempt at dual-rating their motors, and continue to manufacture them to either the NEMA or IEC standards. What all of this means is that there are no definite rules that will always apply to all motors the construction electrician will encounter in the field, so having a basic understanding of both standards is essential.

## Motor Lead Wire Color Codes

The NEMA terminal wire color code found in Figure 5-1 is applicable to all single-phase motors manufactured under the NEMA motor standard, but the IEC terminal wire color code may change between different motor manufacturers. The color code is especially handy to keep as a reference, because the terminal wires themselves may not be marked with identifying labels. Sometimes only the terminal block terminals of single-phase motors are labeled, and once the wires are lifted from the marked terminals that reference is lost; the color code may be the only remaining wire identifier.

The NEMA terminal wire color code for three-phase motors found in Figure 5-2 shows that the actual terminal wire color is undefined, but it is usually black. The actual color is unimportant, because each terminal lead is marked or tagged using the T system of labeling T1 thru however many motor leads there are. The IEC color code is especially handy for the same reason as the NEMA single-phase color code; sometimes only the terminal block terminals of IEC motors are labeled, and once the lead wires are lifted from the terminal block the color code may be the only remaining identifier.

| NEMA single-phase | IEC single-phase |
|---|---|
| T1 - Blue | U1 - Blue |
| T2 - White | U2 - Black |
| T3 - Orange | Z - Brown |
| T4 - Yellow | Ground - Green/yellow |
| T5 - Black | |
| T8 - Red | |
| Ground - Green, green/yellow | |



FIGURE 5-1 NEMA and IEC single-phase motor wire color codes

| NEMA three-phase | IEC three-phase |
|---|---|
| T1 thru T12 - Undefined | U1 - Black |
| Ground - Green, green/yellow | U2 - Green |
| | V1 - Blue |
| | V2 - White |
| | W1 - Brown |
| | W2 - Yellow |
| | Ground - Green/yellow |



FIGURE 5-2 NEMA and IEC three-phase motor wire color codes

## NEMA Conduit Box Electrical Connections

It is most common for NEMA motors to use connection terminal blocks in the end bell opposite of the motor's drive end for single-phase motors, as shown in Figure 5-3. The terminal block will normally have threaded nuts on binding posts for the line leads, in addition to male quick-connect terminals. The motor coil connections for different operating voltage configurations and reversing

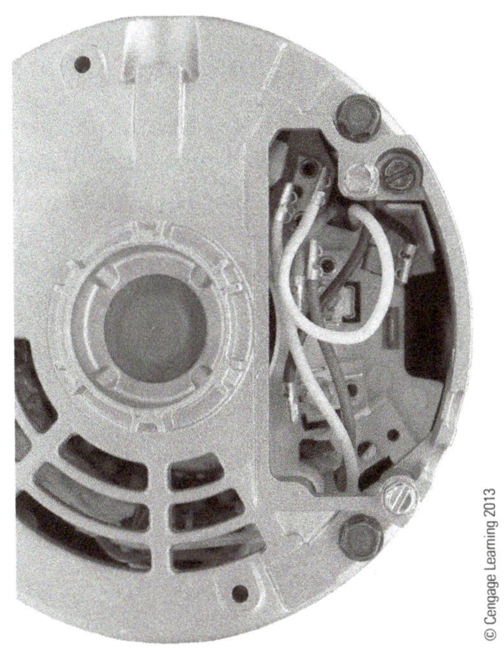



FIGURE 5-3 Front of single-phase motor showing the terminal block

**FIGURE 5-4** NEMA three-phase motor with individual connection leads

connections are made with quick-connect terminals to make them easy to reconfigure.

NEMA three-phase motors normally use open-ended lead wires marked with the T label system in the conduit box, as shown in Figure 5-4. Many older motors may have identifying tags applied to the lead wires, but most new motors have the T label marked directly on the lead wire insulation. Many different methods are used for making the connections to the line conductors, such as split bolts for large motors, to nuts and bolts for smaller motors, and even wire nuts for the smallest motors.

## IEC Conduit Box Connection Terminals

It is most common for IEC motors to use terminal blocks for both single-phase and three-phase motors, as shown in Figure 5-5, which makes the color code important for both types of IEC motors. The motor terminal block is mounted inside the motor conduit box, and for the motor manufacturer shown here the leads are marked for both NEMA and IEC three-phase conventions. The terminal block for single-phase motors would probably look similar, only the labels would be different.

## Reversing Direction of Rotation for Single-Phase Motors

Not all single-phase motors are designed to be reversible, because to be reversible all of the coil leads must be brought out to the motor terminal

**FIGURE 5-5** IEC conduit box terminals

block to reconfigure the connection in the field. It is less expensive, and it simplifies the field connection, if the manufacturer makes the internal coil connections without bringing the coil leads out to the terminal block, but the motor will be limited to a single direction of rotation. For single-phase motors that can be reversed, the electrical connection between the auxiliary start winding and the main run winding must be reversed. Reversing either the auxiliary start winding or the main run winding, but not both, would accomplish the same goal of reversing the motor's direction of rotation. Single-phase motor connection diagrams will always designate exchanging the leads of the auxiliary start winding, T5 and T8, because it is a valid method of reversing the motor regardless of the main run windings configuration for low or high voltage.

## Standard Direction of Rotation

NEMA and IEC motors use different terminologies for discussing the direction of rotation for electric motors. NEMA uses the terms "clockwise" and "counterclockwise," and IEC uses the terms "clockwise" and "anticlockwise." The standard direction of rotation for both NEMA and IEC motors is the same, but they also designate it differently. NEMA motors define the standard direction of rotation as counterclockwise from the front, or non-drive shaft, end of the motor. IEC motors define the standard direction of rotation as clockwise from the drive shaft end of the motor. For NEMA motors, when the odd-numbered terminal of the auxiliary start winding is connected to the even-numbered terminal of the main run winding, the rotor will turn in the standard direction of rotation. For IEC motors, when the even-numbered terminal of the auxiliary start winding is connected to the even-numbered terminal of the main run winding, the motor will turn in the standard direction of rotation.

## Electrical Connections

In order to better understand the actual terminal connections provided by the manufacturer on the motor nameplate, this section will explain how motors are connected internally, and how the stator coil leads are brought out to the motor conduit box for connecting. Understanding the internal connections of motors will not only help make sense of the motor operation, and aid in troubleshooting a malfunctioning motor, it also will help the technician reason out the connection in the all-too-familiar situation where the manufacturer's connection information is missing. The single-phase, two-wire motor connection will not be discussed, because there are no field alterations that can be made to affect the motor operation.

**NEMA Single-Phase, Dual Voltage, Capacitor Start, Reversible Motor Connection.** The capacitor start motor is the most common single-phase motor the construction electrician will encounter in the field (not necessarily dual voltage), so it is the first motor covered here and will be given the most detail. The diagram in Figure 5-6 shows the connection diagram for a NEMA single-phase, dual voltage, capacitor start, reversible motor. The motor attains the dual voltage capability by

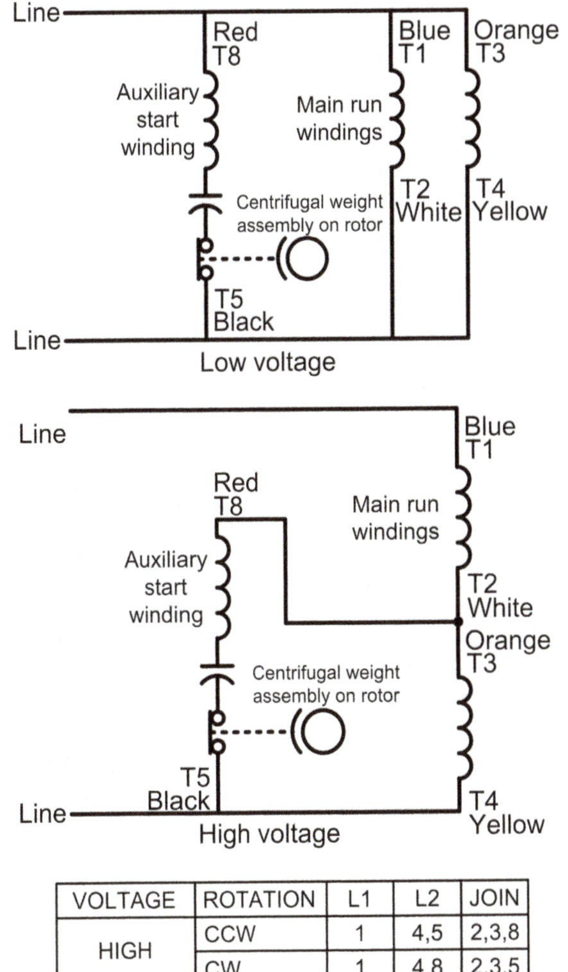


| VOLTAGE | ROTATION | L1 | L2 | JOIN |
|---|---|---|---|---|
| HIGH | CCW | 1 | 4,5 | 2,3,8 |
| | CW | 1 | 4,8 | 2,3,5 |
| LOW | CCW | 1,3,8 | 2,4,5 | -- |
| | CW | 1,3,5 | 2,4,8 | -- |



**FIGURE 5-6** NEMA single-phase, dual voltage, capacitor start, reversible connection diagram

providing two main run windings, each rated for the same voltage, which will be connected in parallel for low voltage operation, and in series for high voltage operation. Figure 5-7 shows a sketch of a connection diagram by lead wire color, similar to one that would be found on the motor nameplate, and Figure 5-8 shows a sketch of the motor terminal block connections.

Look at the low voltage connection diagram provided in Figure 5-6 and it will become apparent that all three windings are rated for the same voltage: the lower voltage rating of the motor. Notice that for the low voltage connection, all three windings are connected in parallel so that each receives the full phase voltage of the power supply.

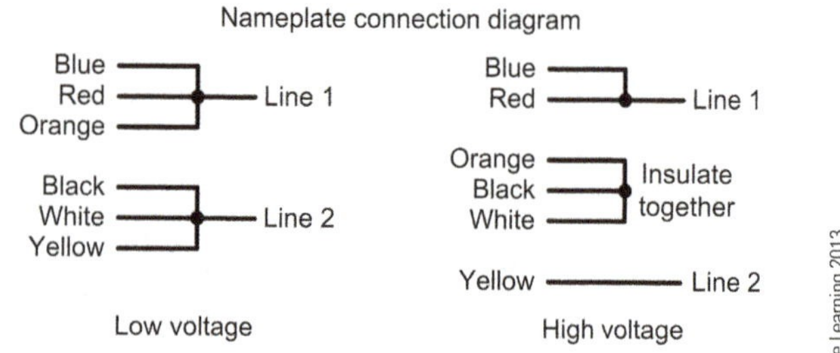


FIGURE 5-7 NEMA single-phase, dual voltage, capacitor start, reversible nameplate diagram

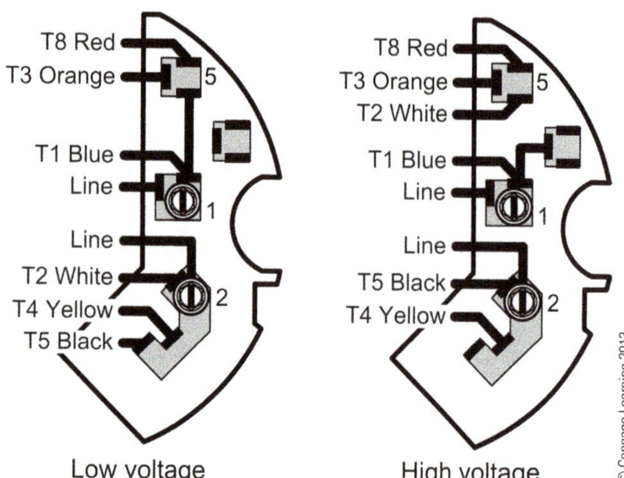


FIGURE 5-8 NEMA single-phase motor terminal block

For the high voltage connection, the two main run windings are connected in series, so that each winding receives half of the higher power supply voltage, and the auxiliary start winding is connected in parallel with only one of the main run windings. During the start of the motor when all three coils are energized, the voltages across each coil will not be exactly the same, because the parallel combination of the auxiliary start winding and one of the main run windings will cause less voltage to drop across the parallel pair than the voltage that will drop across the other main run winding that is in series with the parallel pair. The voltage unbalance will last for only a short time, however, because when the centrifugal switch opens the auxiliary start winding as the motor approaches full speed, the two main run windings will be left by themselves connected in series, with half of the applied supply voltage dropped across each of them.

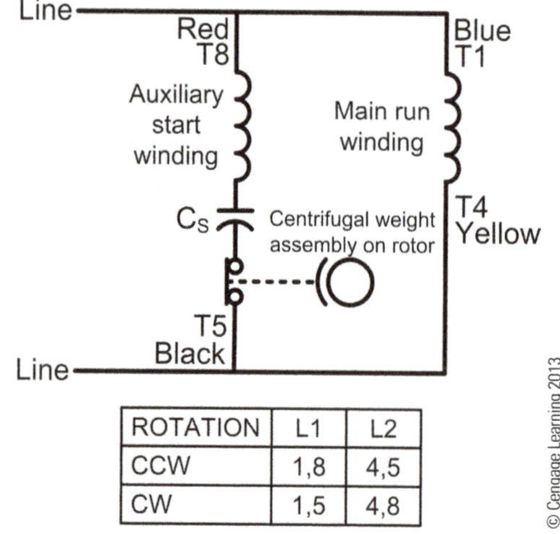


| ROTATION | L1 | L2 |
|---|---|---|
| CCW | 1,8 | 4,5 |
| CW | 1,5 | 4,8 |

FIGURE 5-9 NEMA single-phase, single voltage, capacitor start, reversible connection diagram

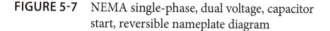

**NEMA Single-Phase, Single Voltage, Capacitor Start, Reversible Motor Connection.** The diagram in Figure 5-9 shows the connection diagram for a single-phase, capacitor start, single voltage, reversible, split-phase motor. The only difference between the single voltage and dual voltage single-phase designs is that the single voltage has only one main run winding. The auxiliary start winding is labeled T5 and T8 as always, but the main run winding may be labeled T1 and T2, or T1 and T4, depending on the manufacturer. Both windings of the motor are rated for the full motor voltage, so they are connected in parallel to receive the full supply voltage.

**IEC Single-Phase, Single Voltage, Capacitor Start, Reversible Motor Connection.** In the United States, single-phase, 115/230-volt, dual voltage motors are common, because the normal single-phase utilization voltages are dual 120/240 volts, 60 Hz. Single-phase IEC motors are normally a single voltage, 220 volts, because the normal single-phase utilization voltage in the counties where IEC motors are found is a 220-volt, 50-Hz power source.

IEC motor manufacturers document connection information with different drawings than NEMA motor manufacturers, but the motors are functionally similar. To demonstrate the two different methods of documenting the connection diagrams between NEMA and IEC, the connection diagram in Figure 5-10 is drawn in the delta configuration used by one IEC motor manufacturer, and Figure 5-11 shows the diagram type used by one NEMA motor manufacturer with the IEC labels transposed in the drawing. Close examination will reveal that the two diagrams are the same. Figure 5-12 demonstrates how the connection would look on the terminal block of an IEC motor.

**NEMA Single-Phase, Single Voltage, Capacitor Run, Reversible Motor Connection.** Capacitor run motors are different from capacitor start motors in two ways. First, both the run capacitor and the auxiliary winding are designed to be connected in energized circuits permanently without causing damage to either. Second, because the run capacitor is connected in the circuit permanently, there is no need for a wear-prone centrifugal switch to disconnect it when the motor approaches operating speed. The addition of the capacitor in the auxiliary winding will cause a greater phase-shift between the main run winding and the auxiliary winding, which will improve operating torque. The run capacitor will be a low microfarad value, paper and foil type. The NEMA capacitor run motor diagram and connection information are shown in Figure 5-13.

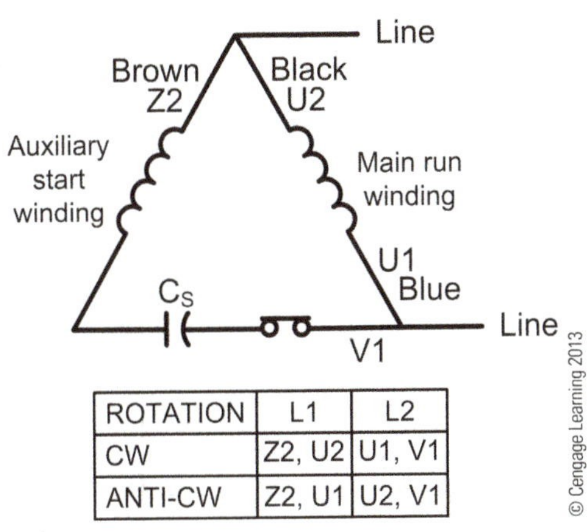


| ROTATION | L1 | L2 |
|---|---|---|
| CW | Z2, U2 | U1, V1 |
| ANTI-CW | Z2, U1 | U2, V1 |

**FIGURE 5-10** IEC connection diagram for a single-phase, single voltage, capacitor start, reversing motor

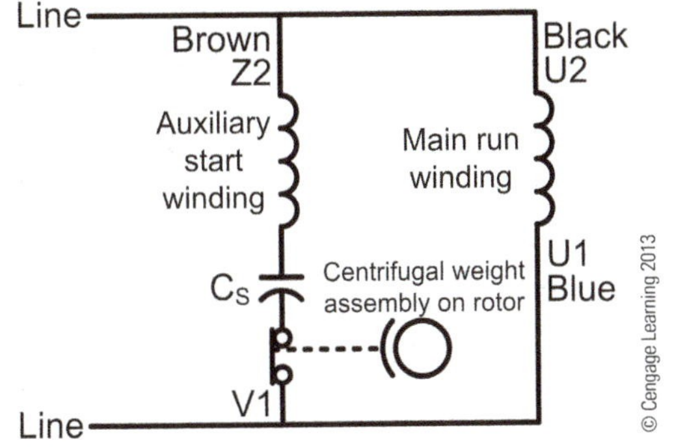


**FIGURE 5-11** IEC-type labels on a NEMA-type coil drawing

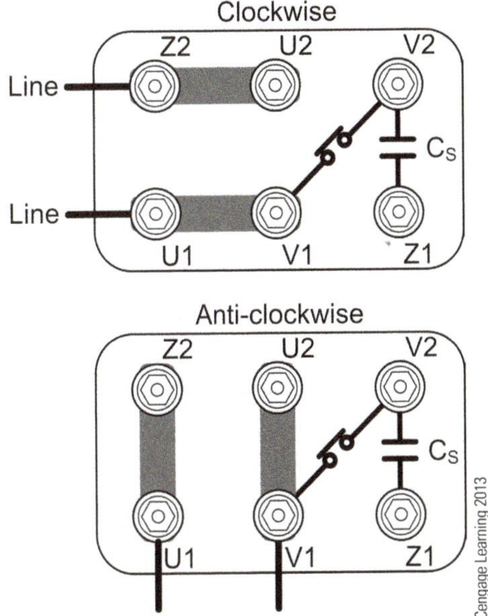


**FIGURE 5-12** IEC single-phase, single voltage, capacitor start, reversing motor terminal block

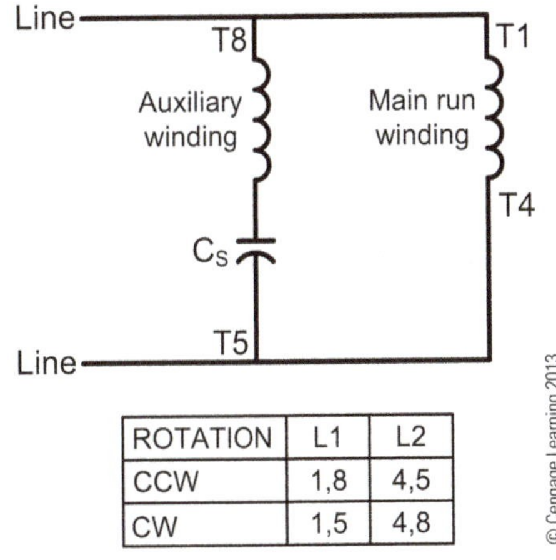


| ROTATION | L1 | L2 |
|---|---|---|
| CCW | 1,8 | 4,5 |
| CW | 1,5 | 4,8 |

**FIGURE 5-13** NEMA single-phase, single voltage, capacitor run, reversible connection diagram

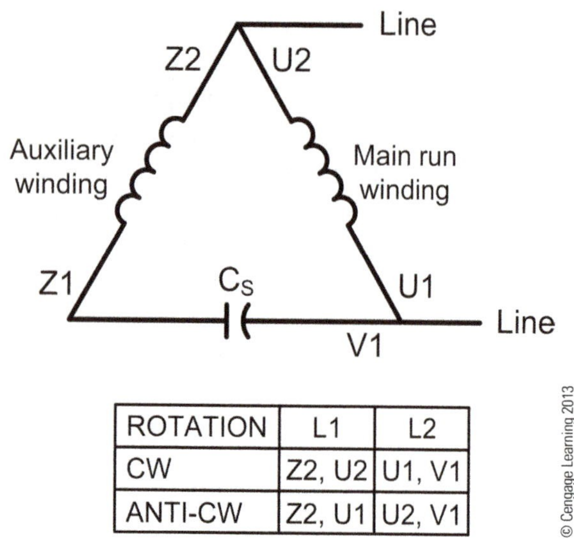


| ROTATION | L1 | L2 |
|---|---|---|
| CW | Z2, U2 | U1, V1 |
| ANTI-CW | Z2, U1 | U2, V1 |



**FIGURE 5-14** IEC single-phase, single voltage, capacitor run, reversing motor

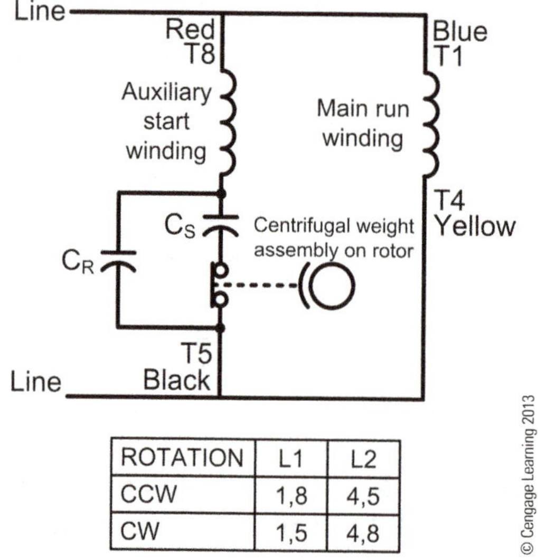


| ROTATION | L1 | L2 |
|---|---|---|
| CCW | 1,8 | 4,5 |
| CW | 1,5 | 4,8 |



**FIGURE 5-15** NEMA single-phase, capacitor start, capacitor run connection diagram

**IEC Capacitor Run Motor.** The IEC single-phase, single voltage, capacitor run motor diagram and connection information are shown in Figure 5-14.

**NEMA Capacitor Start and Run, Reversing Motor.** Single-phase capacitor start and run motors combine the increased locked rotor starting torque of capacitor start motors, with the increased operating torque of capacitor run motors. These motors require two separate capacitors, a high microfarad value AC electrolytic type capacitor for starting, and a low microfarad value paper and foil-type capacitor for running. The AC electrolytic starting capacitor must be removed from the circuit with a centrifugal switch when the motor approaches full speed the same as the capacitor start motor, but the run capacitor remains in the circuit whenever the motor is energized. The NEMA capacitor start and run motor diagram and connection information are shown in Figure 5-15.

**IEC Capacitor Start and Run Motor.** The IEC capacitor start and run motor diagram and connection information are shown in Figure 5-16.

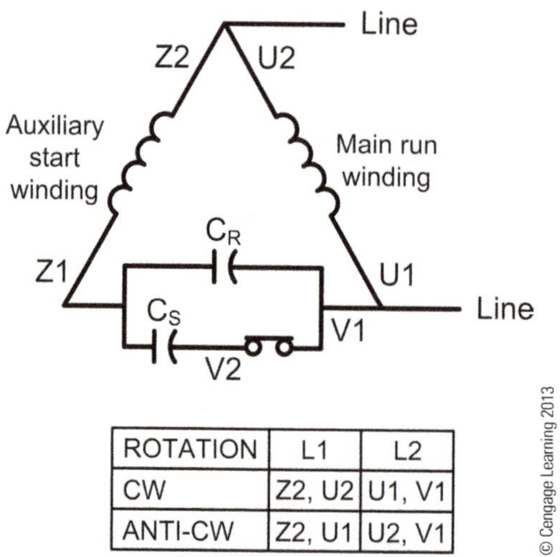


| ROTATION | L1 | L2 |
|---|---|---|
| CW | Z2, U2 | U1, V1 |
| ANTI-CW | Z2, U1 | U2, V1 |



**FIGURE 5-16** IEC single-phase, capacitor start and run coil drawing

## Three-Phase Motor Connections

There are two different three-phase connection configurations, wye and delta. Wye is the coil configuration where one end of each of the three individual phase coils of the motor is connected to a common point to form a Y shape, representing that the coils are separated by a phase rotation of 120 degrees. Delta is the coil configuration where each of the three individual phase coils are connected in series with the next to form a triangular shape, representing that the coils are separated by a phase rotation of 120 degrees. The electrical characteristics of these connection configurations is discussed in other parts of this book, but a shape recognition of each configuration is adequate for now.

**Three-Phase, Three-Lead Motor Connections.** Three-phase, three-lead induction motors are manufactured in either a wye or delta configuration, and are rated for only a single voltage. Figure 5-17 shows the connection diagrams for three-phase, three-lead motors. These motors are reversed in the same way as any other three-phase motor, by exchanging any two power supply leads, although the accepted standard practice is to exchange T1 and T3 at the motor starter. These motors are not very common, because they are not as versatile for being connected in different supply voltage situations as are other three-phase motors with two coil windings for each phase.

**Three-Phase Terminology.** For many years the wye configuration drawing was rotated 60° (the first drawing in Figure 5-17) to show one phase coil going straight up, with the other two phase coils going down, still separated by 120°. When drawn this way, it was called the star configuration. Many older electricians, and a lot of older electrical equipment such as motors and transformers still use the star terminology. If ever encountered in the field, the wye and star terms are interchangeable, but the star terminology will not be used here.

Also important to note here are the terms "line" and "phase." When talking about a three-phase configuration, the voltage measured from one power supply line to another power supply line is called the line voltage. The voltage measured from one power supply line to the common point with another coil winding is called the phase voltage. In the wye configuration line and phase voltage are two different values, but in the delta configuration they are the same, as shown in the second two drawings of Figure 5-17. Three-phase induction motors are a balanced three-phase load (no neutral conductor is required), so they are connected with only the three line leads from the electrical power supply. Three-phase induction motors are connected for the line voltage of the power supply, but the phase voltage is identified here to help explain the dual voltage coil configurations in each phase of the motor later in this chapter.

**NEMA Six-Lead, Three-Phase Motors.** Six-lead, three-phase motors have only a single coil per phase, similar to the three-phase, three-lead motor mentioned earlier, except that both lead ends of all coils are brought out to the motor conduit box. It may seem that having only one coil for each phase would be a detriment, because of the limited configuration possibilities like the three-phase, three-lead motor, but having both lead wires of each coil brought out to the motor conduit box makes the motor more versatile than the three-lead motor. Look at the three coils in Figure 5-18, and notice that making all six of the coil leads accessible in the motor conduit box allows this motor to be connected in either the wye or delta configurations to meet the needs of available utilization voltages, which is not true of the three-lead, three-phase motor. This versatility, coupled with the time savings and simplification of connecting only

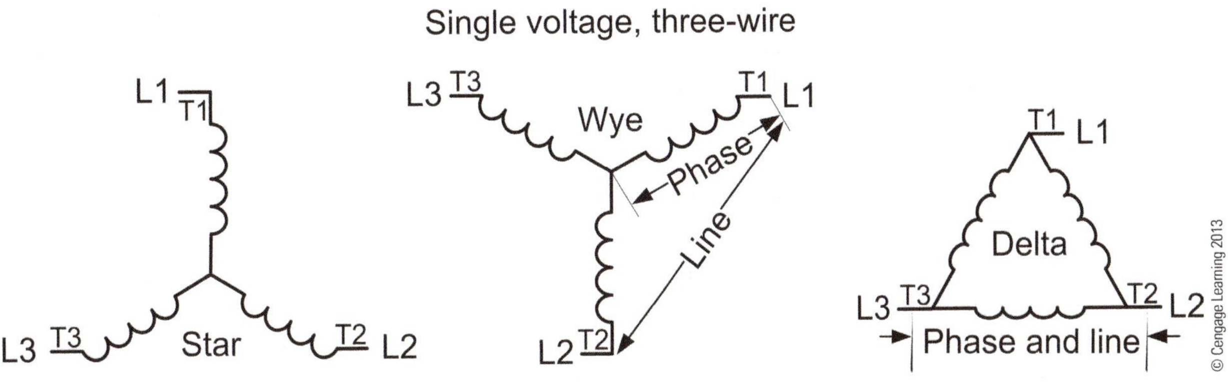


**FIGURE 5-17** Three-phase definitions

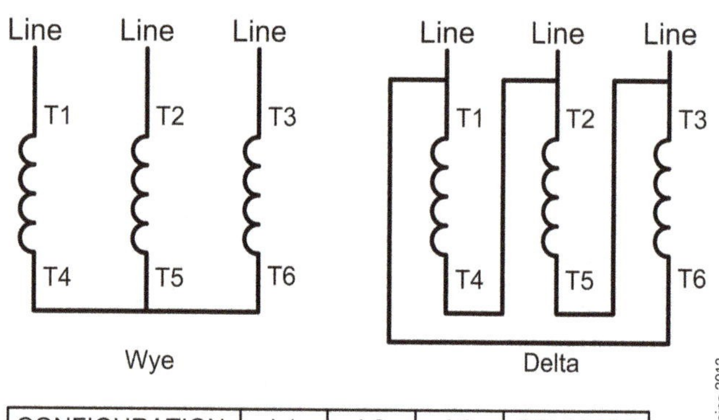

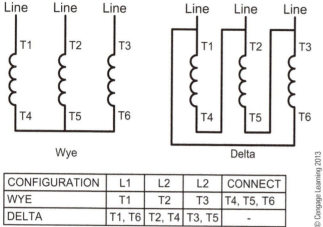

| CONFIGURATION | L1 | L2 | L2 | CONNECT |
|---|---|---|---|---|
| WYE | T1 | T2 | T3 | T4, T5, T6 |
| DELTA | T1, T6 | T2, T4 | T3, T5 | - |



**FIGURE 5-18** IEC six-lead, three-phase motor connected in either wye or delta configurations

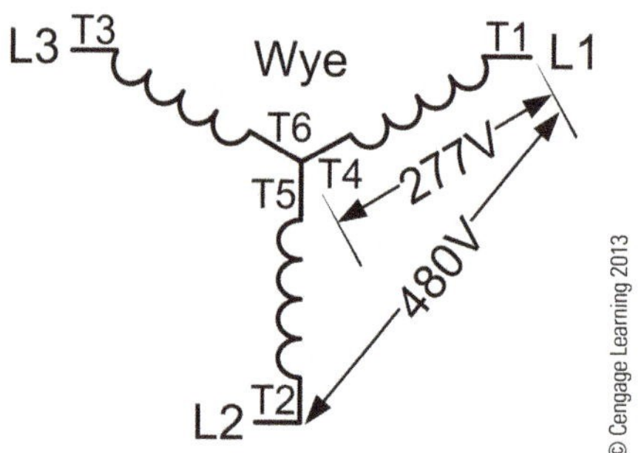




**FIGURE 5-19** NEMA reduced voltage on wye motor coils drawing

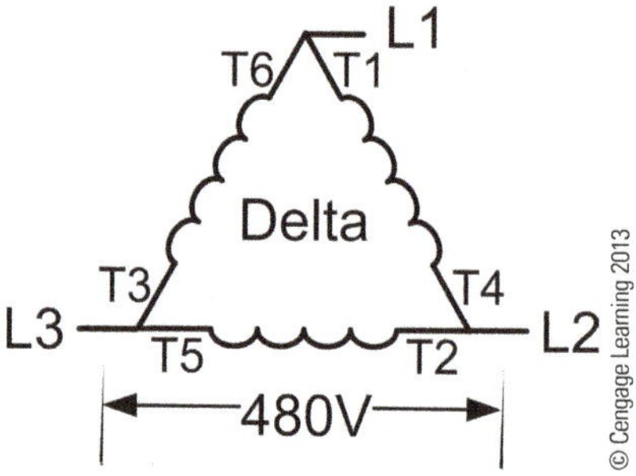




**FIGURE 5-20** NEMA full-rated voltage on delta motor coils drawing

six leads, rather than nine or twelve leads, makes this motor a good choice in some situations.

One application of six-lead, three-phase motors involves a motor starting method for large horsepower motors called wye-delta reduced voltage starting. A detailed explanation of the wye-delta reduced voltage motor starting scheme is too advanced for this study, but a short discussion will help explain the unique flexibility of the six-lead, three-phase motor for this application. The purpose of wye-delta starting is to limit the locked-rotor inrush starting current of the motor by reducing the voltage applied to each motor coil, which will reduce the current draw of each coil.

The term "reduced voltage starting" can appear misleading, because the power supply line voltage supplied to the motor does not change; the reduced voltage pertains to the phase voltage impressed on each of the motor coils. The reduced phase voltage to the motor coils is accomplished by the electrical characteristics between the wye and delta configurations. When the motor is first started, the motor starter will connect the motor coils in the wye configuration, and the voltage to each coil will be reduced to 57.7% of the power supply line voltage.

To better understand how the wye configuration reduces the applied voltage to each motor coil, review the three-phase phase and line definitions mentioned earlier in this chapter, and diagramed in Figures 5-17 and 5-19. With less voltage on the motor coils, the locked rotor inrush starting current of the motor will be less than if the full power supply line voltage is applied to the motor coils at startup. Once the motor is turning at a predetermined speed, the motor coil connection is changed over to the delta configuration, shown in Figure 5-20, by the motor starter. In the delta configuration the full power supply line voltage is applied to each motor coil, which is the rated voltage of the motor, and the motor will produce its rated horsepower and torque.

There are many methods of accomplishing the wye-delta reduced voltage start, and one of the more easily understood diagrams is shown in Figure 5-21. Most wye-delta reduced voltage starters will have three contactors: in the example the contactors are labeled M, for main; S, for start; and R, for run; along with one overload unit. To start the motor the main contacts are closed with the start contacts to connect the six-lead, three-phase motor into the wye configuration to start it turning from the locked rotor condition. When the motor reaches the predetermined speed the start contacts

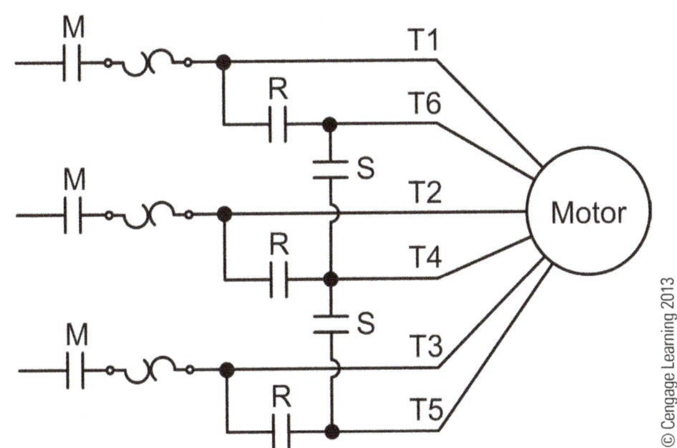


FIGURE 5-21 NEMA six-lead, three-phase, wye-delta scheme, reduced voltage start

are opened, and the run contacts are closed to connect the motor in the delta configuration for full voltage operation.

**IEC Six-Lead, Three-Phase Motors.** In countries that use IEC motors, six-lead, three-phase motors are probably more popular than their NEMA counterparts are in the United States, but for a different reason. Rather than being used for wye-delta, low voltage starting schemes, the IEC six-lead motor is popular because of its dual voltage rating capability. The dual voltage rating of the IEC six-lead motor is more than anything else the result of the differences in utilization voltages between the United States and countries that conform to the IEC standard. In the United States utilization voltages often follow a ratio of 2 to 1, such as 120/240 V, or 240/480 V. In countries that conform to the IEC standard, however, it is more common to find utilization voltages that follow a ratio closer to 1.732 (the square root of 3) to 1, such as 220/380 V.

In the IEC scheme, the motor coils are rated for the lower voltage of the available power supply. In the example utilization voltage of 220/380 V, the motor can be connected for 220-volt, low voltage operation if it is connected in the delta configuration, as shown in Figure 5-22, because the line voltage is the same as the phase voltage.

The same motor can be connected for 380-volt, high voltage operation if it is connected in the wye configuration, as shown in Figure 5-23. When 380 volts is applied to the lines of a wye-connected

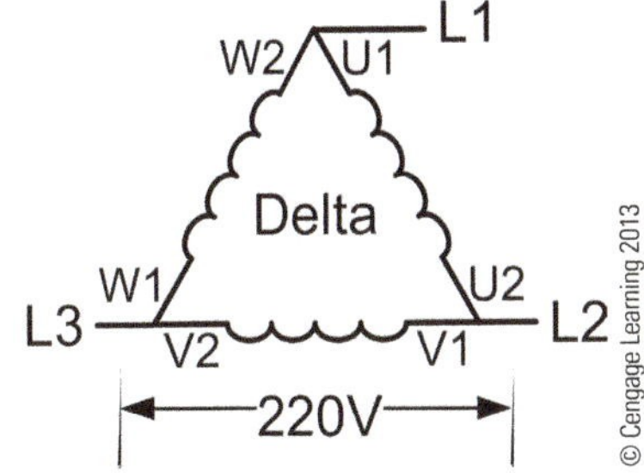


FIGURE 5-22 IEC low voltage delta motor configuration drawing

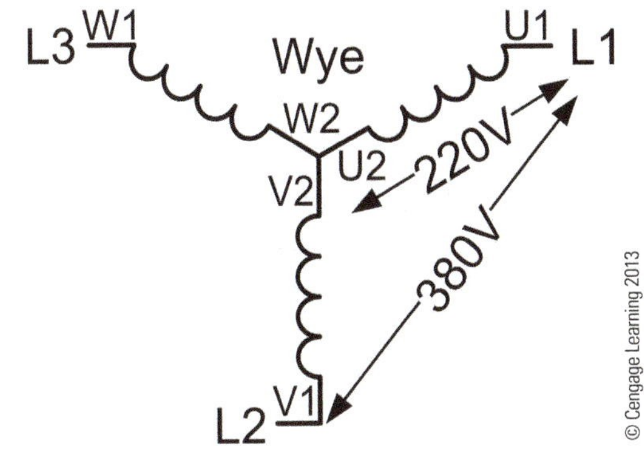


FIGURE 5-23 IEC high voltage wye motor configuration drawing

motor, each of the coils will be 380 V/1.732, or approximately 220 V. An IEC motor rated for 220/380 volts would not make a suitable dual voltage motor on the electrical distribution system found in the United States, because of the differences in voltage levels.

## Nine-Lead, Three-Phase Motors

**Dual Voltage, Three-Phase Motors.** Dual voltage, three-phase motors have two coils per phase, and each coil is rated for the same voltage. When connecting the motor for the lower of its rated voltages, referred to as the low voltage connection, the two coils are connected in parallel so that each coil receives the full phase voltage. When connecting the motor for the higher of its rated voltages, referred to as the high voltage connection, the two coils are connected in series so that each coil receives only half of the phase voltage. There is no difference between the nine- and twelve-lead NEMA and IEC motors, so only the NEMA connections will be drawn here.

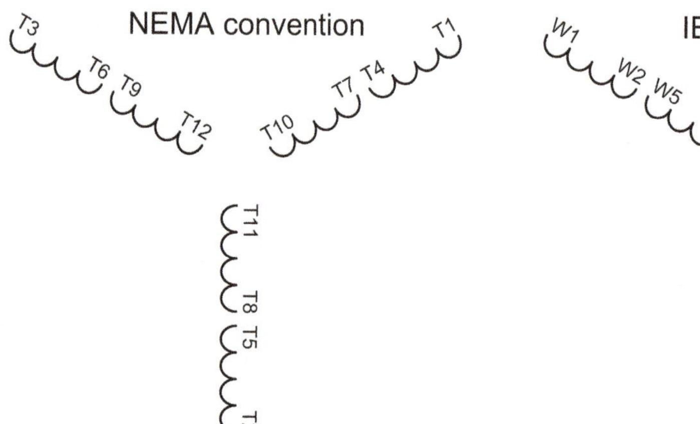


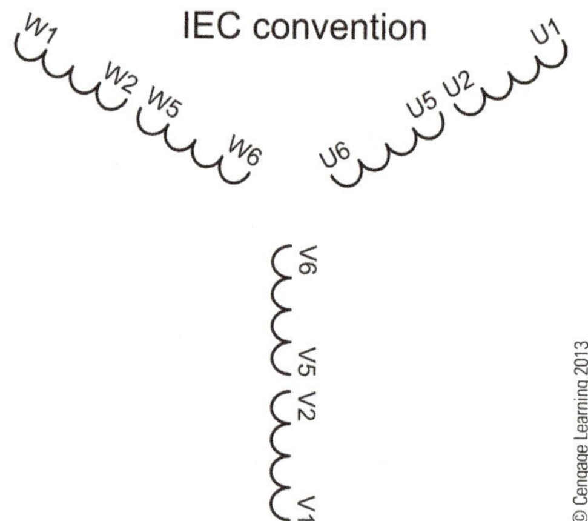


FIGURE 5-24 NEMA and IEC coil labels for wye

**Three-Phase Wye Configuration.** Figure 5-24 shows the three-phase wye coil configuration with two windings per phase, with both the NEMA and IEC convention labels.

**Labeling the Wye Configuration Coil Winding Leads.** Dual voltage, wye configuration, three-phase motors have six separate coil windings inside them, with a total of twelve leads. Each of the twelve leads requires a unique label to differentiate each coil winding from all the others, and a scheme to indicate each coil winding's magnetic polarity in relation to all the others. NEMA and IEC have different labeling conventions, but they both hold to patterns that are easy to remember.

IEC designates the labels 1 and 2 for the first coil winding of each phase, and labels 5 and 6 for the second coil winding. For each phase, the magnetic polarity of each coil winding from the 1 label to the 2 label, and the 5 label to the 6 label, is going to be the same. The coil windings must be further differentiated by phase, or there would be three leads with the label 1, three with the label 2, etc., in each motor conduit box. To differentiate the phases, the IEC convention uses the letters U, V, and W. Whether the motor is connected for high voltage or low voltage operation, the 1 label coil windings of each phase will connect to lines L1, L2, and L3 from the power supply.

NEMA designates all of the coil winding leads with the label T, and then further designates each of the leads with the numbers 1 thru 12 to differentiate individual coil windings. At first it may appear that the NEMA coil winding leads are numbered randomly, but they are not. The NEMA convention for labeling the leads is derived from a pattern, which is demonstrated in Figure 5-25.

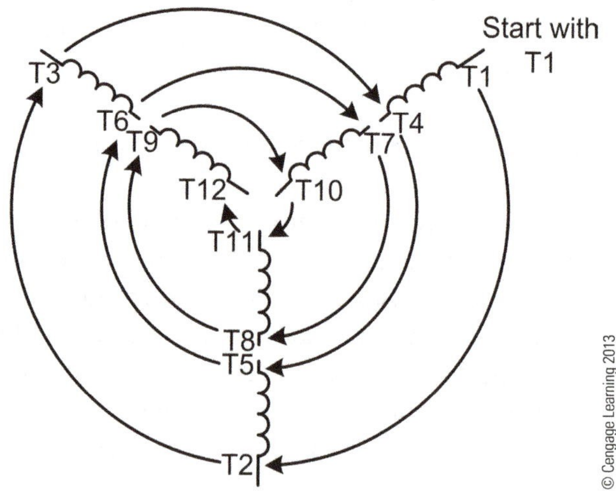


FIGURE 5-25 Labeling the leads of a wye-connected motor

**Wye Configuration Lead Labels.** The wye coil winding leads are numbered by starting at one of the outside points of the wye, and rotating around the outside points in a clockwise direction, numbering the beginning point of the first coil in each phase with the labels T1, T2, and T3. After numbering the beginning point of the first coil in each phase, continue rotating around in the same clockwise direction, and label the ends of the first coils in each phase, continuing with labels T4, T5, and T6. Repeat the same pattern for the beginning point of the second coil in each phase, with the labels T7, T8, and T9. Finally, label the ends of the second coil in each phase with the labels T10, T11, and T12.

**Three-Phase Delta Configuration.** Figure 5-26 shows the three-phase delta coil configuration with two windings per phase, with both the NEMA and IEC convention labels.

**Labeling the Delta Configuration Coil Winding Leads.** Dual voltage, delta configuration, three-phase motors, like dual voltage, wye configuration, three-phase motors, have six separate coil windings inside them, with a total of twelve leads. And, like the wye configuration, the twelve leads require a unique label to differentiate each coil winding from all the others, and a scheme to indicate each coil winding's magnetic polarity in relation to all the others. Again, NEMA and IEC have different labeling conventions, but they both hold to patterns that are easy to remember.

IEC designates the labels 1 and 2 for the first coil winding of each phase, and labels 5 and 6 for the second coil winding. For each phase, the magnetic polarity of each coil winding from the 1 label to the 2 label, and the 5 label to the 6 label, is going to be the same. The coil windings must be differentiated further by phase or there would be three leads with the label 1, three with the label 2, etc. in each motor conduit box. To differentiate the phases, the IEC convention uses the letters U, V, and W. Whether the motor is connected for high voltage or low voltage operation, the 1 label coil windings of each phase will connect to lines L1, L2, and L3 from the power supply.

Again, NEMA designates all of the coil winding leads with the label T, and then further designates each of the leads with the numbers 1 thru 12 to differentiate individual coil windings. The NEMA convention for labeling the leads is derived from a pattern, which is demonstrated in Figure 5-27. Figure 5-27 is drawn in two stages, because trying to show the entire circular pattern in one drawing is too cluttered.

**Delta Configuration Lead Labels.** The delta coil winding leads are numbered by starting at the beginning of the first coil winding at a corner of the delta diagram and rotating around the outside points in a clockwise direction, numbering the beginning point of the first coil in each phase with the labels T1, T2, and T3. After numbering the beginning point of the first coil in each phase, continue rotating around in the same clockwise direction and label the ends of the first coils in each phase, continuing with labels T4, T5, and T6. Repeat the same pattern for the beginning point of the second coil in each phase with the labels T7, T8, and T9. Finally, label the ends of the second coil in each phase with the labels T10, T11, and T12.

**Drawing Coil Winding Leads Crossed.** From this point on, the individual leads between the two coil windings of each phase will be drawn longer, and crossed, as shown in Figure 5-28. When the leads are crossed, it can appear that the T lead

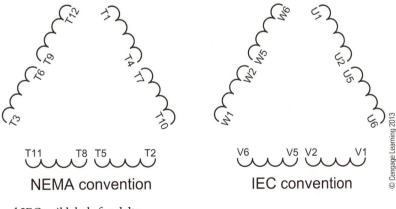

**FIGURE 5-26** NEMA and IEC coil labels for delta

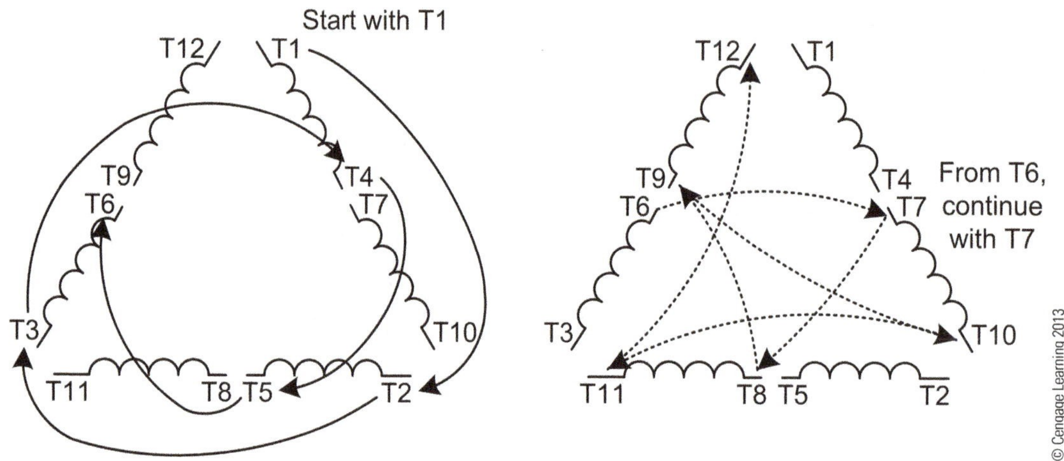


FIGURE 5-27 Labeling the leads of a delta-connected motor

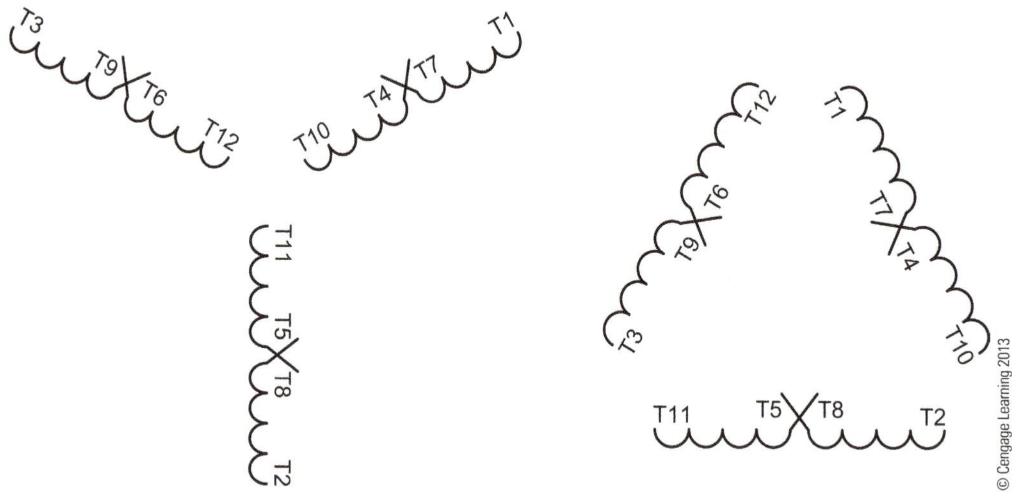


FIGURE 5-28 Wye and delta coils crossing

number labels have been changed, but they have not. Rather than showing the T lead number label at the end of the coil, they are now moved to the end of the coil lead. The reason for drawing them crossed is to help simplify the connection drawings when showing them connected in series and parallel for high voltage and low voltage configurations.

## The Three-Phase, Twelve-Lead Motor

The three-phase, twelve-lead motor has both lead ends of all six of the individual coil windings brought out to the motor conduit box. This motor configuration has the advantage that it may be connected for either wye or delta, and high voltage or low voltage. Bringing all twelve lead ends out to the conduit box also allows the motor to be used for specialized motor starting schemes, such as the wye-delta reduced voltage starting scheme, or part winding motor starting where only some of the motor windings are energized for starting purposes. These types of specialized motor starting schemes are reserved for only the largest motors that represent a small fraction of the induction motors installed. For the vast majority of three-phase induction motors, twelve-lead motors will not be the preferred type, because of the additional time it takes to connect twelve motor leads in the field, rather than only nine leads, and the added opportunity there is to make mistakes because of the additional connections.

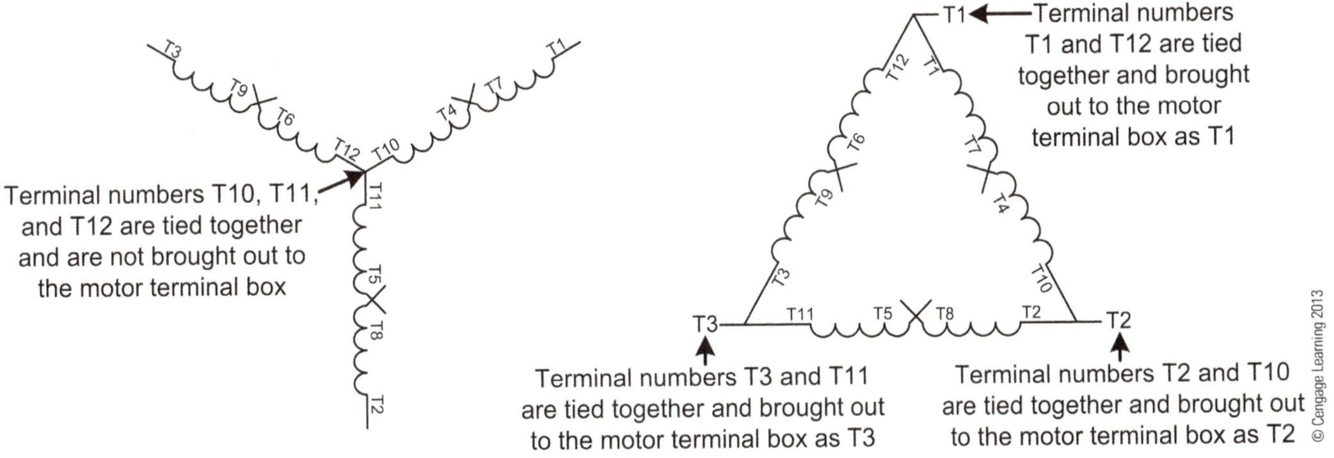


FIGURE 5-29 NEMA internal stator coil connections of wye and delta configurations

## Three-Phase, Nine-Lead Motor Internal Stator Coil Connections

The three-phase, nine-lead motor is far and away the most common motor found in new commercial and industrial construction work. The motor leads are numbered T1 through T9. If dual voltage three-phase motors actually have twelve leads as mentioned above, how is it that only nine of those leads are brought out to the conduit box to be connected to the line? Figure 5-29 shows that when the internal winding connection is made for wye, terminal numbers T10, T11, and T12 all are connected internally at the time of manufacture, and cannot be accessed from the motor's terminal box. When the internal winding connection is made for delta, T1 and T12 are connected and brought out to the terminal box as T1; T10 and T2 are connected and brought out to the terminal box as T2; and T11 and T3 are connected and brought out to the terminal box as T3. For either the wye or delta configurations, nine-lead motors will not have the terminal numbers T10, T11, or T12 brought out to the terminal box of the motor.

**Nine-Lead Wye Low Voltage Connection Diagram.** When a nine-lead, wye-connected motor is wired for low voltage, the phase coils must be connected in parallel so that each coil has its rated voltage applied. This presents a dilemma, because the point where terminal numbers T10, T11, and T12 were connected at the time of manufacture cannot be accessed to complete the parallel connection. To solve this problem, the low voltage wye connection ends up being connected as two separate wye connections within the motor, as shown in Figure 5-30. By connecting T1 with T7, T2 with T8, and T3 with T9, the internal T10, T11, and T12 connection creates the first wye connection. Then, by connecting terminal numbers T4, T5, and T6 in the terminal box, a second wye connection is made. This connection in effect connects each of the phase coils in parallel, which means that the applied voltage is impressed on each of the individual coils.

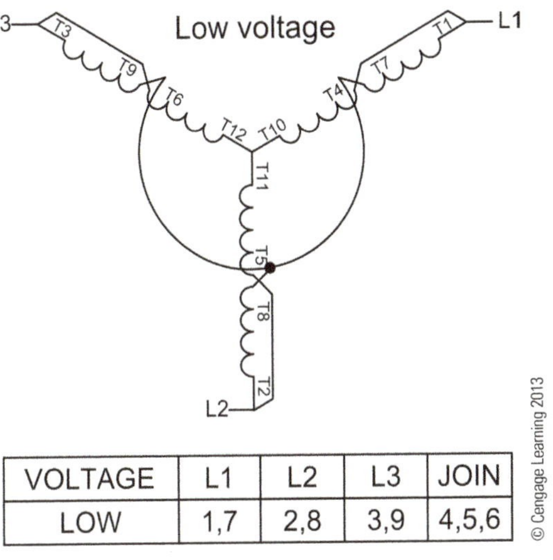


| VOLTAGE | L1 | L2 | L3 | JOIN |
|---|---|---|---|---|
| LOW | 1,7 | 2,8 | 3,9 | 4,5,6 |

FIGURE 5-30 NEMA nine-lead, low voltage, wye-connected motor

**Nine-Lead Wye High Voltage Connection Diagram.** When a nine-lead, wye-connected motor is wired for high voltage, the phase coils must be connected in series so that only half of the applied phase voltage is impressed across each individual coil, as shown in Figure 5-31. By connecting T4 and T7 together, T5 and T8 together, and T6 and T9 together, the phase coils are connected in series; and the internal connection of T10, T11, and T12 completes the wye connection.

**Nine-Lead Delta Low Voltage Connection Diagram.** When a nine-lead, delta-connected motor is wired for low voltage, the phase coils are connected in parallel, as shown in Figure 5-32. At this point it is important to remember that although motor terminal numbers T10, T11, and T12 are not brought out directly to the motor terminal box, they still are accessible via T1, T2, and T3. Remember that T12 and T1 are connected internally, as are T10 and T2, and T11 and T3. Therefore, any connection to T1 is also a connection to T12, and the same with T2 and T10, and T3 and T11. T1, then, connects to both T6 and T7, T2 connects to T4 and T8, and T3 connects to T5 and T9. This arrangement connects the phase coils in parallel, and the supply voltage is impressed on each individual coil.

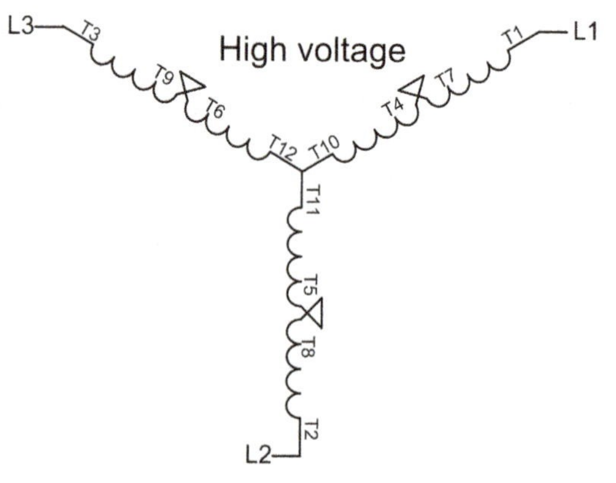


| VOLTAGE | L1 | L2 | L3 | JOIN |
|---|---|---|---|---|
| HIGH | 1 | 2 | 3 | 4,7 |
| | | | | 5,8 |
| | | | | 6,9 |

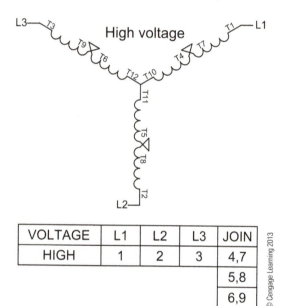

**FIGURE 5-31** NEMA nine-lead, high voltage, wye-connected motor

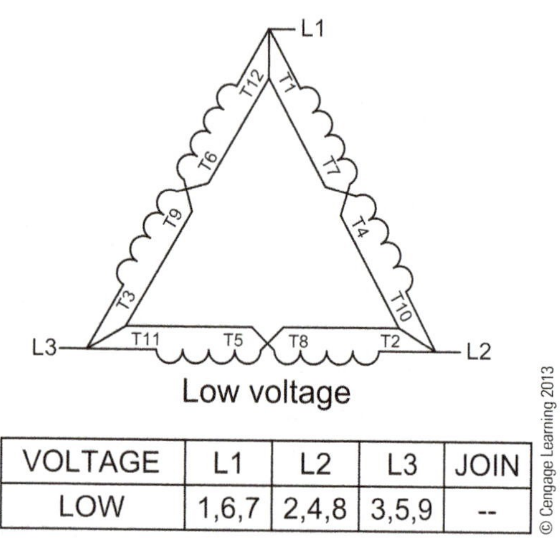


| VOLTAGE | L1 | L2 | L3 | JOIN |
|---|---|---|---|---|
| LOW | 1,6,7 | 2,4,8 | 3,5,9 | -- |



**FIGURE 5-32** NEMA nine-lead, low voltage, delta-connected motor

**Nine-Lead Delta High Voltage Connection Diagram.** When a nine-lead, delta-connected motor is wired for high voltage, the phase coils are connected in series, as shown in Figure 5-33, so that half of the supply voltage is impressed on each individual coil. By connecting T4 and T7 together, T5 and T8 together, and T6 and T9 together, the phase coils are connected in series; and the internal connections of T1 and T12, T2 and T10, and T3 and T11 complete the delta connection.

## Common High Voltage Connections

It is important to observe that although the low voltage connections for the wye and delta configurations are completely different, the high voltage

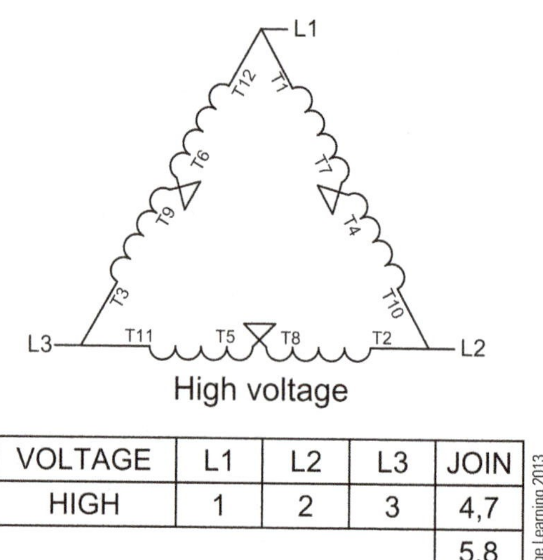


| VOLTAGE | L1 | L2 | L3 | JOIN |
|---|---|---|---|---|
| HIGH | 1 | 2 | 3 | 4,7 |
| | | | | 5,8 |
| | | | | 6,9 |



**FIGURE 5-33** NEMA nine-lead, high voltage, delta-connected motor

connections are exactly the same. This is important to note, because the majority of the times you connect the nine-lead motor it will be for the high voltage connection. The common misconception is that we connect motors for their higher-rated voltage, because they are less expensive to operate. This is completely false; any given dual voltage motor will cost the same to operate, regardless if it is operating at the lower or higher voltage.

**Money Savings for High Voltage Connection.** To understand why a motor costs the same to operate on both the low and high voltages, a little series and parallel theory is necessary. To keep the math simple, let's say that each individual coil of the motor is rated for 100 volts and draws 2 amperes at that voltage. When the coils are connected in parallel, low voltage, the voltage will be 100 volts for both coils, but the current will be the sum of the two coils, or 4 amperes (100 volts times 4 amperes equals 400 watts, what we pay for). When the coils are connected in series, high voltage, the voltage will be the sum of the two coils, 200 volts, but the current will remain the same 2 amperes through both coils; 200 volts times 2 amperes equals 400 watts. With no difference in power, the operating cost of the motor would be exactly the same, regardless of the voltage on which it is operated.

So, where is the savings in connecting the motor for the higher of the rated voltages? The initial cost of installing the motor is less expensive. When the motor is connected for the higher of its rated voltages, it will draw the lower of its rated currents. This means that the electrical materials and equipment required to install the motor can use a smaller and less expensive components, from the disconnect, starter, wire, pipe, etc., right down the line. The lower current draw also would tax the facility electrical system less, leaving more current capacity for other applications.

**Predicting the Motor Direction of Rotation.** Three-phase motors are not like single-phase motors, where the direction of rotation can be determined ahead of time by the physical relationship between the auxiliary start and main run windings. Three-phase motors do not have auxiliary start windings. They create their own rotating magnetic field in the stator, as each of the three phases, physically placed 120° out of phase with each other, are energized and de-energized in sequence. The three phases of an electrical supply are completely random, and it is only their relationship to each other that matters. Therefore, even though motors all may be manufactured with all their coils wound in the same way, every three-phase facility electrical power supply to which they will be connected is random and cannot be predicted.

The chances are 50/50 that a three-phase motor is going to turn in the intended direction of rotation the first time it is first energized. When a three-phase motor is connected initially, it is normally "bumped" by starting and stopping it quickly and noting the direction of rotation. If the direction of rotation is correct, the installation is complete. If the direction of rotation is wrong, any two of the motor leads may be exchanged to reverse the rotation of the motor. As mentioned earlier in this part, the standard is to exchange motor leads T1 and T3 at the bottom of the motor starter.

**Determining the Motor Direction of Rotation.** Some motor installations where the motor turns in the wrong direction would be unacceptable, possibly even damaging, even to bump the motor in the wrong direction. For these situations, there are motor rotation testers that can be used to compare the rotation of the three-phase power supply against the actual intended rotation of the motor. The purpose of this tester is to assure that the motor will turn in the correct direction on the first try. It will take a little three-phase theory and a little magnetic theory to understand how low the tester works, and how to apply it, but this is how it works.

First of all, every three-phase power supply has a rotation that describes how the three phases are ordered in relation to each other, and there are only two possibilities: clockwise or counterclockwise. It is all based on the motor stator, as shown in the diagram of Figure 5-34. If one phase is chosen to be phase A, phases B and C can follow in only one of two orders: either BC or CB. One of these orders would constitute a clockwise rotation, and the other would constitute a counterclockwise rotation.

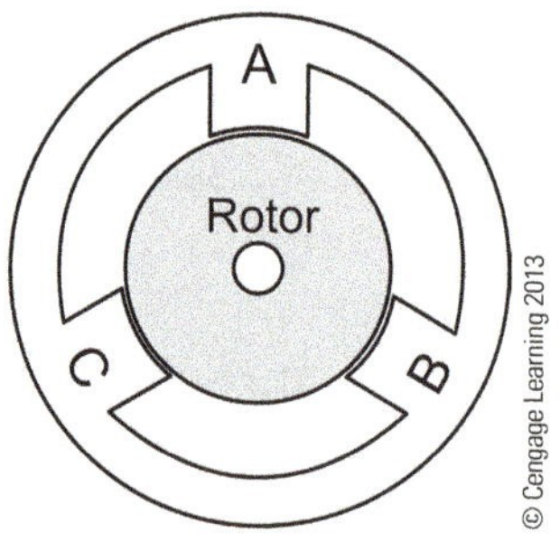


FIGURE 5-34 Pole relationships for CW or CCW rotation diagram

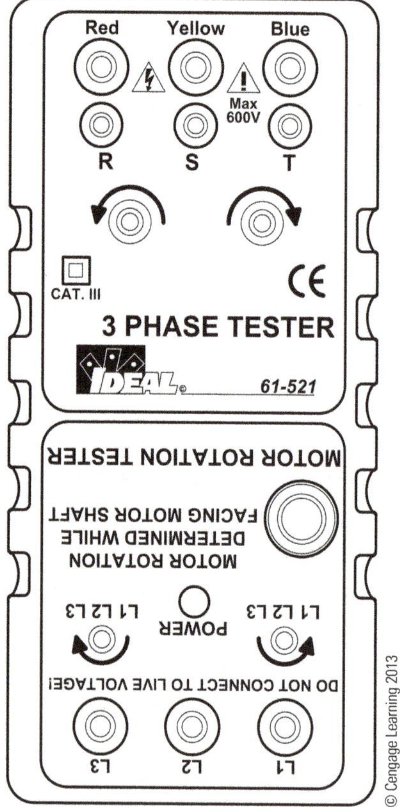


FIGURE 5-35 Ideal three-phase tester drawing

Stated another way, if the motor in Figure 5-35 was turned on and the direction of rotation was found to be clockwise, it would mean that the phases were following an ABC pattern. If the direction of rotation had to be changed, theoretically it would not matter which two phases were interchanged. If phases A and B were interchanged, the phase pattern would become BAC, which is counterclockwise. If instead phases A and C were interchanged, the phase pattern would become ACB, which also is counterclockwise.

**Three-Phase Rotation Testers.** The rotation of the three-phase supply is read with the phase rotation tester, such as the Ideal manufacturer's model 61-521, called a three-phase tester/motor rotation tester, shown in Figure 5-35. By connecting the three leads of the three-phase tester side to the three-phase power supply, the order of each phase is read in relation to the other phases. The three leads of the tester are color-coded red, yellow, and blue. The meter leads initially may be connected in any order, but that same order must be maintained for all testing with that meter. Rather than trying to remember some arbitrary order, most people connect up the leads from left to right on the tester: red, yellow, blue, which is the same sequence they appear on the meter. Sometimes when you look inside a three-phase electrical panel you may see where someone has written "RYB-CW" with a marker. This means that someone before you had a need to know the rotation of that panel, and so they tested it by connecting a three-phase meter red, yellow, blue, and found the rotation to be clockwise for that connection order.

**Motor Rotation Tester.** As shown in Figure 5-36, the opposite end of the Ideal 61-521 tester is a motor rotation tester, which will indicate if the induction motor it is connected to is turning clockwise or counterclockwise. To understand how the meter can sense the electrical rotation of a motor, it is important to understand the residual magnetism concept of magnetic theory. Once a piece of iron has been magnetized, it will never become completely demagnetized after the magnetizing force has been removed. With soft iron, the kind used in electric motors, the residual magnetism is very low, but it can be detected by sensitive instruments. The ideal motor rotation tester is such an instrument. The soft iron rotor of an induction motor becomes an electromagnet during operation, and therefore retains some residual magnetism thereafter. By connecting the motor rotation tester to the motor leads, and spinning the shaft of the motor in the direction you want it to turn, the residual magnetism in the rotor iron laminations will cut through the conductors of the stator coils and induce a voltage. Motor rotation testers

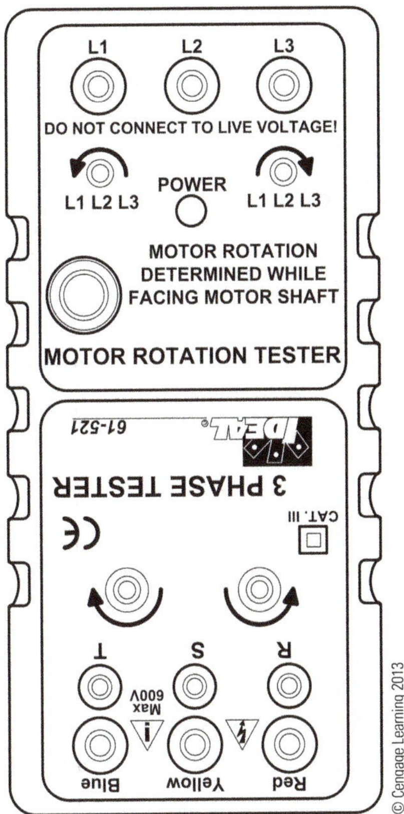


**FIGURE 5-36** Ideal motor rotation tester drawing

can detect if the motor coils are electrically related to one another in a clockwise or counterclockwise sequence for the direction that the motor shaft was turned while the rotation tester was connected.

**Using the Tester.** Notice that the tester has two different ends to it; it is actually two different testers in one package. Using the tester lead color sequence red, yellow, blue, carefully connect the "Three-phase Tester" side of the meter (not the side that states "Do not connect to live voltage!") to the three-phase power supply: L1, L2, and L3. It is safest, but not always possible, to de-energize the three-phase supply before connecting the tester, and then reenergize the circuit to make the meter reading. Once the meter is connected and the circuit is energized, note if the tester indicates a clockwise or counterclockwise rotation for that power source.

Disconnect the tester from the line side of the motor starter, and remove the test leads from the tester. Turn the tester 180° to the side labeled "Motor Rotation Tester." Plug the test leads into this side of the tester in the same red, yellow, and blue order, and connect the tester leads to the T1, T2, and T3 terminals of the motor, connected for either high or low voltage. Hold down the tester test button and spin the motor shaft in the direction you wish the motor to turn. Note if the tester determines that the direction that you spun the motor was electrically a clockwise or counterclockwise rotation for that motor. If the rotation direction of the power supply line agrees with the rotation direction of the motor load terminals, the motor can be connected and it will turn the desired direction when energized without any wiring changes. If the two rotation readings were different, exchange the T1 and T3 leads of the motor with the line leads of the power supply, and the motor will turn in the desired direction when energized.

## CHAPTER SUMMARY

- Whenever connecting any motor, it is always best to use the nameplate information provided by the manufacturer.
- It is impossible to create standardized connection diagrams and tables that would apply across different manufacturers of IEC motors in all cases.
- It is most common for NEMA motors to use terminal blocks in the front end bell for single-phase motors.
- NEMA three-phase motors normally use open-ended lead wires marked with the T label system in the conduit box.
- IEC motors use terminal blocks for both single-phase and three-phase motors.
- The main run windings of a split-phase motor may be labeled either T1 and T2, or T1 and T4, depending on the manufacturer.
- The auxiliary start winding of a split-phase motor is labeled T5 and T8.
- The direction of rotation may be changed for single-phase motors by exchanging the leads for either the main run winding or the auxiliary start winding, but not both.
- NEMA motors define the standard direction of rotation as counterclockwise from the front, or non-shaft end of the motor.
- IEC motors define the standard direction of rotation as clockwise from the shaft end of the motor.
- Six-lead, three-phase motors have the versatility that they can be connected in either the wye or delta configurations.
- The wye-delta reduced voltage start scheme is a method of limiting the locked rotor starting current of a motor by reducing the voltage applied to each motor coil when the motor is first energized.
- Dual voltage rated motors have two coils, which are connected in parallel for low voltage operation, or in series for high voltage operation.
- Three-phase induction motors are manufactured in either a wye or delta configuration, and single or dual voltage.
- The voltage measured from one line to another line is called the line voltage. The voltage measured from one line to the common point with another coil winding is called the phase voltage.
- The three-phase, nine-lead motor is the most common motor found in new commercial and industrial construction work.
- The low voltage connections for the wye and delta configurations are completely different, but the high voltage connections are exactly the same.
- Any given dual voltage motor will cost the same to operate, regardless if it is operating at the lower or higher voltage.
- The cost savings associate with running a motor at the higher of its rated voltages comes from the initial installation. The electrical materials and equipment can use a smaller and less expensive disconnect, starter, wire, pipe, etc.
- The three-phase, twelve-lead motor brings all twelve stator coil leads out to the motor terminal box, but is less common than nine-lead motors.
- Both wye and delta coil configurations are numbered the same way on a diagram. Starting at one of the outside points of either of the wye or delta drawing, start with the label T1, and rotate around the coils in a clockwise direction, moving in.
- The direction of rotation for a three-phase motor cannot be predicted without a tester. The chances are 50/50 that the motor will turn in the correct direction the first time it is connected.

- The direction of rotation for a three-phase motor may be reversed by exchanging any two of the three power supply conductors, but the accepted standard is to exchange T1 and T3 on the bottom of the motor starter.
- For the installations where it is unacceptable for a three-phase induction motor to turn in the wrong direction, a motor rotation tester is used to predetermine the correct direction of rotation.

## REVIEW QUESTIONS

1. What is the best source for motor connection information?
2. With what motor standards are original equipment manufacturer motors required to comply?
3. When are the motor lead wire color codes especially handy?
4. What terminal labels are given to the auxiliary start windings of NEMA single-phase motors?
5. What terminal labels are given to the main run windings of NEMA single-phase motors?
6. What is the standard direction of rotation for NEMA and IEC single-phase motors?
7. How is the direction of rotation reversed for single-phase motors?
8. How are the two main run windings of a dual voltage, single-phase motor connected for the low voltage connection?
9. How are the two run windings of a dual voltage, single-phase motor connected for the high voltage connection?
10. What are the two different configurations in which three-phase motors are manufactured?
11. What terminal labels are given to the leads of a three-phase, three-lead motor?
12. Why are three-phase, three-lead motors not very common?
13. What advantage does a six-lead, three-phase motor have over a three-lead, three-phase motor?
14. What savings does the text associate with six-lead motors over nine- and twelve-lead motors?
15. What is the advantage of manufacturing three-phase motors with nine leads brought out to the motor terminal box?
16. What terminal labels are given to the leads of three-phase, nine-lead motors?
17. What is significant to note about the high voltage connection for both the wye- and delta-configured three-phase motors?
18. What is the operating cost difference between operating a given dual voltage, three-phase motor on the higher-rated voltage, as compared to operating it on the lower-rated voltage?
19. What are the cost savings associated with installing a given dual voltage, three-phase motor on the higher-rated voltage, as compared to installing it on the lower-rated voltage?
20. What are the chances that a three-phase motor is going to turn in the intended direction of rotation the first time it is energized?
21. For motor installations where having the motor turn in the wrong direction would be unacceptable, or possibly even damaging, what type of tester can be used to assure the correct direction of rotation the first time?
22. Number each of the coil winding shown on page 74 leads with the correct NEMA T number labels, and IEC U, V, and W labels, for both the wye and delta configurations.

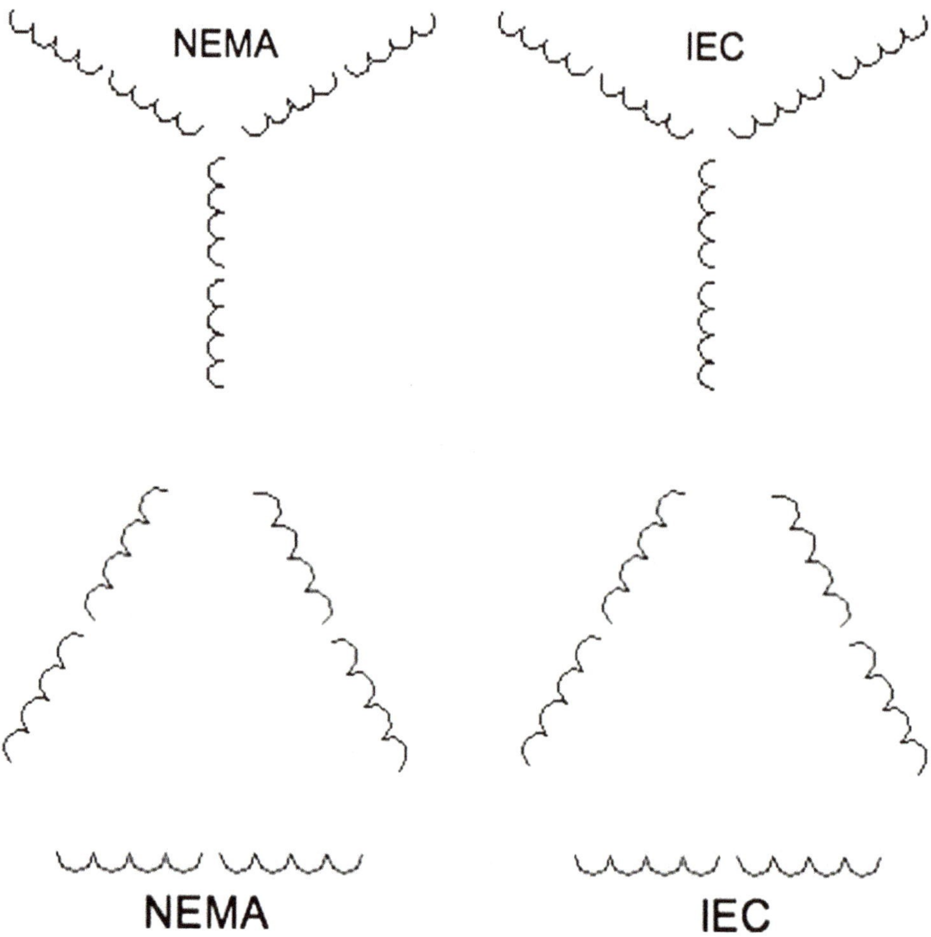
NEMA
IEC
NEMA
IEC

CHAPTER

6

# Motor Nameplates

**PURPOSE**

To familiarize the learner with the information found on induction motor nameplates.

**OBJECTIVES**

After studying this chapter on induction motor nameplates, the learner will be able to:

- Explain the purpose of motor nameplate information
- Identify the two common motor standards for electricians
- Explain common motor nameplate ratings

## MOTOR NAMEPLATE INFORMATION

The motor nameplate pictured in Figure 6-1 most likely does not exist anywhere, but is provided here to contrast how NEMA and IEC motors document nameplate information. The left-hand side of the nameplate shows the NEMA conventions, and the right-hand side shows the same motor attributes with the IEC conventions. The bottom portion of the drawing shows the nameplate information that is the same for both conventions.

### Standards Agencies

Standards agencies such as NEMA and IEC specify certain information that is to be listed on the motor's nameplate describing the physical and operating characteristics for each type of motor design. The purpose of standardizing this information is to ensure the correct interchangeability of motors between different motor manufacturers. The NEMA standard, MG-1 (Motors and Generators 1), sets the standard for the construction and manufacture of AC and DC motors and generators, and will be the main standard referenced.

The IEC standard, 60034-1, pertains to all rotating electrical machines, and will be referenced for comparison purposes with the NEMA standard where applicable. Both NEMA and IEC motors and motor control components are common in the electrical industry today, because electrical equipment is manufactured in the global economy under both standards and may be shipped anywhere in the world. Also, electrical contractors in the United States, trying to cut costs, often purchase the less expensive imported IEC manufactured components for their electrical installations.

### Nameplate Information

The following list is not intended to be an all-inclusive list of all information that can be found on nameplates, but it is a good start to understanding what types of information may be provided. As each of these information items are discussed in the remainder of this chapter, some will require more explanation that will be covered in other chapters, and some that are not covered in other parts of the book will be elaborated on here. Not every motor nameplate will have every item covered here.

1. Manufacturer's name and address
2. Rated voltage (V, or VOLT)
3. Rated full load amperes (AMPS, or FLA)
4. Rated frequency (F, Hz, or FREQ)
5. Number of phases (P, PH, or PHASE)
6. Rated full load speed (RPM)
7. Rated horsepower (HP)
8. Locked rotor kVA code letter (CODE)
9. Rated duty or time rating (DUTY)
10. Insulation class temperature (INS, or INSUL CLASS)
11. Temperature rise (RISE)

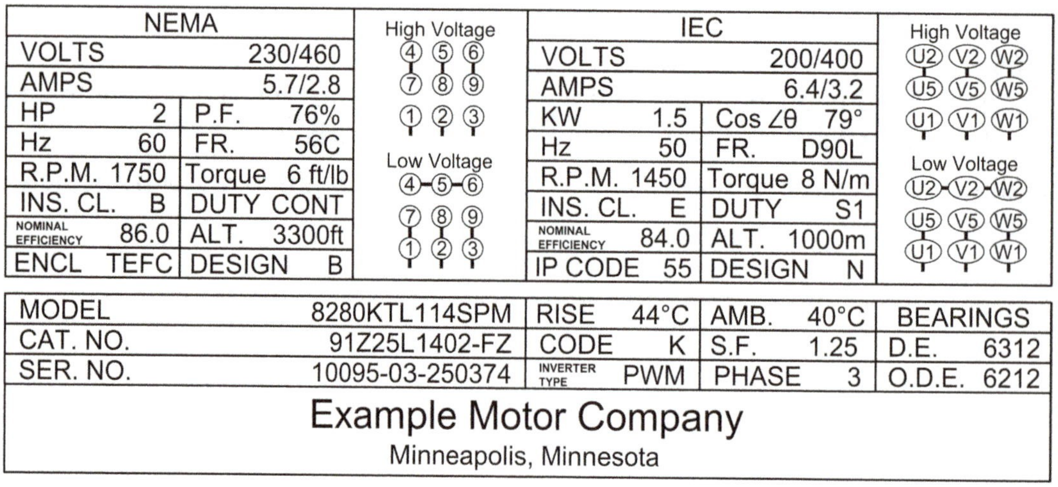

| NEMA | | | |
|---|---|---|---|
| VOLTS | 230/460 | | |
| AMPS | 5.7/2.8 | | |
| HP | 2 | P.F. | 76% |
| Hz | 60 | FR. | 56C |
| R.P.M. | 1750 | Torque | 6 ft/lb |
| INS. CL. | B | DUTY | CONT |
| NOMINAL EFFICIENCY | 86.0 | ALT. | 3300ft |
| ENCL | TEFC | DESIGN | B |

High Voltage: ④ ⑤ ⑥ / ⑦ ⑧ ⑨ / ① ② ③
Low Voltage: ④-⑤-⑥ / ⑦ ⑧ ⑨ / ① ② ③

| IEC | | | |
|---|---|---|---|
| VOLTS | 200/400 | | |
| AMPS | 6.4/3.2 | | |
| KW | 1.5 | Cos ∠θ | 79° |
| Hz | 50 | FR. | D90L |
| R.P.M. | 1450 | Torque | 8 N/m |
| INS. CL. | E | DUTY | S1 |
| NOMINAL EFFICIENCY | 84.0 | ALT. | 1000m |
| IP CODE | 55 | DESIGN | N |

High Voltage: U2 V2 W2 / U5 V5 W5 / U1 V1 W1
Low Voltage: U2-V2-W2 / U5 V5 W5 / U1 V1 W1

| | | | | | | | | |
|---|---|---|---|---|---|---|---|---|
| MODEL | 8280KTL114SPM | RISE | 44°C | AMB. | 40°C | BEARINGS | |
| CAT. NO. | 91Z25L1402-FZ | CODE | K | S.F. | 1.25 | D.E. | 6312 |
| SER. NO. | 10095-03-250374 | INVERTER TYPE | PWM | PHASE | 3 | O.D.E. | 6212 |

Example Motor Company
Minneapolis, Minnesota



**FIGURE 6-1** Example motor nameplate drawing

12. Ambient temperature (AMB, or TEMP)
13. Starts-per-hour (S/HR, or STARTS/HR)
14. Type (TYPE)
15. Rated torque (TORQUE)
16. Altitude (ALT)
17. Power factor (PF, or cos $\angle\theta$)
18. Service factor (SF)
19. Efficiency (EFF, or NOM EFF)
20. Frame (FR, or FRAME)
21. Design code letter (CODE, DES, or DESIGN)
22. Enclosure type (ENCL)
23. Thermal protection (PROT)
24. Bearings (BRGS, or BRG NO)
25. Shaft type (SHAFT)
26. Mounts (MOUNT)
27. Power factor correction (MAX CORR kVAR)
28. Model (MOD, MODEL)
29. Serial number (SN)
30. Catalog number (CAT)
31. Special instructions
32. Special markings
33. Connection diagrams
34. Inverter type
35. Unique or special features

Additional information may also appear on motor nameplates.

**Manufacturer's Name and Address.** This is the name and address of the manufacturer that made the motor, but the address is usually only a country, or possibly a state if the motor was manufactured in the United States. There is no requirement as to how detailed the address information on a motor nameplate has to be.

**Rated Voltage (V, or VOLTS).** The rated voltage of a motor listed on the nameplate is called the terminal voltage. This indicates the actual voltage on the motor's terminals at which the manufacturer designed the motor to operate. Terminal voltage is not the same as the electrical distribution system's nominal voltage. Nominal voltage is the design or configuration voltage of the electrical distribution system: for example, 120/208-V wye, or a 240/480-V delta. However, the actual voltages present in different parts of the electrical distribution system are different because of voltage drops of electrical feeder and branch circuit conductors. Motor terminal voltages are rated slightly less than the nominal system voltages to compensate for the electrical distribution system voltage drops. This way, when the voltage drops of the feeders and branch circuits are deducted from the nominal supply voltage, the terminal voltage rating listed on the motor nameplate will be closer to the actual voltage levels present in the electrical distribution system where the motor is installed.

NEMA standard motors are designed to operate within a 10% voltage deviation tolerance from the nameplate rated voltage. A motor designed to operate on a 480-V electrical distribution system would be rated at 460 V (terminal voltage). The motor will perform best at the rated voltage of 460 V, but it could tolerate a voltage swing of 10% either side of 460 V (414 V to 506 V), which should accommodate any electrical distribution system voltage fluctuations. Operating outside the 10% voltage limitation can severely affect the life expectancy of the motor, even if the motor horsepower rating is derated to compensate.

IEC standard motors are designed to operate within 5% voltage tolerance from the nameplate rated voltage.

**Rated Full Load Amperes (AMPS, or FLA).** When a motor design is tested by the manufacturer, it is loaded to the full load horsepower using the motor's rated frequency and voltage, and the measured current becomes the nameplate full load amperes (FLA) for that motor. The motor will draw less than full load amperes when it is less than fully loaded, and more than full load amperes when it is overloaded. On the Example Motor Nameplate shown in Figure 6-1, notice that if the motor is multiple-voltage rated, the current for each rated voltage is listed.

The full load amperes current of the motor is not used when sizing motor circuit starters, disconnects, or conductors; that current value comes from the tables at the end of Article 430 in the *National Electrical Code® (NEC)®*. The only motor starter circuit component sized from the motor's full load amperes listed on the nameplate is the motor starter overload unit heater element.

**Rated Frequency (F, FREQ, or Hz).** This is the rated frequency at which the motor is designed to operate. All of North America is standardized on 60 Hz, but other parts of the world are standardized on 50 Hz. A motor designed to operate on one of these frequencies cannot be operated satisfactorily on the other frequency, because motors are only designed to tolerate a frequency variation of 5%. However, there are motors that are rated to operate on either 50 Hz or 60 Hz, and they are becoming more common.

**Number of Phases (P, PH, or PHASE).** Phase identifies the type of power source the motor is designed to operate on, either single-phase or three-phase.

**Rated Full Load Speed (RPM).** Full load speed is measured in revolutions per minute (RPM), and indicates the rotor speed when the motor is producing its full rated horsepower. The motor speed will be slightly higher if the motor is lightly loaded or unloaded, but the rotor still will not turn at synchronous speed. Remember that synchronous speed is the rotating speed of the magnetic field of the stator, which is determined by the line frequency of the source voltage and the number of stator magnetic pole coil windings of the motor.

**Rated Horsepower (HP).** This is the horsepower output the motor is designed to deliver on a continuous basis when the motor is energized with its rated voltage and frequency. Motors below one horsepower are referred to as fractional horsepower motors, and motors 1 horsepower or more are called integral horsepower motors.

IEC motors are rated in kW rather than horsepower. The conversion from horsepower to wattage is 1 horsepower = 746 watts, so the Example Motor Nameplate shown in Figure 6-1 lists the 2 horsepower NEMA rating as 1.5kW for the IEC rating. The conversion will usually not be as clean as the Example Motor Nameplate, in which case the next largest motor of the rating being converted to would be substituted. For example, an IEC 2 KW motor that needs to be replaced would be equal to approximately 2.7 horsepower, which is not a standard NEMA horsepower, so the next higher standard motor of three horsepower would be substituted.

**Locked Rotor kVA Code Letter (CODE).** When AC induction motors are started with their full rated frequency and voltage applied, they draw locked rotor current for a short period of time, which is an inrush current that is many times higher than the value of the full load running current of the motor. The intensity of the locked rotor current is calculated using the motor's design code letter. The code letter represents a multiplier based on the design characteristics of the motor that will determine the maximum kVA per horsepower that the motor is capable of drawing under locked rotor conditions. The chart in Figure 6-2 shows the locked rotor kVA per horsepower multiplier for each code letter. Code letter F is highlighted because it is considered average. Code letters below F indicate a low locked rotor current, and code letters above F indicates a high locked rotor current.

The following example explains how the kVA per horsepower multiplier is applied. The locked rotor amperes (LRA) for a three-phase, 460-V, 29.5 FLA, 25-HP motor with a code letter F is found as follows: The multiplier range on the chart for design Code F = 5.0–5.59; approximately half way between the two is 5.3, so that will be used in the

| CODE | kVA/HP | CODE | kVA/HP |
|---|---|---|---|
| A | 0 – 3.14 | L | 9.0 – 9.99 |
| B | 3.15 – 3.54 | M | 10.0 – 11.19 |
| C | 3.55 – 3.99 | N | 11.2 – 12.49 |
| D | 4.0 – 4.49 | P | 12.5 – 13.99 |
| E | 4.5 – 4.99 | R | 14.0 – 15.99 |
| F | 5.0 – 5.59 | S | 16.0 – 17.99 |
| G | 5.6 – 6.29 | T | 18.0 – 19.99 |
| H | 6.3 – 7.09 | U | 20.0 – 22.39 |
| J | 7.1 – 7.99 | V | 22.4 and up |
| K | 8.0 – 8.99 | | |

Low ← → High



**FIGURE 6-2** Design code letter kVA-per-HP chart

formula. The formula for calculating locked rotor amperes is:

LRA = (kVA per HP multiplier * 1000 * HP) ÷ (Voltage * 1.732)

The first part of the formula, (kVA per HP multiplier * 1000 * HP), multiplies the code letter multiplier from the table by 1000 to convert kVA into volt-amperes (VA). The volt-amperes are then multiplied by the motor's horsepower to calculate the total volt-amperes the motor is capable of drawing under locked rotor conditions. The desired result of the calculation, however, is current, so the total volt-amperes must be divided by voltage, which is the second part of the formula.

Three-phase kVA is calculated by multiplying the voltage of the line, times the current of the line, times the 1.732 three-phase multiplier. In order to calculate the line current from this formula, the kVA must be divided by both the line voltage and the 1.732 multiplier. Rather than dividing the kVA by the line voltage and then the 1.732 multiplier, this formula multiplies the line voltage by 1.732 before performing the division operation, as shown here:

LRA = (5.3 kVA per HP multiplier * 1000 * 25 HP) ÷ (460 V * 1.732)

LRA = 132,500 VA ÷ 796 V

LRA = 166 amperes; 5.6 times higher than the motor's FLA of 29.5 amperes

When applying this formula to a single-phase motor, the 1.732 three-phase multiplier is omitted.

**Duty or Time Rating (DUTY).** This specifies the maximum length of time the motor can operate at its rated horsepower and be able to dissipate the heat generated inside the motor effectively, without exceeding the motor's temperature rise. The temperature that a motor stabilizes at after a long period of operation is called the equilibrium temperature, because the motor is capable of dissipating any additional generated heat above that temperature on a continuous basis. A motor that is operating as intermittent duty never reaches equilibrium temperature, because it is never run long enough before being shutdown and allowed to cool again.

The NEMA duty ratings for motors are intermittent, continuous duty, and special duty. If the motor is rated for intermittent duty, the maximum time it may run would be expressed as a ratio of how many minutes it may run, out of a total number of minutes of the run/rest cycle; as an example, 10 minutes per hour. Continuous duty means that the motor is capable of continuously dissipating the heat from normal operation of the motor, and has no limit to the length of time the motor may run.

IEC duty ratings are a little more complicated, with a range of duty ratings from S1, which is continuous duty, to S8. IEC duty ratings not only take into consideration the run time of the motor for the duty rating as they move through the range, they also include intermittent loading, electric breaking, and load speed changes as part of the duty rating criterion.

**Insulation Class (INS, or INSUL CLASS).** A good rule of thumb to remember is that for every 10 degrees C the operating temperature of a motor increases over rated temperature, the motor life will be cut in half, because excessive heat will cause the type of electrical insulation used in motors to fail more quickly. The insulation class identifies the maximum temperature that the insulation around the stator coil wires of the motor can tolerate before breaking down. The thermal capacity of the motor insulation temperature ratings is determined by the ambient temperature and the temperature rise of the motor. The thermal capacity of the motor insulation is classified by NEMA letter designations A, B, F, and H. Designation A means that the motor can safely operate at 105°C, B = 130°C, F = 155°C, and H = 180°C, respectively.

These same insulation classes are used for IEC ratings, except that IEC adds a Class E, which has a temperature rating of 120°C that NEMA does not have.

**Temperature Rise (RISE).** Temperature rise is the increase in the motor's internal operating temperature caused by the motor's operation. The majority of the electrical energy drawn by a motor is converted into mechanical energy. However, some of the electrical energy is converted into heat, mainly $I^2R$ copper losses from current flowing through the electrical resistance of the wire windings. The manufacturer has determined that the motor may be run at the rated voltage, frequency, and horsepower on a continuous basis, and the motor temperature will not increase more than this rated amount above the ambient temperature of the motor. The temperature rise of the motor, plus the ambient temperature of the motor, is what determines the insulation class of the motor.

**Ambient Temperature (AMB, or TEMP).** The ambient temperature rating of the motor determines the maximum ambient temperature at which the motor may be operated. Ambient temperature is the temperature of the existing air surrounding the motor without the motor running. In general, maximum ambient temperature for motors is 40°C. Operation of a motor at ambient temperatures exceeding the motor's rating may reduce the life of the motor, depending on whether the motor is operating at or near its rated horsepower. The ambient temperature of the motor, plus the temperature rise of the motor, is what determines the insulation class of the motor.

**Starts per Hour (S/HR, or STARTS/HR).** Also closely related to the duty rating of the motor is the number of starts the motor can sustain per hour without causing thermal damage to the motor coil windings. The life of a motor is closely related to the number of starts it makes per hour, because the motor must dissipate the additional heat generated at startup. Remember the discussion at the end of Chapter 3 about motor losses; locked rotor losses can be 49 times higher than motor losses when running at full load. Every time a motor is energized, it experiences locked rotor current for a short period of time until the rotor approaches full load speed. If the motor is not allowed to run long enough to dissipate the additional heat caused by the locked rotor starting current before it is de-energized and started again, the stator coils may develop hot spots that may exceed the temperature rating of their insulation.

It can be complicated to determine how many times a motor can be started in a given period of time without causing damage to the coil windings. The size of the motor, the load it is driving, the code letter of the motor, any external cooling methods, and other factors all enter into the equation. There is no magic safe number for how many times a motor may be started in a given time, but the NEMA MG-1 standard does provide a formula that takes into consideration the various criteria listed here. Using this formula generically for a 1 horsepower motor, one result is six starts per hour; a 50 horsepower motor is 2.5 starts per hour; and the number of starts per hour goes down as motor horsepower continues to go up. It is best to consult the motor manufacturer if it becomes necessary to determine a safe number of times a given motor may be started per hour.

**Type (TYPE).** Type identifies the kind of motor it is: general purpose, capacitor start, permanent split, permanent split capacitor run, and so forth. More detailed descriptions and operating characteristics for each of these motor types is covered in Chapters 4 and 5.

**Rated torque (TORQUE).** This is the maximum twisting force supplied by the motor at its rated horsepower. Torque is normally measured in foot-pounds (ft-lb) for NEMA, and newton-meters (N-m) for IEC. The formula for torque in foot-pounds is: torque = 5252 * horsepower / RPM.

**Altitude (ALT).** Altitude indicates the maximum height above sea level that the motor will have dense enough air to dissipate the heat it generates during normal operation at its rated horsepower and remain within its design temperature rise. If the motor is used above this altitude, the air will be too thin to dissipate the heat generated during operation, and the motor may exceed its rated temperature rise. For higher altitudes, higher grades of insulation, additional exterior cooling provisions, or motor derating may be required.

**Power Factor (PF for NEMA, or cos ∠θ for IEC).** As discussed in Chapter 1, power factor is the ratio of the motor's watts divided by the volt-amperes, expressed as a percentage. The best possible power factor is 100%, called unity power factor, where watts equal volt-amperes, and any power factor less than unity is a less desirable condition. The induction motor power factor changes with the mechanical load on the motor. It is the lowest at no load, and it becomes higher closer to unity, as additional load is applied to the motor. Power factor is usually the highest for induction motors at full-load, but does not drop sharply until the motor falls below about 60% loading. If the load is relatively constant, the power factor for induction motors can be corrected with capacitors.

**Service Factor (SF).** Motor service factor is the percentage of overloading above the rated horsepower the motor can handle for short periods of time, such as a mechanical load surge. The service factor is expressed as a percentage, which can be multiplied by the motor's rated horsepower to determine the maximum horsepower output the motor is capable of. For example, a motor with a 1.15 service factor can produce 15% greater horsepower for a short period of time than a motor with a 1.0 service factor rating. It is not a good practice to operate motors in the service factor area above the rated horsepower for long periods of time. The service factor rating will appear on the nameplate only when it is higher than 1.0.

Motors manufactured under the IEC standard are not rated with a service factor. If a particular application requires more horsepower (kW) above the horsepower (kW) rating of the motor for short periods of time, a larger horsepower (kW) motor would have to be used.

**Efficiency (EFF, or NOM EFF).** Efficiency is the ratio of the power output divided by the volt-ampere input, and multiplied by 100 to express it as a percentage. Efficiency indicates how well the motor converts electrical energy into mechanical energy. Motors that list an efficiency rating on the nameplate normally give a nominal efficiency, which is an average at full load. Generally, larger motors are more efficient than smaller motors, and newer large motors may have efficiencies as high as the mid-90s, but mid-80s is more common for smaller motors. Induction motors are most efficient at about 75% of full load, but the efficiency only changes a small amount between 50% loading and full load. Efficiency does not drop sharply until the motor falls below about 25% loading.

**Frame (FR, or FRAME).** The frame is such an important element of the motor nameplate information that it justifies an extended explanation here. The purpose of specifying standard motor mounting dimensions with a system of frame sizes is to facilitate substituting one manufacturer's motor for another. The frame designation insures that motors will be interchangeable to the point that they will have the same mounting requirements, and the shaft will be the same height from the bottom of the motor base. Beyond that, the other dimensions such as overall length, diameter of the motor housing, or overall height may vary significantly. As a frame number becomes higher, the physical size of the motor generally becomes larger.

Today's NEMA motor frame standard for integral horsepower motors is the "T" system, and is denoted in the frame size notation; for example, 182T, 256T, etc. Electricians are sometimes called for service work to replace a motor, and it is important to note that motors manufactured prior to 1964 conformed to the "U" system standard, and are called U-frame motors; and motors manufactured prior to 1952 are called pre-U-frame, or original frame size motors. It is less common as time goes on that these pre-1964 motors need to be replaced, not because they quit working, but because the equipment they were in became obsolete and was removed from service. If it is necessary to replace one of these older motors, it will probably require significant mounting modifications, or in some cases where that is not possible, the motor may have to be re-wound by a motor shop.

Generally, smaller, fractional horsepower, induction motors use a two-digit frame designator system, where the number represents how many sixteenths of an inch the center of the shaft is above the bottom of the motor base, which is dimension D in Figure 6-3. For example, a 56 frame would have a shaft height of 56/16, or 3.5 inches. Some motors with horsepower ratings of 1 horsepower or more may still use the two-digit frame designator, but the three digit T-frame designation is more common.

Larger, integral horsepower motors use a three-digit T-frame designator. In this system, the first two digits represent how many fourths of an inch the center of the shaft is above the bottom of the motor base, which is dimension D in the drawing of Figure 6-3. The third digit represents the number of inches there are between the mounting holes in the base, which is dimension 2F in the drawing of Figure 6-3. For example, a 326T frame would have a shaft height of 32/4, or eight inches, and the distance between the mounting holes would be six inches.

IEC frame sizes are determined a little differently, in that the shaft height is measured in millimeters, rather than sixteenths or quarters of an inch. The millimeter shaft height above the bottom of the motor base determines the frame size; if the center of the shaft is 90 millimeters high, the motor frame is 90. The bolt hole pattern dimensions of IEC motors are also given in millimeters rather than inches. The conversion between millimeters and inches is: one inch = 25.4 millimeters. Even with conversion numbers between inches and millimeters, it is almost impossible to get direct interchangeability between NEMA and IEC frame sizes, because motor manufacturers do not consider interchangeability as part of their design criteria.

The table in Figure 6-3 demonstrates three examples of the interchangeability difficulties between NEMA and IEC motors. The NEMA standard labels motor dimensions in inches, but the conversion to millimeters has been made for this table to highlight the differences. The first example frame shows an IEC Frame 56 motor, which does not have a NEMA frame size that even comes close enough to it to consider it as a replacement. The second example frame shows a NEMA fractional horsepower motor Frame 56, and the closest IEC

| FRAME | | SHAFT HEIGHT NEMA=D IEC=U | MOUNTING HOLES LENGTH NEMA=2F IEC=2B | MOUNTING HOLES WIDTH NEMA=2E IEC=2A |
|---|---|---|---|---|
| N/A | NEMA | N/A | N/A | N/A |
| 56 | IEC | 56 mm | 35.5 mm | 45 mm |
| 56 | NEMA | 88.9 mm | 38.1 mm | 61.9 mm |
| 80 | IEC | 80 mm | 50 mm | 62.5 mm |
| 182T | NEMA | 114.3 mm | 57.2 mm | 95.29 mm |
| 112S | IEC | 112 mm | 57 mm | 95 mm |



**FIGURE 6-3** Sample frame size comparison

frame motor where the mounting could be modified to make it work. The third example frame shows a NEMA integral horsepower motor Frame 182T, and the closest IEC frame motor where the mounting could be modified to make it work. Usually, when interchanging NEMA and IEC design motors, some type of modification, such as shimming or a conversion plate, is necessary to make the new motor fit and align properly.

The dimension diagram shown in Figure 6-4 identifies the frame rating motor measurements that are important for mounting motors. The NEMA letter designations are given, and the equivalent IEC letter designation is given in parentheses. Many more dimensions can be identified on the motor drawings, but if they do not pertain to mounting the motor, the measurements and letter designations become much less standardized from one motor manufacturer to the next and are not shown here.

**NEMA Design Code Letter (CODE, DES, or DESIGN).** Like the frame designation, NEMA has designated specific designs of motor operating characteristics for general purpose motors, to make them interchangeable between motor manufacturers. The motor design code letter describes the standardized locked rotor torque, breakdown torque, slip, locked rotor starting current, and other operating characteristics for each particular letter designation. The four most common NEMA standard designs for induction motors are A, B, C, and D. Refer back to Figure 2-14 in Chapter 2 for the speed-torque curves that are associated with each of the design letters.

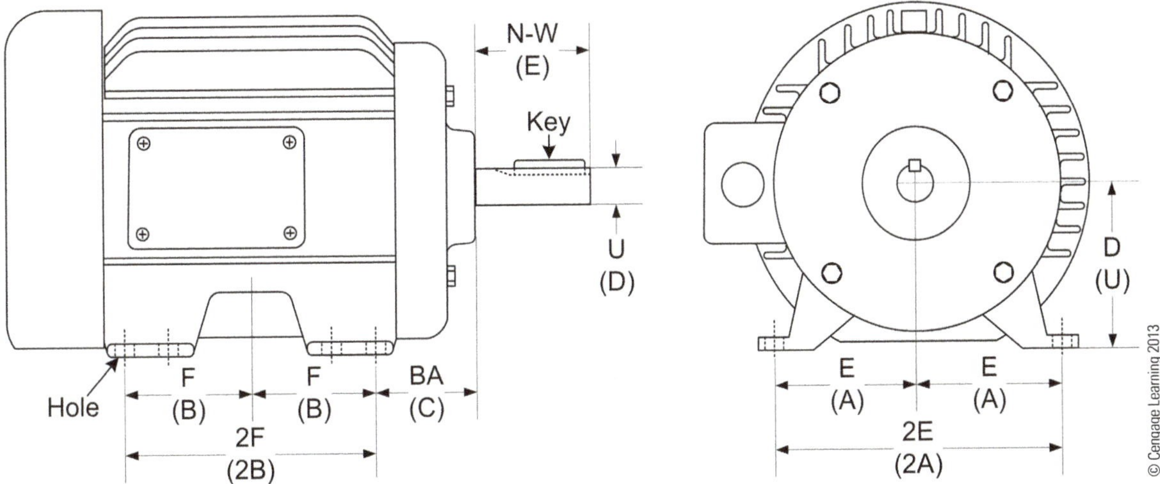


**FIGURE 6-4** Motor dimensioning drawing

The IEC motor has a similar system of design operating characteristics, but the letter designations are different. The IEC design N, which may be thought of as normal torque, is similar to the NEMA design B, which is the general purpose design, considered to be NEMA's normal torque motor. The IEC design H, which may be thought of as high torque, is similar to the NEMA design C, which is considered to be NEMA's high torque motor. Although the design types between NEMA and IEC are similar, they are not exact matches.

**Enclosure Type (ENCL).** The NEMA standard for motors contains two methods for rating the protection provided by motor enclosures regarding their operating environment. The first method is the traditional descriptive method the older electricians may be most familiar with: open (O), open drip proof (ODP), guarded (G), or totally enclosed fan cooled (TEFC) to name a few examples. These descriptive codes are very common and adequate for most motor applications, but they lack a high degree of specificity regarding such aspects as object size.

The second method, which is also used with IEC motors, is the ingress protection (IP) two-digit code, which provides a greater degree of specificity regarding the definition of motor enclosure protections for two areas, people and objects, and water. The first digit of the code ranges from zero through six, and designates the protection of persons or foreign objects coming in contact with live or moving parts of the motor. The second digit ranges from zero through eight, and designates the motor against the intrusion of water. Generally, as the numbers become larger, the protection becomes better.

Both the NEMA and IEC standards also specify a two-digit international cooling (IC) rating, which identifies the method of cooling for the motor. The first digit signifies the cooling circuit arrangement and the second digit the method of supplying power to circulate the coolant. The two ratings are closely related, so care must be taken so that the IC rating is not confused with, or substituted for, the IP rating.

Although some equivalent enclosure ratings can be made between NEMA and IEC, a complete cross-reference would not be possible, because NEMA categories are quite broad and IEC categories are more narrowly defined. For example, an IEC IP22 could be used instead of a NEMA Open Drip Proof, but the reverse may not always be true, because the IEC numeric system is very specific about the level of protection against dust and water infiltration, where the NEMA descriptive system is more general.

**Thermal Protection (PROT).** This designation describes what type of thermal protection the motor has, if any. When a protection method is listed on the nameplate, the protection method is integral to the motor, rather than external protection methods such as the overload unit on a motor starter. Integral thermal protection is most commonly found on single-phase, fractional horsepower motors. Following is a list of thermal protection designations:

- ***Auto (Automatic Reset).*** Automatic protection usually contains a temperature-sensing device in the stator coil windings that disconnects the power supply if the internal temperature of the motor becomes too high due to failure to start or overload. After the motor cools down sufficiently, the thermal protector automatically will restore power and restart the motor. This type of motor should never be used where the unexpected restarting of the motor would be a hazard.
- ***Imp (Impedance).*** Impedance protection is found only on small motors, and is designed so that the motor will not burn out under locked rotor conditions. A good way to think about this type of motor protection is like the doorbell transformer, which has an intrinsically safe design; the internal impedance will prevent it from developing sufficient thermal energy from an overload condition to start a fire.
- ***Man (Manual Reset).*** Manual reset protection usually contains a temperature-sensing device in the stator coil windings that disconnects the power supply if the internal temperature of the motor becomes too high due to failure to start or overload. After the motor cools down sufficiently, an external reset button must be pushed to restore power to the motor. This motor is preferred where the unexpected restarting of the motor would be a hazard.
- ***T-St (Thermostat).*** This type of motor has a temperature sensing device installed inside the motor, with separate leads brought out to be connected into the motor starter control circuit. If the internal temperature of the motor becomes too high because of failure to start or because of overload conditions, the thermostat contacts will open the motor starter control circuit and stop the motor.

**Bearings (BRGS, or BRG NO).** The motor bearings are the physical connection between the stationary stator and the rotating rotor, and are the only wear point on three-phase induction motors. The inside race of the bearing is pressed onto the rotor shaft, and the outside race of the bearing is either friction fit into the motor end bell for smaller motors, or held in the end bell with a bolt or held in the end bell with a bolt plate for larger motors. The main types of bearings on smaller motors are sleeve or ball, but there are many different specialized bearings for larger motors and specific applications. Different designations are used by different manufacturers to identify the bearing on each end of the motor. Some of the most common are DE/NDE (Drive End/Non Drive End), DE/ODE (Drive End/Opposite Drive End), PE/NPE (Pulley End/Non Pulley End), and SE/OSE (Shaft End/Opposite Shaft End); and there are many other variations on the same method.

The reason for documenting the motor bearing numbers on the nameplate is so that the motor does not have to be disassembled to get the bearing numbers if they have to be replaced. This way, the motor may remain in service while the bearings are ordered, and the motor does not have to be removed from service until the bearings are in hand, which will reduce the motor's downtime.

**Shaft Type (SHAFT).** This describes the general type of shaft the motor has. Some types of motor shafts are listed here, but this is certainly not a complete list:

- ***Flat.*** Found on motors up to ½" diameter shaft. The flat portion of the shaft provides an area for a setscrew on the sheave, coupler, or whatever other device is connecting the motor to the mechanical load, to seat and prevent the shaft from slipping.
- ***Key.*** Found on motors with 5/8" and larger shaft diameters. This is the type of motor shaft shown in the dimensions drawings under the Frame section covered earlier in this chapter. The purpose of the key is to prevent the shaft from slipping on the coupling device with the mechanical load.

- ***Round.*** Found on very small shaded pole C-frame motors; the entire length of the shaft is round. This type of shaft design is limited to only very low torque requirement applications, because only the friction of the setscrew of the coupling device with the mechanical load is preventing the shaft from slipping.
- ***Thd (Threaded).*** This type of motor shaft has threads on the end of the shaft, and is used on uni-directional motors for special applications such as driving jet pump impellers. The threads on the end of the motor shaft are of the opposite hand thread as the motor rotation, so the impeller will turn in a tightening direction rather than a loosening direction as the motor runs.

**Mounts.** Unless specified otherwise, smaller motors are designed to be mounted in any position, except when the enclosure considerations, such as open drip-proof, require a certain mounting position to maintain the enclosure rating. Larger motors often have bearing types, cooling schemes, or other considerations that might require a specific mounting position. The three most common mounting methods for fractional horsepower motors are listed below:

- ***Rigid base.*** As shown in Figure 6-5, the rigid mount motor base is bolted, welded, or part of the motor frame cast, and allows motor to be rigidly mounted on equipment.
- ***Resilient base.*** As shown in Figure 6-6, the resilient mount motor base forms a cradle, which is used for mounting purposes and to hold the motor. The motor is isolated from the cradle base with rubber isolation (or resilient) grommets to absorb vibration and noise. Because the rubber grommets will electrically isolate the motor from the base, a bonding conductor must be used to ground the motor to the base.
- ***C face mount.*** As shown in Figure 6-7, the C-face mount motor has a machined face, with the mounting holes in the flange threaded to receive bolts, so that the motor can be mounted directly on equipment. These motors are designed to mount directly to fan housings, pump housings, etc. without a base type of mount. Many motors are designed with both a C-face mount and a rigid base mount to make the motor more versatile.

**Power Factor Correction (MAX CORRECTION kVAR).** This is the maximum value capacitor, in kVARs, to be used with this motor when correcting the power factor for only that one load. Using a higher-value capacitor could result in higher motor terminal voltages, and damage to the motor.

**Model Number (MOD, MODEL).** The model number is important when trying to replace an

FIGURE 6-5 Rigid base motor mount picture

FIGURE 6-6 Resilient base motor mount picture

FIGURE 6-7 C-face motor mount picture

existing motor, because there may be modifications or enhancements that are important to replace.

**Serial Number (SN).** The serial number or identification number may be useful when dealing with the manufacturer, and may be necessary to identify manufacturing changes that could affect replacement parts.

**Catalog Number (CAT).** Catalog numbers may be the most useful when ordering a motor, or finding information, accessories, or replacements for newer motors that are still being manufactured and sold.

**Special Instructions.** Some special instructions might require that only copper conductors be used to terminate the motor, or that a specific overload protection be used.

**Special Markings.** Some motors may have special markings that reflect third-party certification or recognition. Some of the common certifications include:

- CSA (Canadian Standards Association)
- UL (Underwriters Laboratories)
- ASD (Adjustable Speed Drive)
- ANSI (The American National Standards Institute)
- IEEE (The Institute of Electrical and Electronic Engineers)
- NEMA (The National Electrical Manufacturers Association)
- IEC (The International Electrotechnical Commission)

**Connection Diagrams.** The connection diagrams give the specific terminal designations of how to connect the particular motor to the electrical power supply. If the motor is dual voltage rated, the connection diagram will be given for both the high and low voltage connections, and if the motor is reversible that information will be given also.

**Inverter Type.** The most common type of inverter technology for smaller integral horsepower three-phase motors is pulse-width modulation (PWM). The wire coil insulation used in the motor must be rated for the fast switching times used by PWM, or it could be damaged. Older, non-rated motors normally cannot be used with PWM speed controllers without causing damage to the motor. Other inverter technologies may require different ratings.

**Unique or Special Features.** Special features listed on the nameplate might include encapsulated stator windings to protect them from chemical contaminants, or shafts or housings made of special materials for specific uses.

## CHAPTER SUMMARY

- Standards agencies such as NEMA and IEC specify certain information that describes the physical and operating characteristics of each type of motor design, which is listed on the motor's nameplate.
- The purpose of standardizing this information is to ensure the correct interchangeability of motors between different motor manufacturers.
- Both NEMA and IEC motors and motor control components are common in the electrical industry today, because electrical equipment is manufactured in the global economy under both standards.
- The rated voltage of a motor listed on the nameplate is called the terminal voltage, because it is the actual voltage on the motor's terminals at which the manufacturer designed the motor to operate.
- Nominal voltage is the design or configuration voltage of the electrical distribution system.
- Motor terminal voltages are rated slightly less than nominal system voltages to compensate for electrical distribution system voltage drops.
- NEMA standard motors are designed to operate within a 10% voltage deviation tolerance from the nameplate rated voltage.
- The nameplate current rating of a motor is measured when the motor is loaded to its full rated horsepower.
- The nameplate RPM rating of a motor is measured when the motor is loaded to its full rated horsepower.
- Motors below 1 horsepower are referred to as fractional horsepower motors, and motors 1 horsepower or more are called integral horsepower motors.
- NEMA rates motor power in horsepower, and IEC rates motor power in kilowatts.
- The locked rotor kVA per horsepower multiplier is used to determine the maximum locked rotor amperes a motor design can draw.
- The NEMA duty ratings for motors are intermittent, continuous duty, and special duty.
- Temperature rise is the increase in the motor's internal operating temperature caused by the motor's operation.
- Ambient temperature is the temperature of the existing air surrounding the motor.
- The temperature rise of the motor, plus the ambient temperature of the motor, is what determines the insulation class of the motor.
- The insulation class identifies the maximum temperature that the insulation around the stator coil wires of the motor can tolerate before breaking down.
- The life of a motor is closely related to the number of starts it makes per hour, because the motor must dissipate the additional heat generated at startup.
- If a motor is operated above its altitude rating, the air may be too thin to dissipate the heat generated during operation, and the motor may exceed its rated temperature rise and damage the insulation.
- Power factor is the ratio of the motor's watts divided by the volt-amperes, and NEMA documents it on the nameplate as a percentage. IEC documents power factor on the nameplate as the cos $\angle\theta$.
- Service factor is the percentage of mechanical overloading above the rated horsepower the motor can handle for short periods of time.
- IEC motors are not rated with a service factor, and are considered to be a service factor of 1.
- The frame designation insures that motors will be interchangeable to the point that they will have the same mounting requirements, and the shaft will be the same height from the bottom of the motor base.

- Today's NEMA motor frame standard for integral horsepower motors is the "T" system.
- Motors manufactured prior to 1964 conformed to the "U" system standard.
- Fractional horsepower motors use a two-digit frame designator system, where the number represents how many sixteenths of an inch the center of the shaft is above the bottom of the motor base.
- Integral horsepower motors use a three-digit system, where the first two digits represent how many fourths of an inch the center of the shaft is above the bottom of the motor base, and the third digit represents the number of inches between the mounting holes in the base.
- The motor design letter describes the standardized locked rotor torque, breakdown torque, slip, locked rotor starting current, and other operating characteristics for each particular letter designation.
- The enclosure designation classifies the motor as to its degree of protection against accidental contact with dangerous parts, environmental considerations such as dust and water, and methods of cooling.
- When a protection method is listed on the nameplate, the protection method is integral to the motor rather than external protection methods such as the overload unit on a motor starter.
- By documenting the motor bearings on the nameplate, the motor does not have to be disassembled to identify them.
- The three most common mounting methods for fractional horsepower motors are rigid, resilient, and C-face mount.

## REVIEW QUESTIONS

1. What is the name given to the voltage that the manufacturer designed the motor to have at the motor terminals?
2. What is the difference between nominal voltage and terminal voltage?
3. Terminal voltage is lower than nominal voltage to compensate for what?
4. Within what voltage deviation tolerance are NEMA motors rated to operate?
5. The ampere rating of an induction motor listed on the nameplate is determined at what horsepower load on the motor?
6. What is the only motor starter circuit component sized from the motor nameplate current?
7. The RPM rating of an induction motor listed on the nameplate is determined at what horsepower load on the motor?
8. NEMA power ratings are in horsepower. How are IEC motors rated, and what is the conversion?
9. What are the three NEMA duty ratings for induction motors?
10. The temperature that a motor stabilizes at after a long period of operation is called what?
11. Ambient temperature is defined as what?
12. The temperature rise plus the ambient temperature determines what motor rating?
13. Why is the life of a motor related to the number of starts it makes per hour?
14. Why does the altitude a motor is operated at make a difference?
15. NEMA motors document power factor on the nameplate as a percentage of watts/volt-amperes. How does IEC document power factor on the motor nameplate?
16. What does the service factor rating on NEMA induction motors identify?
17. What is the service factor rating of all IEC motors?

18. What does the frame designation of a motor identify?
19. Fractional horsepower NEMA motors normally use an identifier of how many digits, and what do they represent?
20. Integral horsepower NEMA motors normally use an identifier of how many digits, and what do they represent?
21. What does the T denote in the frame rating of integral horsepower motors?
22. Motors manufactured prior to 1964 are what frame standard?
23. Even though NEMA does specify different levels of protection for all electrical enclosures, how do the enclosure ratings for motors differ?
24. How do IEC motors specify different levels of motor enclosure protection?
25. When the thermal protection is documented on the nameplate, where is the protection located?
26. What is a benefit of documenting the bearings on the motor nameplate?
27. What are the three most common mounting methods for fractional horsepower motors?

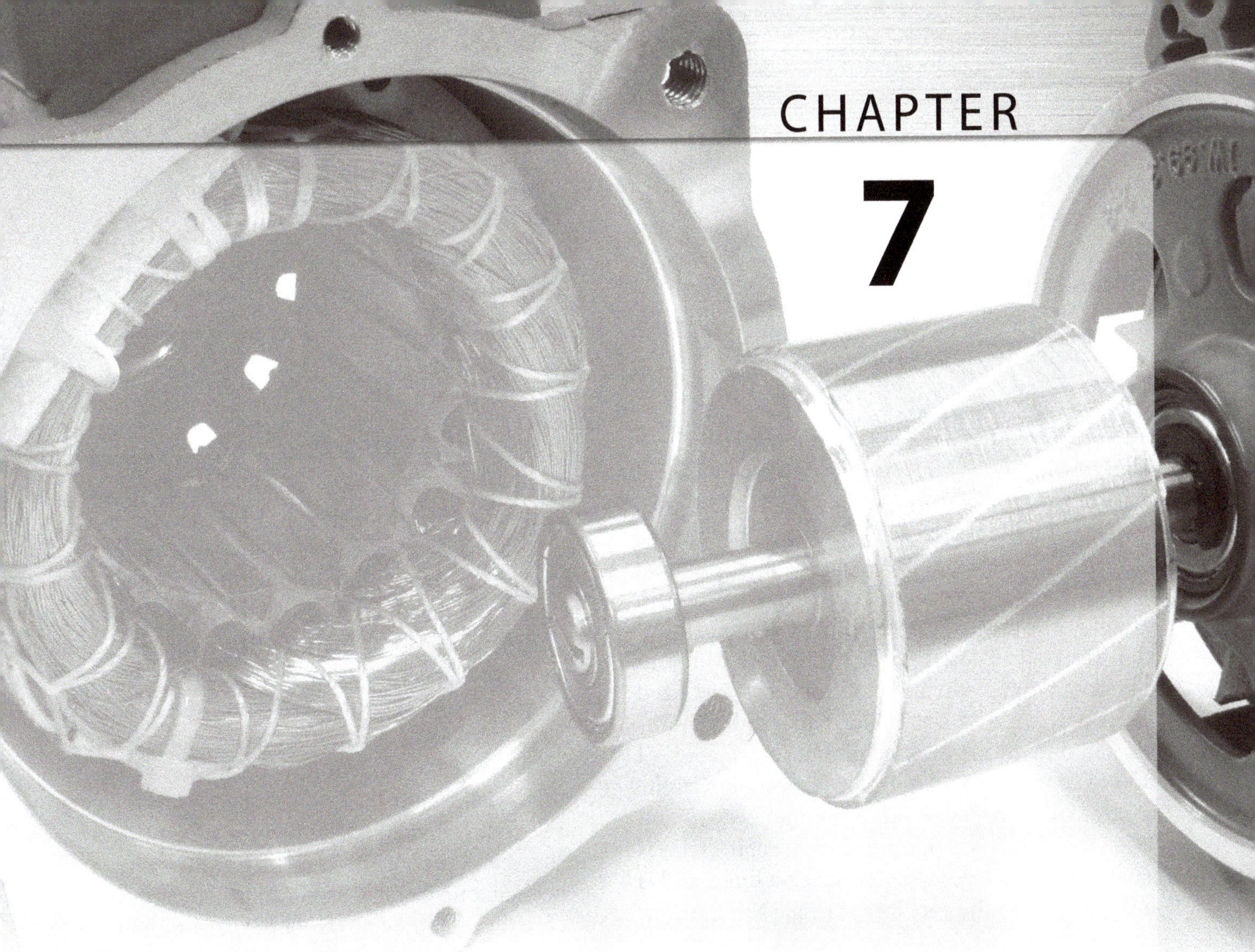


# Magnetic Relays and Contactors

**PURPOSE**

To familiarize the learner with the two most basic types of magnetic switching components used in electric motor control: magnetic relays for control circuits and contactors for energizing and de-energizing electrical loads.

**OBJECTIVES**

After studying this chapter on magnetic relays and contactors the learner will be able to:

- Understand the terminology associated with magnetic relays and contactors
- List, identify, and explain the purpose and function of component parts of relays and contactors
- Understand the function and operating characteristics of the two separate electrical circuits present on each relay and contactor: the control and power circuits
- Explain pull-in and dropout current
- Explain contact configurations and pinout information

- Identify and explain the features of ice cube relays
- Explain the differences between relays and contactors that make them more suited for their different applications
- Explain the purpose of arc chutes
- Identify contactor rating differences between NEMA and IEC
- Explain the versatility of general purpose contactors
- Explain the functionality of definite purpose contactors

## PARTS OF A RELAY

The drawing in Figure 7-1 shows the component parts of a magnetic relay, without the electrical contacts, to simplify the drawing. The yoke provides the physical structure that holds the relay together and is made of iron to provide a low permeability path to guide the magnetic lines of flux of the control coil. The control coil is a coil of wire wound around an iron core, and when the control coil is energized the two become an electromagnet. Terminal numbers 13 and 14 identify the pinout information of how to connect to the control coil, meaning that the control coil is brought out to pin numbers 13 and 14 on the relay base. The movable part of the relay is called the armature, which pivots on a hinge and is held away from the core by a spring when the control coil is de-energized. The armature also either is made of iron, or has an iron component.

### Iron Path

The relay drawing in Figure 7-2 shows how the low permeability of the iron path around the yoke of the device would direct the magnetic flux (dotted line in the diagram) of the electromagnet. Permeability is a measure of how easily a medium can carry magnetic lines of flux and become magnetized. Air has a permeability of one and iron has a permeability greater than one, which makes it a better path to conduct magnetic lines of flux. When the coil is de-energized, spring tension will hold the armature away from the core, creating an air gap between the two. When the control coil is energized, the resulting magnetic lines of flux created by the current flow in the control coil windings will follow the path of best permeability around the iron yoke of the device. When the air gap between the core and armature is encountered by the magnetic flux, the armature will be drawn to the core (against the tension of the spring), to create a path of increased permeability for the magnetic lines of flux to follow. When the control coil is de-energized, again, the magnetic field holding the armature against the core will dissipate, because there is no current flow in the control coil, and the spring tension will return the armature to the de-energized, or relaxed, position, with an air gap between the core and armature.

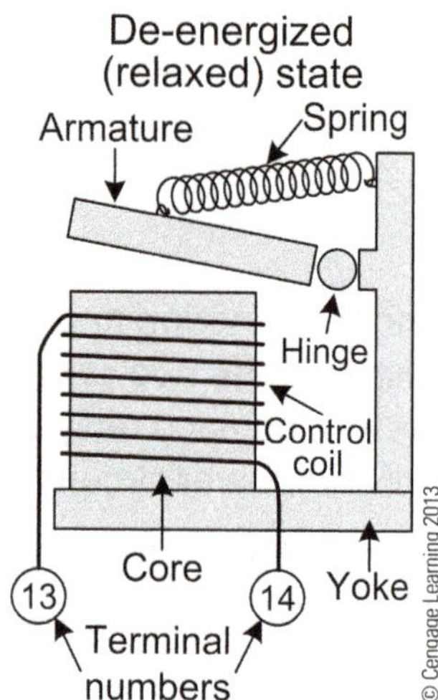


**FIGURE 7-1** De-energized relay without contacts

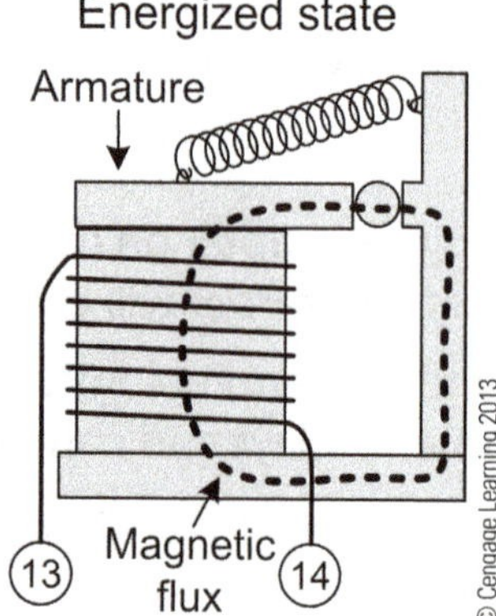


**FIGURE 7-2** Energized relay showing iron path for magnetic lines of flux

## Pull-in and Dropout Current

Relays require a certain minimum level of electrical current to develop the necessary lines of magnetic flux to mechanically draw the armature to the core, against the spring tension holding them apart. Electricians normally talk about controlling voltage rather than controlling current, because for any given electrical load, current is a dependent variable of voltage. However, if the voltage applied to the control coil is too low, there will not be sufficient current to create the necessary lines of magnetic flux to operate magnetically actuated devices like relays and contactors. The minimum current required to actuate a device is called the pull-in current, and is determined by the applied voltage. The pull-in current must be enough to create the magnetic forces necessary for the device to overcome the tension of the spring holding the armature in the de-energized position, and any friction losses in the device. Sometimes pull-in current is called "take-up" current.

Once the armature of the relay is pulled in against the core, less current is needed to hold the armature in the actuated, or energized, position against the core. If the applied voltage is continuously decreased from the pull-in level, a point will be reached where the applied voltage is not sufficient to hold the armature against the core, opposing the spring tension trying to pull them apart. As the applied voltage is decreased, the resulting current flow in the coil windings also will decrease, and eventually the magnetic forces created by the current flow no longer will be sufficient to keep the relay actuated. The current level where the armature returns to the de-energized position is called the dropout current. Dropout current sometimes is called the "let-go" current. Relays and contactors normally operate at a control voltage level considerably higher than the voltage necessary to cause pull-in current, to guarantee positive actuation of the device.

## Control and Power Circuits

When fixed electrical contacts are mounted on the yoke, and movable electrical contacts are mounted on the armature, the device becomes an electrically powered switch, as shown in Figures 7-3 and 7-4. The control circuit usually has a lower voltage, and much lower current rating (as indicated by a single battery and thin wire lines in the figures), which is used to energize the control coil. Energizing the control coil causes the armature, with the movable contacts attached, to draw against the

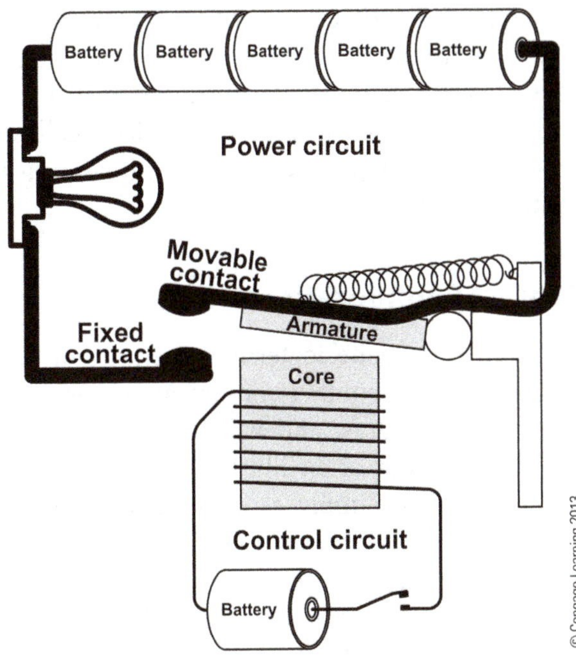



**FIGURE 7-3** De-energized relay with circuits

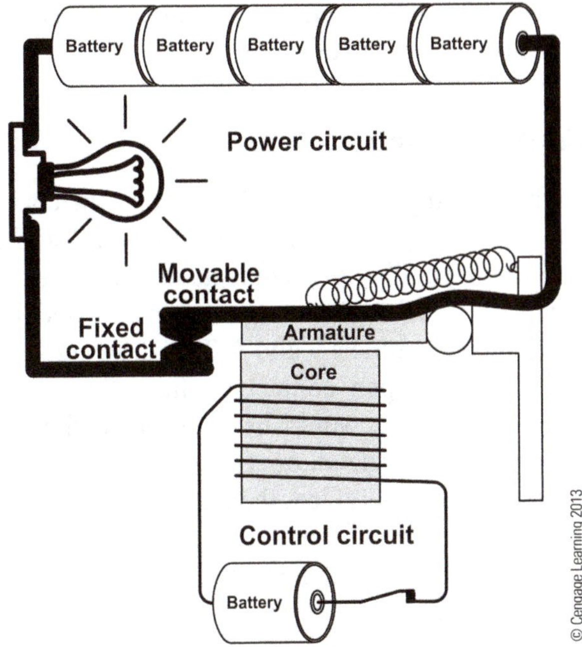



**FIGURE 7-4** Energized relay with circuits

core and close the circuit between the fixed and movable contacts. The fixed and movable electrical contacts that change states to perform the electrical switch function when the control coil is energized are called the power circuit, or main, contacts. The power circuit usually has a higher voltage, and much higher current rating (as indicated by multiple batteries and thick wire lines in the figures), to directly control a larger electrical load. Even though the fixed and movable electrical contacts are mechanically connected, respectively, to the yoke and armature, they are electrically isolated from other parts of the relay, including the control circuit.

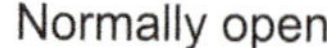


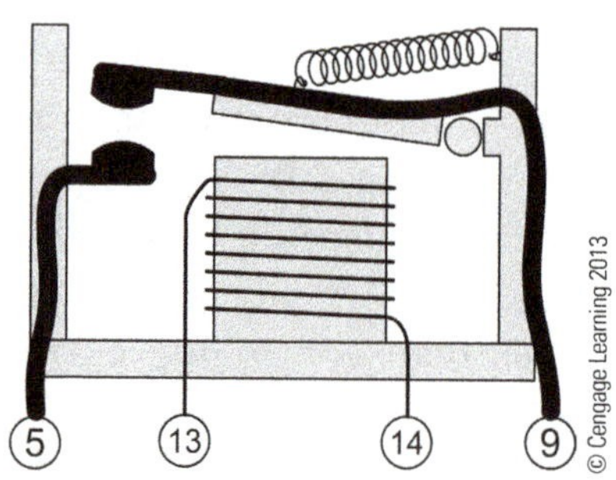


**FIGURE 7-5** Form A relay contacts

Normally closed

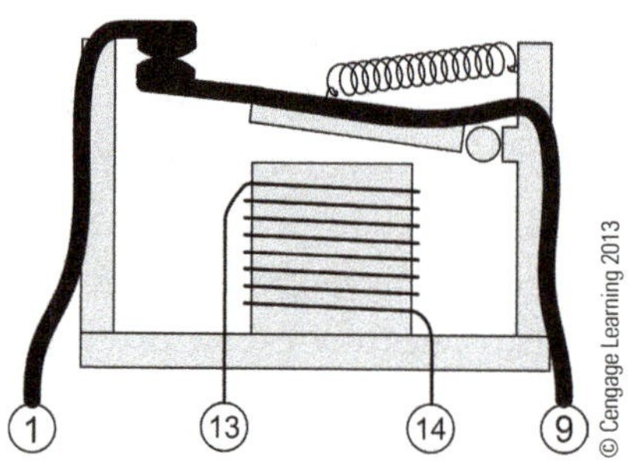


**FIGURE 7-6** Form B relay contacts

## Contact Configuration Callout

"Callout" is a generic term sometimes used to document relay contact configurations on ladder and print diagrams, and will be used here. Like any other callout of an illustration, layout, drawing, or print, it simply explains, defines, details, or describes something about the content that may not be readily apparent by looking at it. It is very important to note that when designating relay contacts as normally open or normally closed, the contact configuration is always called out for the contact position when the relay is in the de-energized or relaxed state, regardless of the contact function in the circuit.

## Contact Configurations

In the drawing of Figure 7-5, note that the control coil is connected between relay terminals 13 and 14, and the normally open contacts of the power circuit, the switching contacts, are connected between relay terminals 5 and 9. This drawing shows the normally open contact configuration, where the power circuit would be open when the relay is in the de-energized, or relaxed state. When the control coil is energized, the armature will be mechanically drawn to the core, causing the normally open power circuit contacts to mechanically close and complete the power circuit. The normally open contact configuration is sometimes called the enable function, because closing the contacts completes the power circuit, and enables something else to operate. In some areas of the electrical industry the normally open contact configuration is also called the Form A configuration.

In Figure 7-6, note that the normally closed contacts of the power circuit, the switching contacts, are connected between relay terminals 1 and 9. This drawing shows the normally closed contact configuration, where the power circuit would be closed, or completed, when the device is in the de-energized, or relaxed, state. When the control coil is energized, the armature will be drawn to the core, causing the normally closed power circuit contacts to mechanically open, and open the power circuit. The normally closed contact configurations is sometimes called the "inhibit" function, because opening the contacts of the power circuit prevents, or inhibits, something else from operating. In some areas of the electrical industry the normally closed contact configuration is also called the Form B configuration.

In Figure 7-7, both the normally open and normally closed contacts are combined on a single

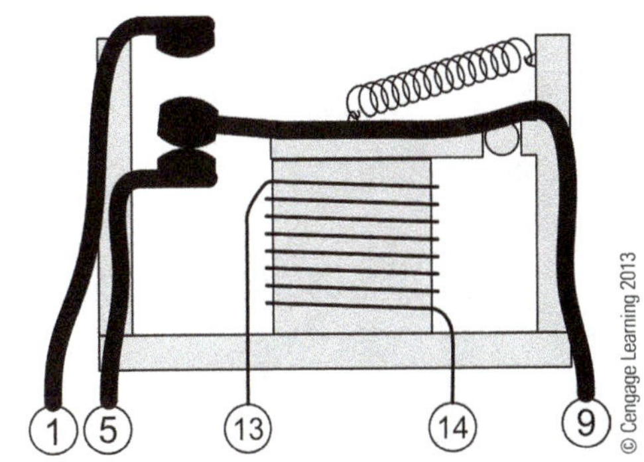

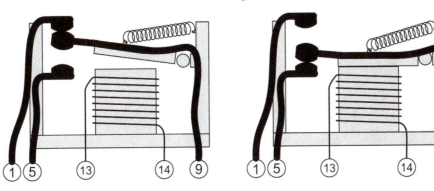

**FIGURE 7-7** Form C relay contacts

pole to form what is known as the changeover configuration, sometimes called the transfer configuration. Notice that relay terminal number 9 is the movable contact, which is common to both the normally open and normally closed fixed contact configurations. In the changeover configuration, the movable contact is called the common, similar to the common, or point, of a three-way switch, because it can make contact with it can make contact with both fixed contacts. Changeover relay contacts reverse states, toggling between the normally open and normally closed contacts when the control coil is energized; the normally open contacts mechanically close, and the normally closed contacts mechanically open. In some areas of the electrical industry the changeover contact configuration is also called the Form C configuration.

## Ice Cube Relays

Figure 7-8 shows the front of an Automation Direct, 14-pin, miniature square base, blade-type, ice cube relay, which has four sets of changeover contacts. Ice cube relays are small, completely self-contained, and plug into bases (see Figure 7-10) with pins that protrude out the back (see Figure 7-9). Before ice cube relays were used, the relays themselves had the wire terminals on them (see Figure 7-11), and they were permanently wired into electrical circuits, which made them more difficult to troubleshoot and change out. With ice cube relays, only the base is permanently wired into the electrical circuit, and the relay is easily inserted and removed for troubleshooting

**FIGURE 7-8** Front of automation direct relay

and replacement purposes. The term "ice cube" comes from the resemblance of the small squared relay to the small squared appearance of what ice cubes used to look like when these relays were first manufactured.

Figure 12 shows the ice cube relay with the cover removed, and all the parts labeled.

Figure 13 shows the same ice cube relay with the armature removed from the yoke to observe the relay construction.

FIGURE 7-9 Back of automation direct relay

FIGURE 7-10 Automation direct relay base

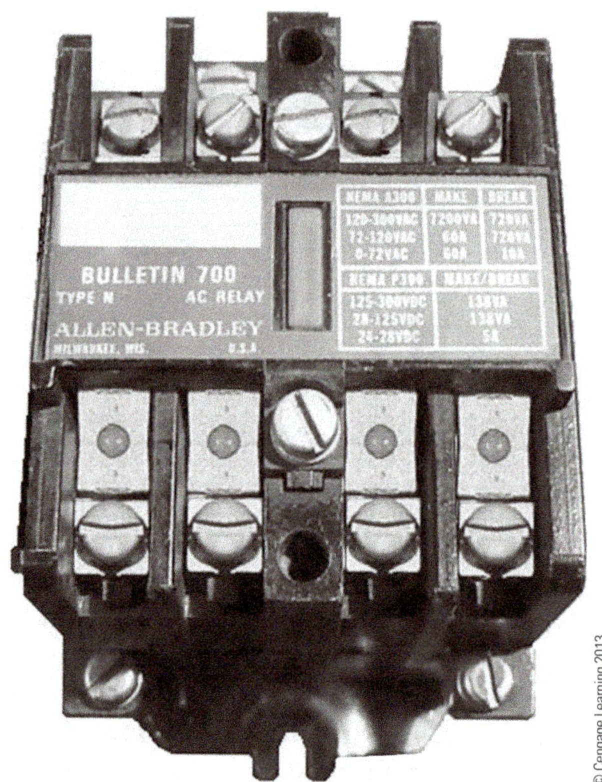




FIGURE 7-11 Allen-Bradley control relay

## Relay Contact Characteristics

The physical characteristics of relay electrical contacts are different from contactor electrical contacts. Specifically, relay contact faces are sometimes domed, as shown in Figure 7-14, and they make and break contact at only one end of the armature; the other end of the armature is hinged. At first thought, having the contact faces domed would seem to be counterproductive to the purpose of the relay contact, because the surface area where the contacts "seat" is reduced. The term "seat" is used to describe where the contact faces meet and connect both mechanically and electrically. Relay contacts are domed to improve the contact resistance by mechanically cleaning the contact faces of oxidation, dirt, and corrosion every time they make contact.

Relay contacts are made from a harder copper, silver-cadmium oxide alloy (usually less silver and more copper) so they do not wear out for a very long time. The main problem with using the harder alloys that contain more copper is that

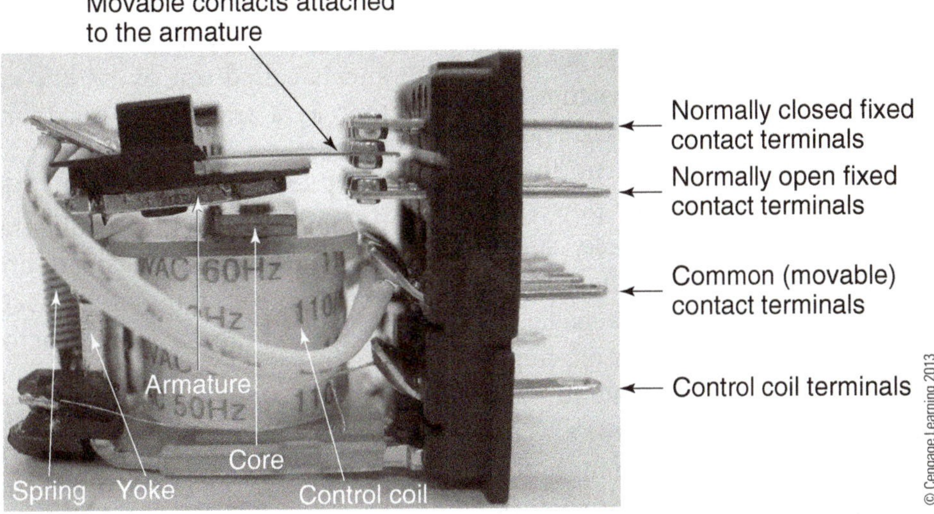


FIGURE 7-12 Automation direct relay without cover

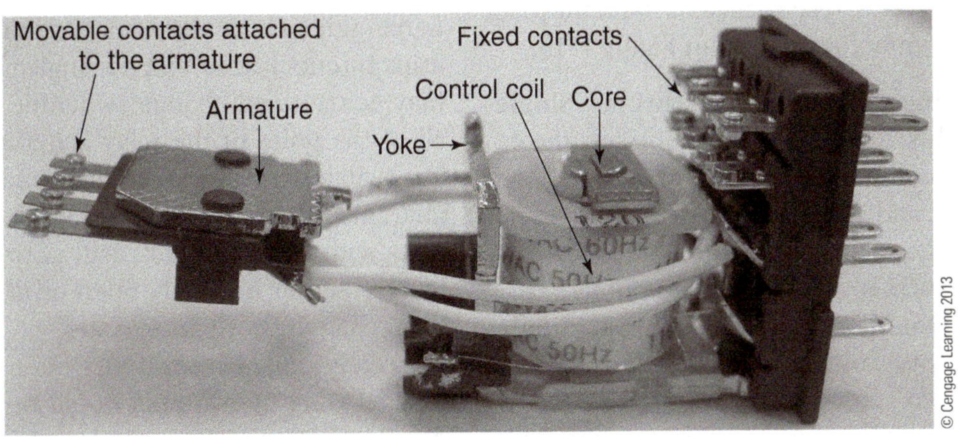


FIGURE 7-13 Automation direct relay laid open

FIGURE 7-14 Relay domed contacts

any oxidation that occurs on the contact face will interfere with a good electrical connection and contribute to contact resistance, because copper oxidation is not a good conductor. Contact resistance is a problem for any electrical contact, but it is especially significant in relay circuits, given the lower voltage and current levels of control circuits where relays are used. When the domed contacts come together, they actually meet before the armature is completely seated against the core. After the contact faces meet, and the armature continues to travel more, the movable contact will push past the stationary contact, because the movable contact is anchored on the hinge end and cannot move. This causes the contact faces to sweep across each other after making initial contact, mechanically cutting through any oxidation that may have accumulated if the relay had not been exercised for a long period of time.

## Relay Contact Configuration Terminology

The terms "poles" and "throws" are the same for relay contacts as they are for toggle switches. The term "pole" refers to the number of separate power circuits that can be switched, and the term "throw" refers to the number of states that the power circuits can be switched into. Single-pole, single-throw toggle switches are used to turn the light in a room on and off have the capacity for a single circuit: single-pole; and can be either on or off: single-throw. Like toggle switches, relays come in many different configurations of poles and throws, as one relay armature may be used to actuate more than one set of contacts. The configurations can get confusing sometimes, so to help explain the poles and throws terminology for relays, consider the following common types of relays called out with the poles and throws terminology, shown in Figure 7-15:

- **SPST—Single-Pole, Single-Throw, Single-Break.** This contact configuration has the capacity to control a single circuit (single-pole) in the on and off positions or states (single-throw). The contacts may be either normally open or normally closed configurations.
- **DPST—Double-Pole, Single-Throw, Single-Break.** This contact configuration has the capacity to control two different circuits (double-pole) in the on or off positions or states (single-throw). The contacts may be either normally open or normally closed configurations.
- **SPDT—Single-Pole, Double-Throw, Single-Break.** This switching configuration is completely different from the single-throw, on-and-off type of contact configuration, and is called out by its known function, either as a transfer contact or a changeover contact.
- **SPST—Single-Pole, Single-Throw, Double-Break.** This contact configuration functions the same as the SPST single-break design. The double-break type of contact configuration is most common for contactors, rather than relays.
- **DPST—Double-Pole, Single-Throw, Double-Break.** This contact configuration functions the same as the DPST single-break design.

## Force-Guided Relays

Force-guided relays have contacts that are mechanically interlocked so that no single pole of the relay may actuate and change switching states without all of the poles of the relay actuating and changing switching states together. Force-guided relays guarantee that even just one welded contact in a relay with many contacts will keep all other contacts, whether normally open or normally closed, from changing switching states.

Force-guided contacts are often a requirement in safety circuits when it would be dangerous for different actions to operate together, or for different actions to not operate together. Force-guided contacts also are called positive safety contacts, forced contacts, linked contacts, positive-guided, or positively driven contacts.

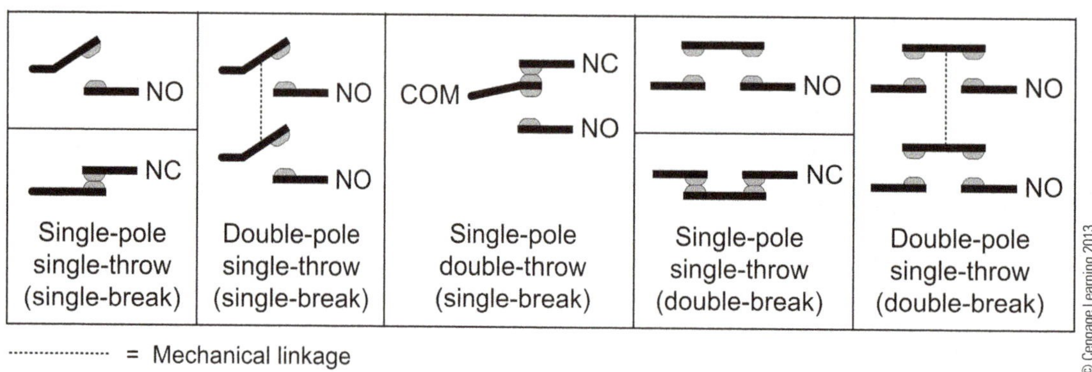


FIGURE 7-15 Relay contact configurations

## Timing Relays

Timing relays, sometimes called time-delay relays, provide a delayed action, changing operating states of the contacts associated with their control coil. The timing relay contacts may be either normally open or normally closed, and change states after the specified time delay after the relay control coil is energized, which is called "on delay." Or, the timing relay contacts may be either normally open or normally closed, and change states after the specified time delay after the relay control coil is de-energized, which is called "off delay." Timing relays are a basic element of motor control circuits, and deserve an extended explanation here.

To make sense of timing relay contacts, it is important to note both the contact type, normally open or normally closed, and the direction of the arrow on the stem. If the arrow is pointing up, it designates that the timing function to change states of the contacts is going to happen after the control coil is energized. If the arrow is pointing down, it designates that the timing function to change states of the contacts is going to happen after the control coil is de-energized. It is important to note that when the timing function is associated with de-energizing the control circuit, the timing contacts switch states when the timer control coil first is energized. The following four circuits, with a timing chart for each, should help make sense of timing relays:

- **NOTC—Normally Open, Timed-Closed.** Normally the NOTC notation would not be included in the ladder diagram; you would be expected to know the function by the symbol. The stem arrow is pointing up, so the timing function happens when the timing relay control coil is energized. Notice by the timing chart on the right of the ladder diagram shown in Figure 7-16 that the timing contacts for S2 do not immediately change states when the timing relay control coil first is energized, the timing contacts remain open. Instead, the relay times out 5 seconds, and then closes the timing contacts to energize S2. When the timing relay control coil is de-energized, the timing contacts open immediately.
- **NCTO—Normally Closed, Timed-Open.** The stem arrow of the timing contact in Figure 7-17 is pointing up, so the timing function happens when the timing relay control coil is energized. Notice by the timing chart on the right of the ladder diagram that the timing contacts for S2 do not immediately change states when the timing relay control coil first is energized; the timing contacts remain closed. Instead, the relay times out for 5 seconds, and then opens the timing contacts to de-energize S2. When the timing relay control coil is de-energized, the timing contacts close immediately.
- **NOTO—Normally Open, Timed-Open.** The stem arrow of the timing contact in Figure 7-18 is pointing down, so the timing function happens when the timing relay control coil is de-energized. Notice by the timing chart on the right of the ladder diagram that the timing contacts for S2 immediately change states when the timing relay control coil first is energized, so that the contacts may revert back to their normally open state after a time delay when the timing relay control coil is de-energized.

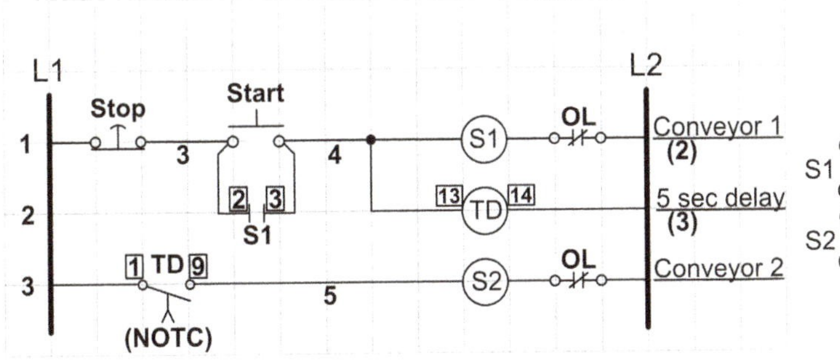


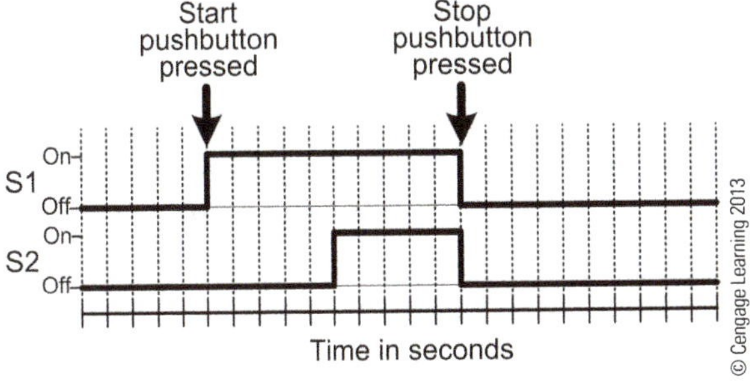




FIGURE 7-16 Normally open, timed-closed time delay

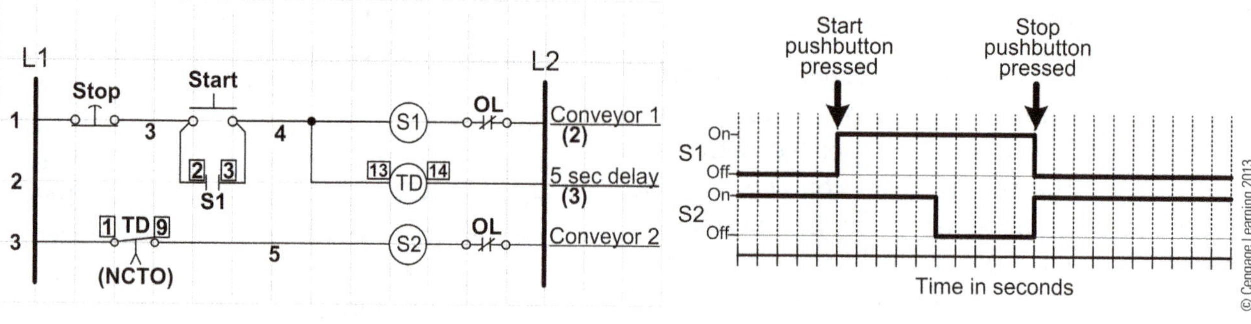


FIGURE 7-17 Normally closed, timed-open time delay

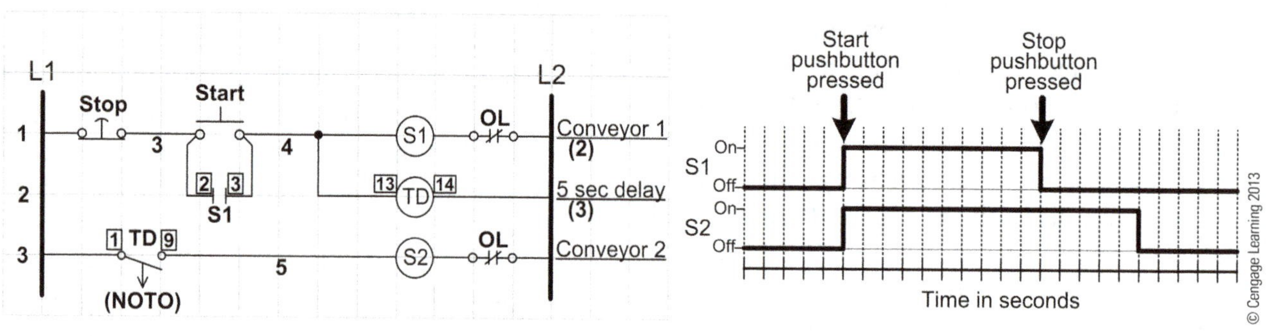


FIGURE 7-18 Normally open, timed-open time delay

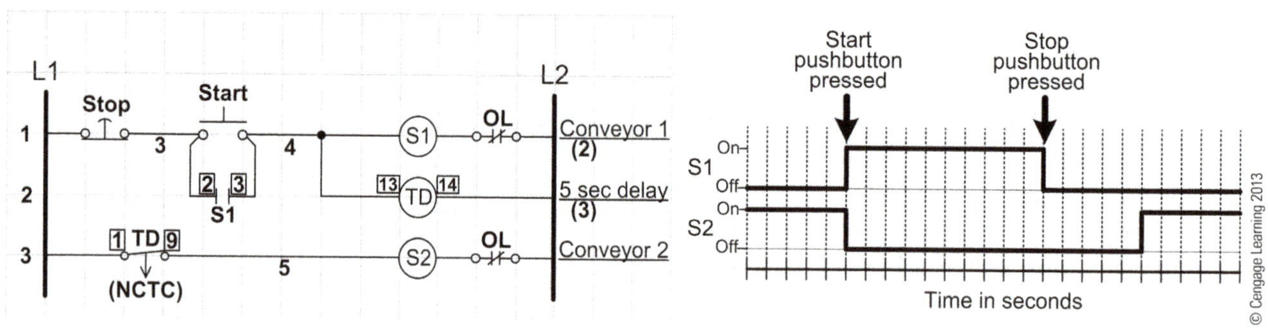


FIGURE 7-19 Normally closed, timed-closed time delay

- **NCTC—Normally Closed, Timed-Closed.** The stem arrow of the timing contact in Figure 7-19 is pointing down, so the timing function happens when the timing relay control coil is de-energized. Notice by the timing chart on the right of the ladder diagram that the timing contacts for S2 immediately change states when the timing relay control coil first is energized, so that the contacts may revert back to their normally closed state after a time delay when the timing relay control coil is de-energized.

## MAGNETIC CONTACTORS

### Purpose

Contactors, like relays, are electrically powered switches that utilize a lower voltage and current control circuit to energize and de-energize the control coil. The power circuit utilizes large, heavy-duty electrical contacts to control directly the higher voltages and currents associated with large electrical loads. Notice in Figure 7-20 that for some contactor units there is no spring to return the armature to

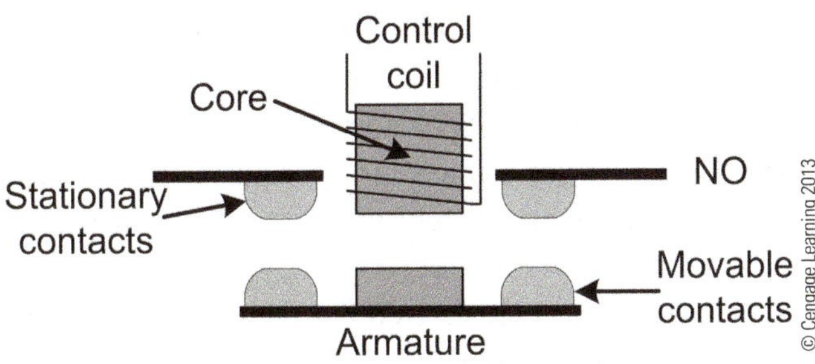


FIGURE 7-20 Contactor contacts drawing

the relaxed (de-energized) position. Instead, gravity is used for that purpose. Depending on gravity to return the armature results in fewer parts that can break and interfere with the operation of the contactor. There are also standards concerning safety that require a means, other than spring tension alone, to open the power circuit contacts when the contactor is de-energized in some applications. It is not uncommon to find contactors that use springs to keep the armature and core apart when the control coil is de-energized, but when the contactor is part of a large industrial motor starter, it is most common to use a gravity-type contactor unit.

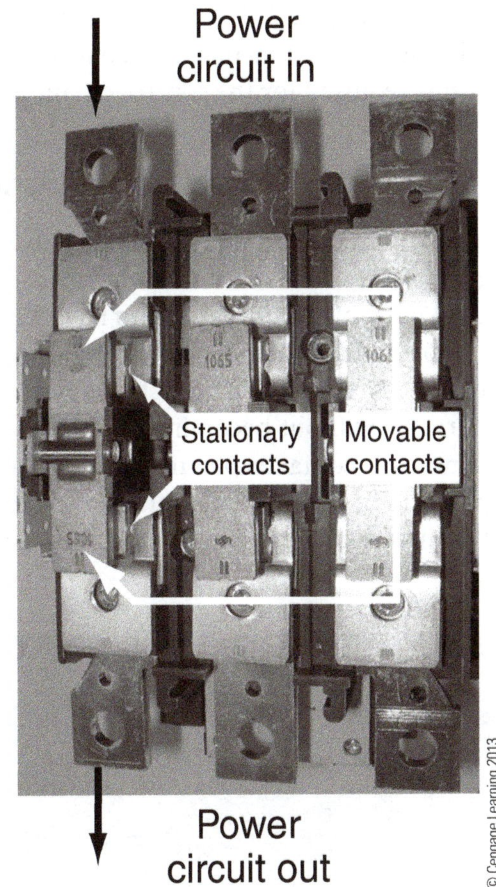


FIGURE 7-21 Contactor stationary and movable contacts

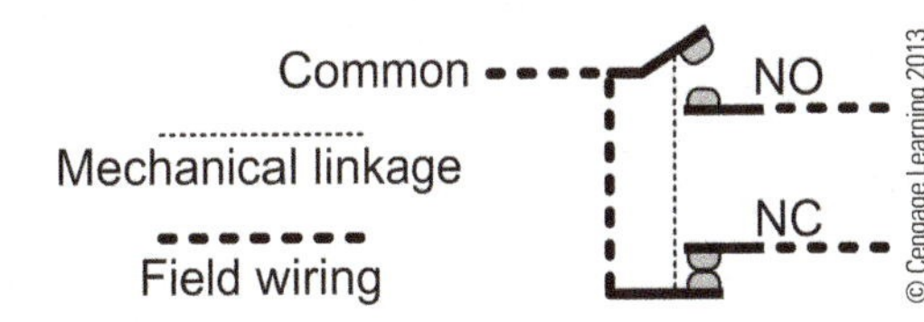


FIGURE 7-22 Class C contacts from NO and NO contacts

## Contactors in the Power Circuit

Whereas relays are found in the control circuit, contactors are found in the power circuit, connected directly between the electrical mains and the load to be powered. Contactors may be used to control electric lighting, electric heating, or other high-power electrical loads, but for the purposes of this study the load will be an electric motor. As shown in Figure 7-21, contactors share the same terminology with relays. Contactors have control and power circuits, stationary and movable contacts, an armature, coil, and core, but the main difference is the physical size: contactors are larger than relays.

Even though contactors are associated with the power circuit because they directly switch electrical loads on and off through their main contacts, contactors may also have auxiliary contacts. Auxiliary contacts are smaller, light duty, normally open or normally closed contacts, also actuated by the contactor armature. Auxiliary contacts, however, are more like relay contacts in that they have lower voltage and current operating ratings, which are designed to be connected in the control circuit to examine or report the operating status of the contactor power circuit contacts. Whereas the contactor power circuit main contacts are normally open and close when the control coil is energized, the auxiliary contacts may be either normally open or normally closed, and change state when the control coil is energized. Some contactor manufacturer's auxiliary contacts may not be manufactured in the changeover configuration. If this is the case, by using a normally open auxiliary contact in conjunction with a normally closed auxiliary contact, as shown in Figure 7-22, the changeover configuration and function may be replicated if that function is required for the control circuit logic.

## Contactor Main Contact Characteristics

Contactor main contact faces must be able to seat a large cross-sectional area in order to carry the high load current. It is necessary to have flat main contacts because domed contacts could not attain the large seating area necessary to carry the large load currents without becoming excessively large. The contactor contacts pictured in Figure 7-23 are approximately 0.5 inches square, and are rated for 200 amperes. These contacts have the surface area and mass to safely dissipate the arc heat of a 200-ampere current, without causing contact welding or pitting. It is also necessary for contactor contacts to have a mass large enough to carry the high currents and safely dissipate the heat generated by the load current ($P=I^2R$), and the inevitable electric arcs that are caused from closing and opening large current loads.

Figure 7-24 shows a magnified view of the flat stationary and movable contacts in both the open (a) and closed (b) positions.

**FIGURE 7-23** Contactor contacts with ruler measuring

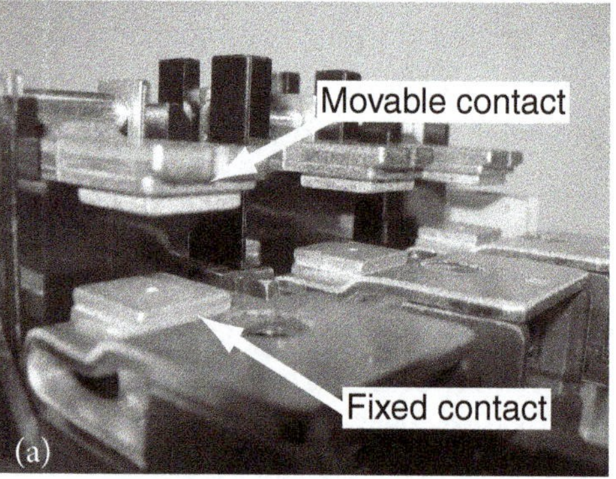

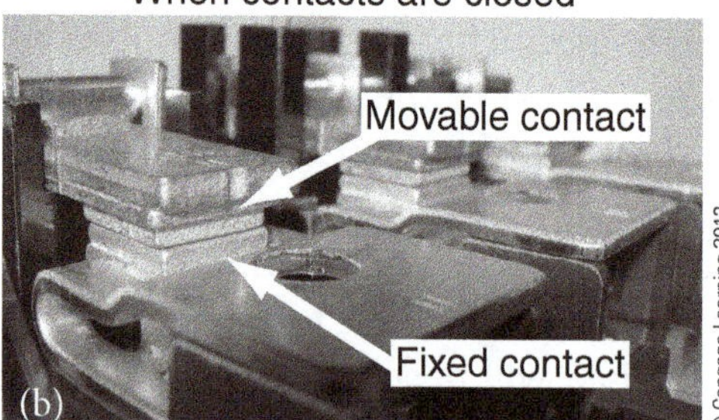




**FIGURE 7-24** Contactor fixed and movable contacts open and closed

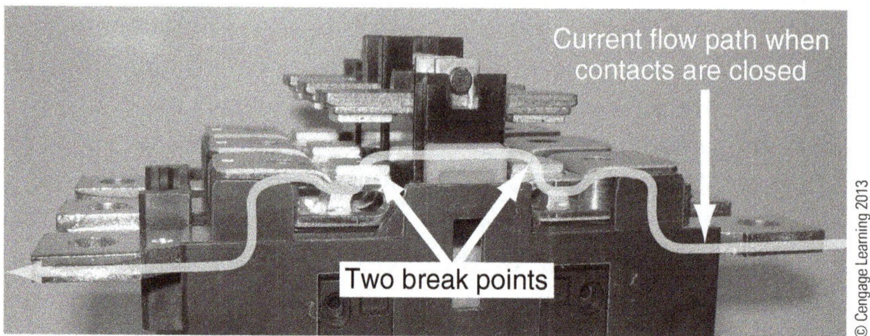




**FIGURE 7-25** Contactor from side showing current path through

Contactor power circuit contacts also utilize the double-break form design to realize a greater open break distance between the fixed and movable contacts than the contacts actually physically moved. The opening distance of the two break points is additive, so the movable contact only has to travel one-half the required opening distance to open the load. Figure 7-25 shows the double-break form design, and the current path through the contactor power circuit contacts when the contacts are closed.

Double-break flat electrical contact faces have one huge disadvantage from single-break domed contact faces, and that is that they cannot effectively perform that cleaning sweep when they close. The single-break domed contact design, which has a small seating area for the contact size that performs the cleaning sweep, has been traded off for the double-break flat contact design that can carry more current for a given size contact because of its greater seat area for the contact size and the greater realized open break distance of the double-break design. To compensate for the loss of the cleaning sweep feature of the domed contacts, the flat contactor power circuit contacts are made with a higher content of silver-cadmium oxide, almost 90%. The high silver content for the contacts is necessary because, unlike copper oxidation, silver oxidation is still a relatively good electrical conductor. When the large flat surfaces of contactor contacts oxidize, it does not cause a high resistance contact for the load current. The high silver content of the contacts does, however, cause two problems: increased expense of using silver, and the fact that silver is very soft and will exhibit poor wear characteristics.

## Arc Chutes

Electrical arcs are caused whenever the main contacts in the power circuit of a contactor are opened under load. When the electrical contacts first move apart, breaking the large electrical load current will cause a small electric arc to form immediately in the air between the contacts, before the contacts move far enough apart to self-extinguish the electric arc. Arc chute plates are placed in close proximity to

**FIGURE 7-26** Arc chutes

where the contacts open, and they use both mass and space properties to help extinguish the electrical arc before it becomes too large. The arc chute plates are made of iron, as shown in Figure 7-26, so they have the mass to absorb the majority of the generated heat of the electric arc so that the contacts do not have to dissipate all of the heat themselves. If the contacts had to dissipate the total heat generated by very large electrical arcs, the contact faces would become pitted, burned, and uneven, which would contribute to contact resistance and cause the heating problem to become worse. The spacing between the arc chute plates draws the arc in and breaks it into smaller pieces, because cooling and extinguishing several small electric arcs is easier than extinguishing one large electric arc. Even with the use of arc chutes and other arc extinguishing methods, the electrical contact faces still may become pitted and burned over their service life. Contactors, unlike relays, are designed to replace the electrical contacts as they become worn, to extend the useful life of the contactor.

## Replaceable Electrical Contacts

The power circuit contacts of electrical contactors are designed to be replaced as they become worn or damaged from use. The power circuit contacts are not the only wear parts of contactors, but they are the most wearing parts. Under normal use, the rest of the contactor will last many times longer than the power circuit contacts. Having the ability to replace the contacts easily is simply a necessity for contactors. Most contactor manufacturers do sell individual power circuit contact replacements, but if the contactor is going to be dismantled to replace one contact, it is just as easy to replace them all. Usually kits are available from the manufacturer to rebuild completely both the fixed and movable contact assemblies for the entire contactor, as shown in Figure 7-27, which will make the contactor as good as new.

## General Purpose Contactors

General purpose contactors are contactors that are designed and manufactured with voltage, current, and horsepower ratings to meet a wide variety of electrical load operating characteristics. The general purpose contactor shown in Figure 7-28 is designed to be expandable to meet the demands of a large range of control circuits and diverse electrical loads. The same general purpose contactor shown in Figure 7-29 demonstrates how auxiliary contacts may be added to expand the function of the contactor.

Generally, NEMA contactors are rated by voltage and horsepower, and once the motor horsepower is known, the contactor horsepower rating is

FIGURE 7-28 General purpose contactor bare

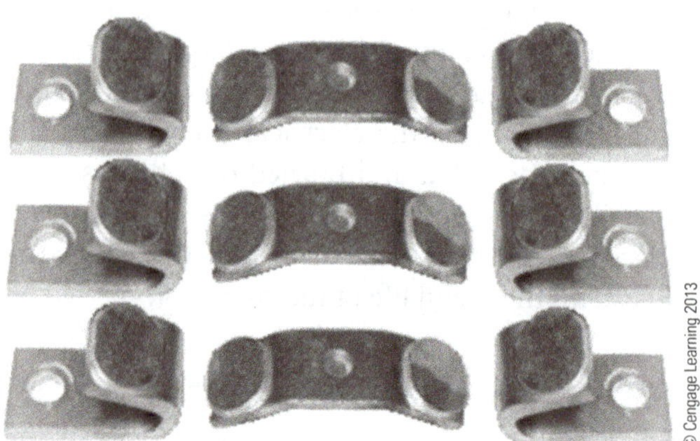


FIGURE 7-27 Replaceable contact set

FIGURE 7-29 General purpose contactor expanded

simply matched to that of the motor, and no other considerations are necessary. NEMA contactors and starters are designed in eleven standardized sizes from 00 to 9, and each of these sizes has a strictly defined current, voltage, and frequency-dependent horsepower rating. When a NEMA contactor is selected for starting an electric motor rated for the correct horsepower and voltage, normally it will handle the higher current jogging and plugging operations associated with controlling that motor.

**Jogging and Plugging Explanation.** "Jogging" is a motor control function where the motor is energized for only an instant, either to verify the direction of rotation for the motor, or possibly to position or align a machine part. Because the motor is jogged from a stopped condition, and then only energized for a very short period of time, and sometimes repeatedly, the motor will draw locked rotor current much more than usual. "Plugging" is a motor control function to stop an induction motor by temporarily applying the full line voltage of a motor in the reverse direction of rotation of the motor. This action causes extra-high current levels to be drawn from the power supply, and can cause damage to the motor control circuit components if they are not rated for plugging applications.

## IEC Contactor Ratings Are Specified in a Completely Different Manner Than NEMA

IEC contactors are not designed to standard sizes that have strictly defined electrical parameters, rather they are designed to meet a number of defined applications referred to as utilization categories. The two most common IEC utilization categories for motor control are AC-3 and AC-4. Utilization category AC-3 would be used for the most common motor starting applications using a standard squirrel cage induction motor. AC-3 rated contactors are capable of across-the-line motor starting where it is subjected to the high motor inrush starting current, but would not be suitable for stopping the motor under locked rotor current conditions, such as a stalled motor, or for other motor operations such as plugging. Larger AC-4 rated contactors are capable of both across-the-line motor starting, and motor applications such as plugging.

When determining which IEC contactor to use for a motor control application, it would be either a category AC-3 or AC-4; more than likely an AC-3 for normal motor starting and stopping. Beyond the category rating, the manufacturer would have to be specified to determine the current rating of the contactor, because there are no standard sizes for IEC contactors. Each manufacturer makes its own sizes and they rarely cross directly to other manufacturers. The table in Figure 7-30 identifies how NEMA and IEC standards compare for motor starter contactors.

The table in Figure 7-31 provides an example cross-reference between NEMA and IEC contactor sizes for an idea of how they compare. Note: This table is not a dependable cross-reference, because each manufacturer of IEC contactors determines its own sizes and category ratings. The IEC sizes listed in this table are specific to a single manufacturer, which is not identified.

## Definite Purpose Contactors

Definite purpose contactors, as shown in Figure 7-32, are designed for specific applications where the electrical load operating characteristics are defined clearly. The load operating characteristics that must be considered include parameters such as full-load amperes, locked rotor amperes, the number of poles, duty cycle, and the total number of expected operations. Definite purpose contactors will not have the provisions general purpose contactors have for adding auxiliary contacts, mechanical interlocks to interlock them with other contactors, or reversing capabilities. These contactors are designed for the special load operating characteristics of heating, air conditioning, and refrigeration equipment type loads. Definite purpose contactors normally feature quick connect terminals and binder head screws for easy wiring.

| CHARACTERISTIC | NEMA | IEC |
|---|---|---|
| RATINGS | Strict horsepower ratings for each standard size | Each manufacturer determines current and utilization category ratings. There are no standard sizes |
| SIZES | Standardized — 00 to 9 | There are no standard sizes |
| JOGGING/PLUGGING | Standard product is adequate as long as HP rating is observed | Must usually use a larger contactor, higher utilization rating and possibly increased current rating for these control operations |
| FAULT CURRENT | Meets NEC requirements | Usually has less fault current withstand |
| TERMINAL MARKINGS | | |
| POWER — IN/OUT | L1, L2, L3 and T1, T2, T3 | 1, 3, 5 and 2, 4, 6 |
| COIL | No standard | A1, A2 |
| HOLDING CONTACTS | Terminals 2 and 3 | Terminals NO and 14/22 |
| OVERLOAD CONTACTS | NC contacts on overload unit | Terminals 95 and 96 |
| AUXILIARY CONTACTS | No standard | NO is 53 and 54; NC is 61 and 62 |



FIGURE 7-30 NEMA and IEC comparison table

| MAXIMUM HP @ 460 VOLTS | IEC CONTACTOR SIZE | NEMA CONTACTOR SIZE |
|---|---|---|
| 2 HP | AC3 7A | 00 |
| 3 HP | AC3 7A | 0 |
| 5 HP | AC3 10A | 0 |
| 7.5 HP | AC3 12A | 1 |
| 10 HP | AC3 18A | 1 |
| 15 HP | AC3 25A | 2 |
| 20 HP | AC3 32A | 2 |
| 25 HP | AC3 37A | 2 |
| 30 HP | AC3 44A | 3 |
| 40 HP | AC3 60A | 3 |
| 50 HP | AC3 73A | 3 |

| MAXIMUM HP @ 460 VOLTS | IEC CONTACTOR SIZE | NEMA CONTACTOR SIZE |
|---|---|---|
| 60 HP | AC3 85A | 4 |
| 75 HP | AC3 105A | 4 |
| 100 HP | AC3 140A | 4 |
| 125 HP | AC3 170A | 5 |
| 150 HP | AC3 200A | 5 |
| 200 HP | AC3 300A | 5 |
| 350 HP | AC3 420A | 6 |
| 450 HP | AC3 520A | 7 |
| 500 HP | AC3 700A | 7 |
| 600 HP | AC3 810A | 7 |
| 900 HP | AC3 1100A | 8 |



FIGURE 7-31 NEMA and IEC motor starter contactor size comparison

## Relay and Contactor Generalizations

- Relays are designed to control small electrical loads in control circuits, such as contactor control coils, indicator lamps, or alarms; contactors are designed to control large electrical loads in the power circuit with high voltage and current requirements.
- Contactors are on/off devices to control a load directly; relays have many different contact configurations to control the logical operation of the control circuit.

**FIGURE 7-32** Definite purpose contactor picture

- Contactors have both large main contacts with high voltage and current ratings meant for use in the power circuit, as well as smaller auxiliary contacts with lower voltage and current ratings meant for use in the control circuit; relay contacts are all the same on a given device.
- Relays generally utilize spring tension to return the armature to the relaxed state when the control coil is de-energized; contactors may use gravity to return the armature to the relaxed state.
- Relays are most often the "plug-in" type, where they plug into a base. This allows the relay to be removed or exchanged without unwiring any part of the relay circuit, an especially useful feature when troubleshooting a control circuit; contactors normally are hard-wired into the power circuit.
- Relays may have domed contact faces, are made of a harder, lower-silver alloy, and they sweep the faces every time they make contact to keep them clean of oxidation, dirt, and debris; contactors have flat contact faces, and the contacts are made of a softer, higher silver alloy, because silver oxidation is still a good electrical conductor.
- Relays generally have a single break point; contactors generally have double break points.

## CHAPTER SUMMARY

- The pinout information identifies the pin numbers on the relay base to which each contact point of the relay is connected.
- The movable part of the relay, with the movable contacts attached to it, is called the armature.
- The armature is held away from the core by a spring when the control coil is de-energized.
- When the control coil is energized, the resulting magnetic forces will draw the armature to the core.
- The minimum current required to actuate a device is called the pull-in current, and is determined by the applied voltage.
- The current level where the armature returns to the de-energized position is called the dropout current.
- When fixed electrical contacts are mounted on the yoke, and movable electrical contacts are mounted on the armature, the device becomes an electrically powered switch.
- When designating relay contacts as normally open or normally closed, the contact configuration is always called out for the contact position when the relay is in the de-energized or relaxed state, regardless of the contact function in the circuit.
- The normally open contact configuration is sometimes called the enable function, or the Form A configuration.
- The normally closed contact configurations is sometimes called the inhibit function, or the Form B configuration.
- The contact configuration that has both normally open and normally closed contacts, and toggles between the two states, is called the changeover, transfer, or Form C configuration.
- Ice cube relays are versatile, self-contained relay packages that plug into relay bases that are permanently wired in the electrical circuit.
- Relay contact faces may be domed, and they make and break contact at only one end of the armature; the other end of the armature is hinged.
- Relay contacts are made from a harder copper, silver-cadmium oxide alloy so they do not wear out for a very long time.
- The term "pole" refers to the number of separate power circuits that can be switched, and the term "throw" refers to the number of states into which the power circuits can be switched.
- Contactors, like relays, are electrically powered switches that utilize a lower voltage and current control circuit to energize and de-energize the control coil; the power circuit utilizes large, heavy-duty electrical contacts to control directly the higher voltages and currents associated with large electrical loads.
- Contactors are found in the power circuit, connected directly between the electrical mains and the load to be powered.
- Contactor main contact faces must be able to seat a large cross-sectional area in order to carry the high load current.
- Contactor power circuit contacts also utilize the double-break form design to achieve a greater open break distance between the fixed and movable contacts than between the contacts actually physically moved.
- Arc chutes are placed in close proximity to where electrical contacts open, and they use both mass and space properties to help extinguish the electrical arc caused by opening the contacts under load.
- The power circuit contacts of electrical contactors are designed to be replaced as they become worn or damaged from use.
- General purpose contactors are contactors that are designed and manufactured with

voltage, current, and horsepower ratings to meet a wide variety of electrical load operating characteristics.

- NEMA contactors are rated by voltage and horsepower, and once the motor horsepower is known, the contactor horsepower rating is simply matched to that of the motor, and no other considerations are necessary.
- IEC contactors are not designed to standard sizes; each manufacturer decides what ratings to manufacture.
- Definite purpose contactors are designed for specific applications where the electrical load operating characteristics are clearly defined, and they are not expandable like general purpose contactors.

## REVIEW QUESTIONS

1. When a relay control coil is de-energized, or relaxed, what determines the position of the armature?
2. When a relay control coil is energized, what determines the position of the armature?
3. What is the minimum electrical current called that is necessary to create the lines of magnetic flux to mechanically actuate a relay? Give both names.

   1. ______________________________

   2. ______________________________
4. What two forces must the magnetic field overcome to actuate the relay?

   1. ______________________________

   2. ______________________________
5. What is the current called at which the energized relay armature will succumb to either gravity or spring tension and allow the armature to assume the de-energized position?Give both names.

   1. ______________________________

   2. ______________________________
6. Relays are commonly referred to as what types of switches?
7. What is the purpose of the control circuit?
8. What are the names of the two relay electrical circuits?

   1. ______________________________

   2. ______________________________
9. In which of the two relay circuits is the control coil?
10. What is the name of the relay power circuit contact that is mounted on the relay yoke?
11. What is the name of the relay power circuit contact that is mounted on the relay armature?
12. What is the term used to document relay contact configurations on all ladder and print diagrams?
13. In what state of the relay (energized or de-energized) are the contact configurations called out?
14. What are two names sometimes used for the normally open contact configuration function?

    1. ______________________________

    2. ______________________________
15. What are two names sometimes used for the normally closed contact configuration function?

1. ______________________

2. ______________________

16. What is the name of the contact configuration when both normally open and normally closed contacts are combined on a single pole? Give all three names.

 1. ______________________

 2. ______________________

 3. ______________________

17. What is the purpose of having domed relay contacts sweep across the faces of each other every time they seat?
18. Are relay contacts manufactured with harder or softer silver alloys than contactors?
19. What term refers to the number of separate power circuits that can be switched by a relay?
20. What term refers to the number of states in which the power circuits of the relay can be switched?
21. What is the purpose of a contactor?
22. In what three specific ways are contactor contacts different from relay contacts?

 1. ______________________

 2. ______________________

 3. ______________________

23. What is the advantage of using flat contact faces for contactor contacts?
24. What is the biggest disadvantage of using flat contacts with contactors?
25. Why is a softer, higher silver alloy used for contactor contacts?
26. What is the advantage of designing contactor contacts as double-break types, rather than single-break types?
27. When it becomes impractical to make the contacts large enough to dissipate the generated arc heat themselves, or move the contacts far enough apart that the arc would become cooled enough to self-extinguish, what is incorporated to help safely extinguish the electrical arcs?
28. General purpose contactors differ from definite purpose contactors in what two specific ways?

 1. ______________________

 2. ______________________

29. Which rating philosophy, IEC or NEMA, most closely matches the contactor unit with the least amount of contactor to the exact electric operating characteristics of the load being powered?
30. Which rating philosophy, IEC or NEMA, least closely matches the contactor unit, and is robust enough to handle a large range of electrical loads to the exact electric operating characteristics of the load being powered?

CHAPTER

8

# Overload Units

**PURPOSE**

To familiarize the learner with the two circuits found on the overload unit and the unique purpose and operation of the overload unit for electric motor protection.

**OBJECTIVES**

After studying this chapter on overload units, the learner will be able to:

- Explain the purpose and need for an overload unit in an electric motor starter circuit
- Identify where in the motor control circuit the overload unit is found
- Explain the term "inverse time delay trip"
- Explain how the overload unit detects an overload condition from the motor load operating current and opens the overload unit contacts
- Explain the two motor starter circuits found on the overload unit, the control and power circuits, and the purposes of both
- Identify the motor circuit components that are sized from the nameplate current

- Select an overload heater from a table, given the actual motor nameplate current
- Discuss the differences between the three most common types of overload units

Overload units contain the overload contacts and utilize the motor load operating current to mimic the thermal characteristics of the motor being protected, to de-energize the motor before damaging heat can build up from excessive current due to mechanical overload conditions.

## THERMAL OVERLOAD UNITS

The overload unit of the motor starter pictured in Figure 8-1 is located below the magnetic contactor, between the contactor unit and the motor load. The overload unit is operated by heat developed from the motor load operating current passing through the overload heaters. When the motor full load current, or less, flows through the overload unit, the temperature created by the overload heaters is not sufficient to trip the overload unit. When the level of motor current reaches a predetermined value, the increased temperature from the increased current will trip the overload unit and open the overload unit contacts. The overload unit contacts are wired in the contactor control circuit, as shown in Figure 8-2, and when they open, the contactor control coil will become de-energized and drop the contactor out, de-energizing the motor.

### Inverse Time Delay

Overload protection is designed with inverse time delay trip characteristics, meaning the greater the overload current the faster the circuit will open, but a smaller overload would take much longer to open the circuit. The high locked rotor starting current of a motor lasts only for a short period of time during a normal motor start, and does not last long enough to trip the overload unit. If the motor failed to start, however, the higher locked rotor current would trip the overload unit much faster than a small overload condition. The diagram

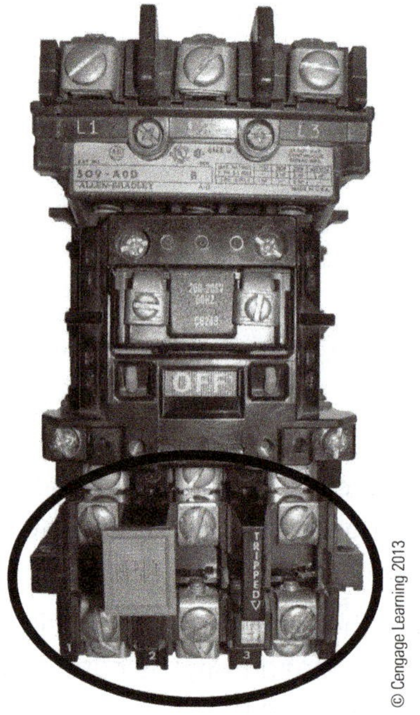


FIGURE 8-1 A-B starter with overload unit

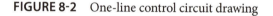

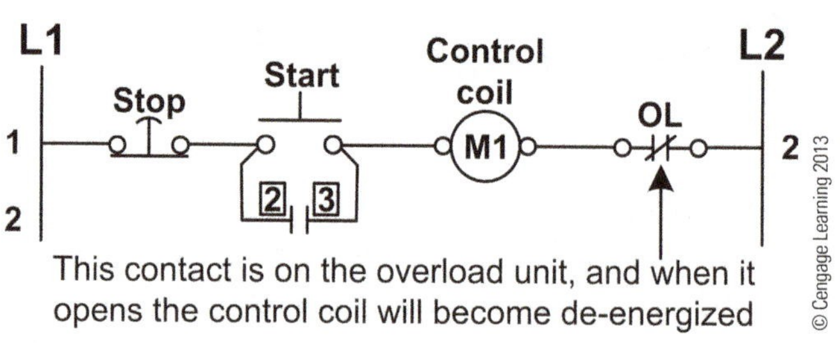


FIGURE 8-2 One-line control circuit drawing

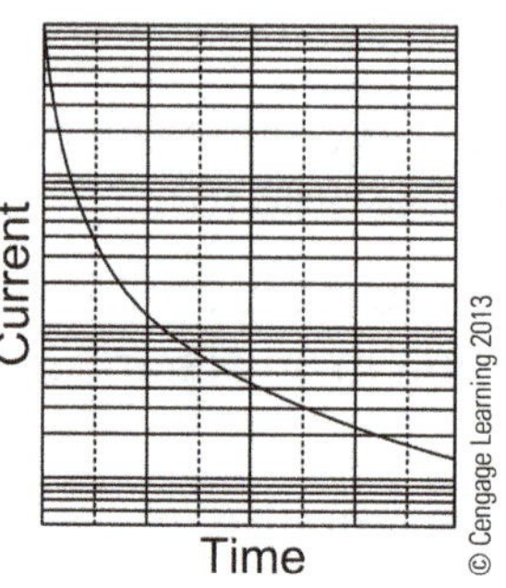


FIGURE 8-3 Inverse time delay drawing

in Figure 8-3 illustrates the inverse time delay trip characteristic, and the current curve represents the motor load current.

## Types of Overload Units

Of the three most common types of overload units, two use overload heaters (melting alloy and bimetallic), and the third is electronic. Melting alloy and bimetallic overload units are the most common types because of their simplicity and durability, but the electronic overload unit offers additional features that are not possible with the others.

## Melting Alloy (Eutectic)

The melting alloy type of overload unit (see Figure 8-4) works by using an overload heater to control the release mechanism. The overload heater contains a heater barrel element. The barrel surrounds a shaft that is connected to a ratchet wheel, which serves as the initial release mechanism. The shaft is held stationary in the barrel by using a solidified eutectic alloy. A eutectic substance, when heated, goes from a solid state to a liquid state without going through an intermediate solid-liquid putty stage.

When the motor current passes through the overload unit, it passes through a heating element that is wrapped around the barrel with the melting alloy inside, as shown in Figure 8-5. Under motor overload conditions, the additional motor current passing through the heating element causes additional heat on the melting alloy.

When the melting alloy is a solid, the shaft of the ratchet wheel is held from turning, as shown in Figure 8-6. The ratchet wheel is holding the overload unit contact closed, under spring tension that would open the contact if the spring were released. When the melting alloy melts, the spring tension on the ratchet wheel causes the shaft to turn in the liquefied alloy and release the spring tension holding the overload unit contact closed, as shown in Figure 8-7. The overload unit contact is connected in the control circuit of the motor

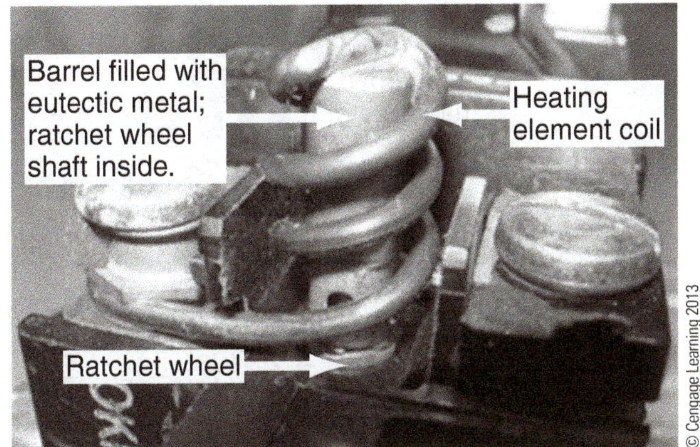


**FIGURE 8-5** Heater element with cover removed

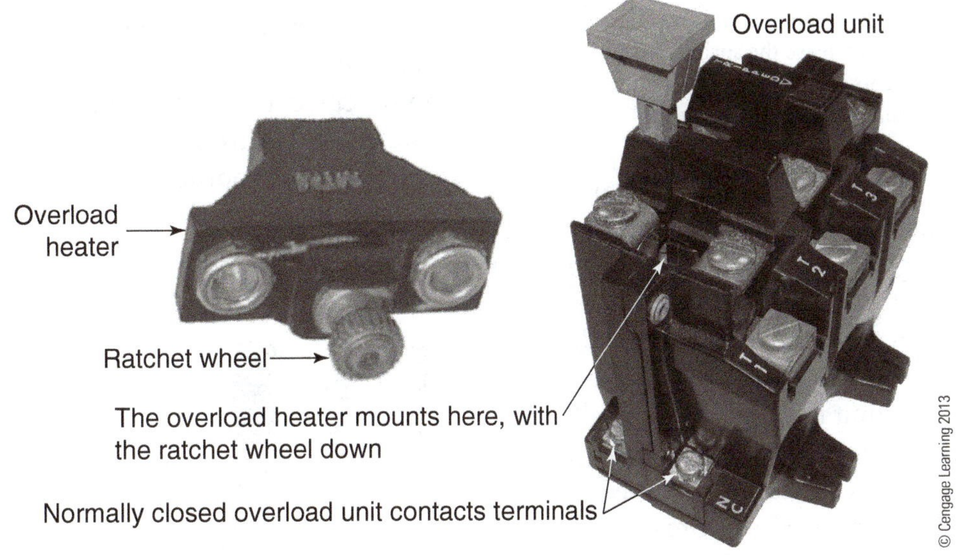


**FIGURE 8-4** Overload unit with heater element removed

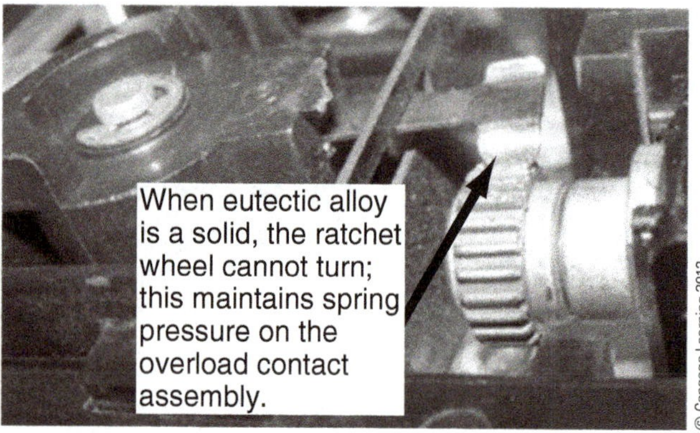


**FIGURE 8-6** Ratchet wheel assembly holding spring loaded arm. Here the eutectic alloy in the heater barrel element is still solid. This prevents the ratchet wheel assembly from turning and releasing the arm.

**FIGURE 8-7** Ratchet wheel assembly with released arm. In an overload condition, the eutectic alloy in the heater barrel element melts and releases the ratchet wheel assembly, which then can turn and release the spring tension on the arm. This allows the overload unit to function and shut down the motor.

starter contactor. When the overload unit contact opens, the contactor control coil will become de-energized and open the contactor main contacts to de-energize the motor. Once the melting alloy has a chance to cool, the alloy will solidify again and prevent the shaft from turning. The overload unit now may be manually reset, and spring tension once again will hold the overload unit contacts in the control circuit closed. Melting alloy overload heaters are sized from the nameplate full load amperes (FLA) of the specific motor to be protected, and are not otherwise adjustable.

## Bimetallic

When the motor load current passes through the overload unit, it passes through a heating element in close proximity to the bimetallic strip that is used as a trip lever. The bimetallic strip is made of two dissimilar metals, with different thermal expansion characteristics bonded together. When the bimetallic strip is heated, the metal with the higher thermal expansion characteristic will become longer, and cause the metal strip to bend. Under motor overload conditions, the additional motor current passing through the heating element causes additional heat on the bimetallic strip. The additional heat causes the bimetallic strip to deflect more and actuate a tripping mechanism, which opens the overload unit contacts, as shown in Figure 8-8. Bimetallic overload heaters are sized from the nameplate full load amperes (FLA) of the specific motor to be protected, but an advantage of most bimetallic overload units is that they are adjustable over a range of approximately 85% to 115% of their value, which allows them to be tailored to each specific motor.

## Electronic

Electronic motor overload units such as the Furnas unit shown in Figure 8-9 protect the motor from overload damage differently than either the melting alloy or the bimetallic overload units. One main difference of electronic overload units is that there are no overload heaters that must carry the motor current to detect an overload condition. Instead, the electronic overload unit contains sensors and a microprocessor that will calculate the intensity of the overload and the duration time to determine when to open the overload unit contact.

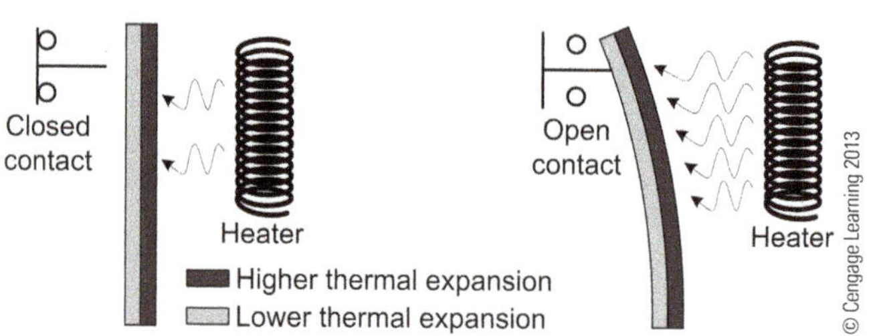

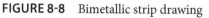

**FIGURE 8-8** Bimetallic strip drawing

FIGURE 8-9 Furnas electronic overload unit

| AMPERAGE | HEATER |
|---|---|
| 20.6 - 23.3 | H1042 |
| 23.4 - 26.0 | H1043 |
| 26.1 - 30.5 | H1044 |



FIGURE 8-10 Overload heater chart

Similar to the bimetallic overload unit, electronic overload units provide some adjustment, and may also perform other protection functions such as ground fault and phase loss protection, which are not offered on the other overload units. Electronic overload units are purchased for a range of motor nameplate full load amperes (FLA), and then the unit is set to the specific nameplate FLA of the specific motor to be protected.

## Sizing Overload Protection

It is important to understand that each different type of overload protection method requires its own procedure, and it is important to keep those procedures separate for each type of overload unit. The best way to assure adequate protection is to consult the cover of the motor starter, the motor control center, or the manufacturer's catalog at the time of installation. Typically, when the overload heater size is selected from the table on the inside of the motor starter cover, the nameplate full load amperes current of the motor is used without applying any multipliers.

The example heater table shown in Figure 8-10 demonstrates how the nameplate full load current of the motor is used to pick the correct size heater from the ranges listed in the table. A motor with a nameplate full load current of 21.2 amperes and a service factor of 1.00 requires three H1042 overload heaters.

Motor nameplate full load current is not used to determine the size or rating of any component in the motor starter circuit other than the overload unit heater size. Article 430 of the *National Electrical Code®* requires that the tables at the end of the article be used to find the motor current based on motor horsepower; to size circuit conductors, branch circuit, short circuit, and ground fault protective devices; and to find ratings of the disconnect switch. The actual full load current for different motors of the same size and type may vary by manufacturer. The table currents are used to ensure that if a motor needs to be replaced, the components of the motor starter circuit will not also need to be replaced if a less efficient replacement motor draws more current.

The heater table example shown in Figure 8-11 is not a valid table for actual use, but is included here to demonstrate how heater tables are used. First, the electrician has to determine for what size motor starter the heaters are needed, because although some heaters may be used in different size starters, the current ranges are different. Once the correct motor starter size column is determined, match the actual motor nameplate full load amperes within one of the current ranges in that column. Notice that some overload heaters are not manufactured for some size motor starters, and some heaters can be used in different size motor starters. When the correct current range is found, follow the row to the far right column and read the heater number. The questions at the end of this chapter will help clarify these differences.

| NEMA SIZES 00, 0, and 1 | NEMA SIZE 2 | NEMA SIZE 3 | HEATER |
|---|---|---|---|
| .960 – 1.07 | | | H1116 |
| 1.08 – 1.21 | | | H1117 |
| 1.22 – 1.35 | | | H1118 |
| 1.36 – 1.52 | | | H1119 |
| 1.53 – 1.70 | | | H1020 |
| 1.71 – 1.90 | | | H1021 |
| 1.91 – 2.10 | | | H1022 |
| 2.11 – 2.33 | | | H1023 |
| 2.34 – 2.62 | | | H1024 |
| 2.63 – 2.93 | | | H1025 |
| 2.94 – 3.27 | 3.72 – 4.10 | | H1026 |
| 3.28 – 3.64 | 4.11 – 4.59 | | H1027 |
| 3.65 – 4.06 | 4.60 – 5.07 | | H1028 |
| 4.07 – 4.55 | 5.08 – 5.65 | | H1029 |
| 4.56 – 5.03 | 5.66 – 6.29 | | H1030 |
| 5.04 – 5.59 | 6.30 – 7.00 | | H1031 |
| 5.60 – 6.25 | 7.01 – 7.82 | | H1032 |
| 6.26 – 6.92 | 7.83 – 8.79 | | H1033 |
| 6.93 – 7.75 | 8.80 – 9.67 | | H1034 |
| 7.76 – 8.63 | 9.68 – 10.8 | | H1035 |
| 8.64 – 9.59 | 10.9 – 12.0 | 11.5 – 12.8 | H1036 |
| 9.60 – 10.6 | 12.1 – 13.4 | 12.9 – 14.3 | H1037 |
| 10.7 – 11.9 | 13.5 – 14.9 | 14.4 – 16.0 | H1038 |
| 12.0 – 13.3 | 15.0 – 17.6 | 16.1 – 17.8 | H1039 |
| 13.4 – 14.7 | 17.7 – 19.0 | 17.9 – 20.3 | H1040 |
| 14.8 – 16.6 | 19.1 – 21.5 | 20.4 – 22.9 | H1041 |
| 16.7 – 18.8 | 21.6 – 24.5 | 23.0 – 26.0 | H1042 |
| 18.9 – 21.2 | 24.6 – 27.9 | 26.1 – 29.5 | H1043 |
| 21.3 – 23.9 | 28.0 – 31.76 | 29.6 – 32.2 | H1044 |
| 24.0 – 27.0 | 31.77 – 36.0 | 32.3 – 37.3 | H1045 |



**FIGURE 8-11** Large overload heater chart

## CHAPTER SUMMARY

- Overload units are designed to mimic the thermal characteristics of the motor load from the motor operating current.
- Overload units are designed with an inverse time delay function, which means that small overload conditions are tolerated for a longer period of time, and large overload conditions are tolerated for a shorter period of time.
- Overload units may be either thermal or electronic.
- Thermal overload units conduct the motor load current through heater elements, which produce heat proportional to the current and open the overload unit contacts if the motor current becomes excessive due to a mechanical overload condition.
- Electronic overload units sense the motor load current, and a microprocessor calculates the intensity of the overload condition to determine when to open the overload unit contacts.
- The overload unit contacts are connected in series with the motor starter contactor unit control coil. When the overload unit contacts open, the contactor unit will become de-energized and stop the motor load.
- Melting-alloy-type thermal overload units use heater elements sized from the full load amperes (FLA) of the specific motor load, and are not otherwise adjustable.
- Bimetallic type of thermal overload units use heater elements sized from the full load amperes (FLA) of the specific motor load, and typically have a 15% adjustment either side of that value to accommodate special conditions.
- Electronic-type thermal overload units are adjustable for a range of motor load currents, and are set for the full load amperes (FLA) of the specific motor load.

## REVIEW QUESTIONS

1. Overload units utilize the motor load current to mimic what?
2. What part of the overload unit does the motor load current (power circuit) pass through?
3. What are the three most common types of overload units?
4. What is meant by the term "inverse time delay trip" for overload units?
5. What is the significance of using a eutectic alloy in the melting alloy type of overload heater?
6. In which electrical circuit are the overload unit contacts wired?
7. How are bimetal strips for bimetallic type overload units made?
8. What is an advantage of the bimetal-strip-type overload unit?
9. What is an advantage of the electronic type of overload unit?
10. What are two types of protection that electronic overload units may offer in addition to overload protection?
11. When choosing overload heaters for a motor starter, what current is used for the lookup table?

**Note:** Please use the table shown in Figure 8-11 for questions 12-16

12. What is the smallest overload heater on the example table that can be installed in a NEMA size 2 starter?

13. What is the smallest overload heater on the example table that can be installed in a NEMA size 3 starter?
14. What size overload heater would be required for a motor with 10.7 full load amperes on the nameplate if a NEMA size 1 motor starter is being used?
15. What size overload heater would be required for the same motor if a NEMA size 2 motor starter is being used?
16. What size overload heater would be required for the same motor if a NEMA size 3 motor starter is being used?

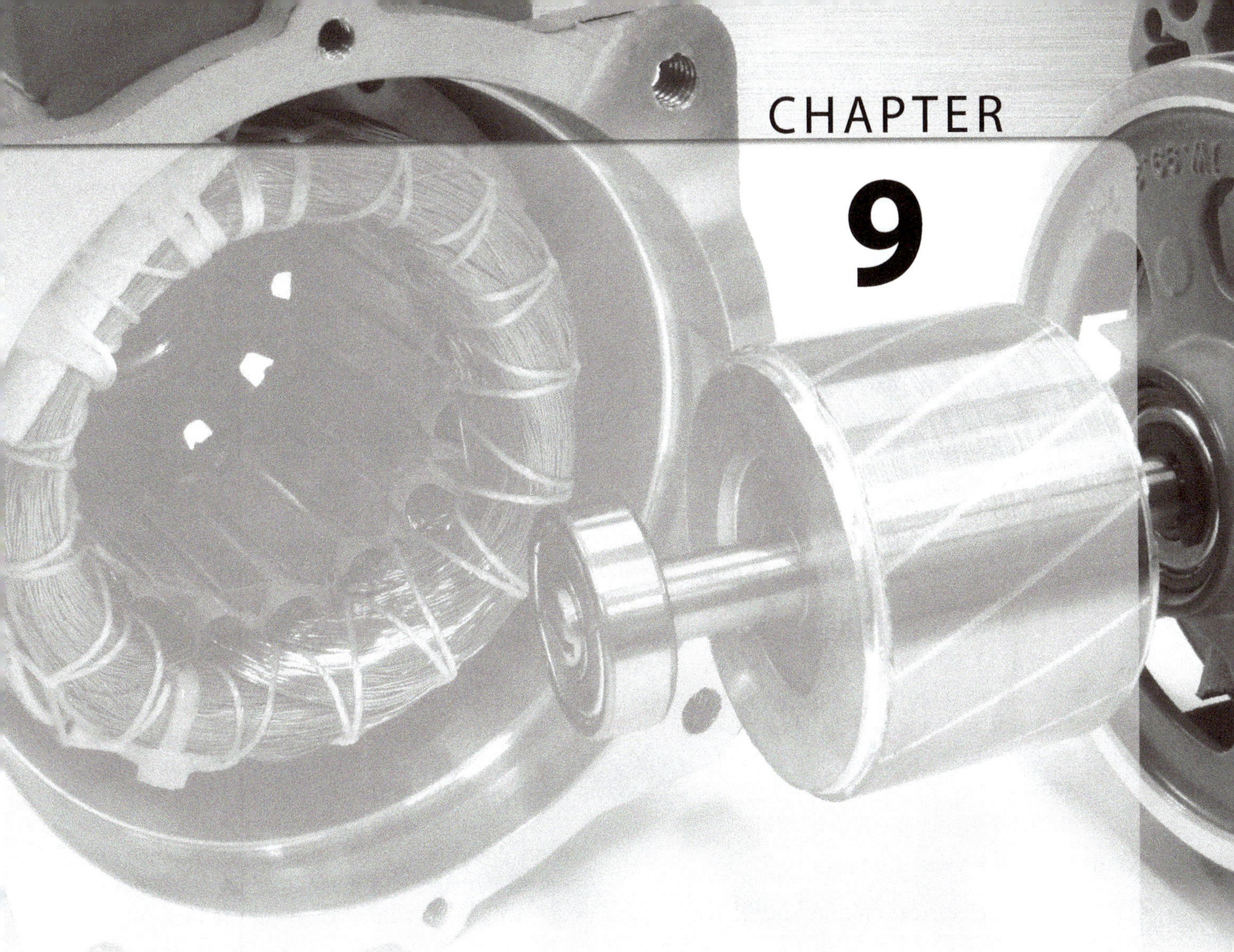

CHAPTER

9

# Magnetic Motor Starters

### PURPOSE

To familiarize the learner with the concept of combining a magnetic contactor with an overload unit to form a motor starter, as well the unique function they perform as a unit.

### OBJECTIVES

After studying this chapter on motor starters, the learner will be able to:

- Explain how the magnetic contactor and overload units function together to create a magnetic motor starter
- Identify the functional parts of a motor starter
- Discuss the differences between magnetic and manual motor starters
- Discuss some of the configurations in which motor starters are found

## DEFINITION

Motor starters are completely unique devices, even though they are simply the combination of two other common electrical components: the contactor and the overload unit. The magnetic contactor unit of the motor starter is designed to close and open the large motor load current, to start and stop the motor under the control of the contactor control coil. The overload unit is designed to measure the running load current of the motor, and open the control coil circuit of the contactor unit to stop the motor in the event the motor draws current levels that could cause damage from excessive heat. The picture in Figure 9-1 shows the contactor and overload unit separated, but they would mount together on the common backplate shown. Figure 9-2 shows this same motor starter assembled with all parts labeled, but altered to show the front and both sides of the unit at the same time.

### Control and Power Circuits

Two separate electrical circuits on all magnetic motor starters pass through both the magnetic contactor and the overload unit: the control and power circuits. The magnetic contactor has both a control circuit and a power circuit, and the overload unit has both a control circuit and a power circuit. For the motor starter application, both of these independent circuits are interconnected to function as a single unit. The motor load current passes through both the contactor unit and the overload unit, but is actually directly controlled by only the contactor unit. The heavy-duty fixed and movable main contacts in the contactor unit are designed to make, carry, and break the motor load current safely and without damage.

### Motor Controllers

Every electric motor must have a controller. For a very small motor, the controller may be a cord-and-plug method, where the motor is plugged in to a power outlet to start it, and unplugged to stop it. Other smaller motors, like a garbage disposal unit, may have a simple toggle switch to start and stop the motor. In these two cases the overcurrent and overload protection are provided by the branch circuit overcurrent device (the circuit breaker in the electric panel). As the motor applications become larger, at some point the branch circuit overcurrent device no longer will be sufficient to provide the necessary overcurrent protection to protect the motor from overload damage, and the simple toggle switch no longer will be capable of carrying the locked rotor starting current of the motor. At that point a more sophisticated system of starting and protecting the motor will be required: a motor starter. Sometimes because of a high motor operating voltage, or the wish for more precise overcurrent protection, motor starters are even used on smaller fractional horsepower motors.

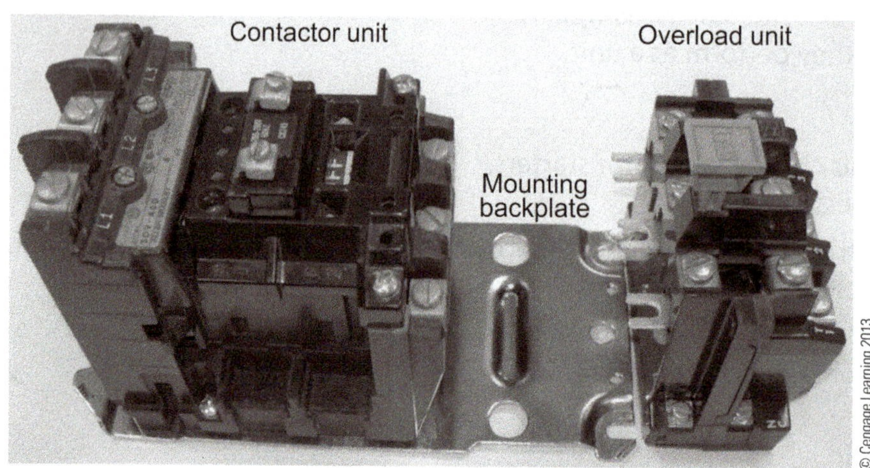



FIGURE 9-1 A-B starter with overload unit separated

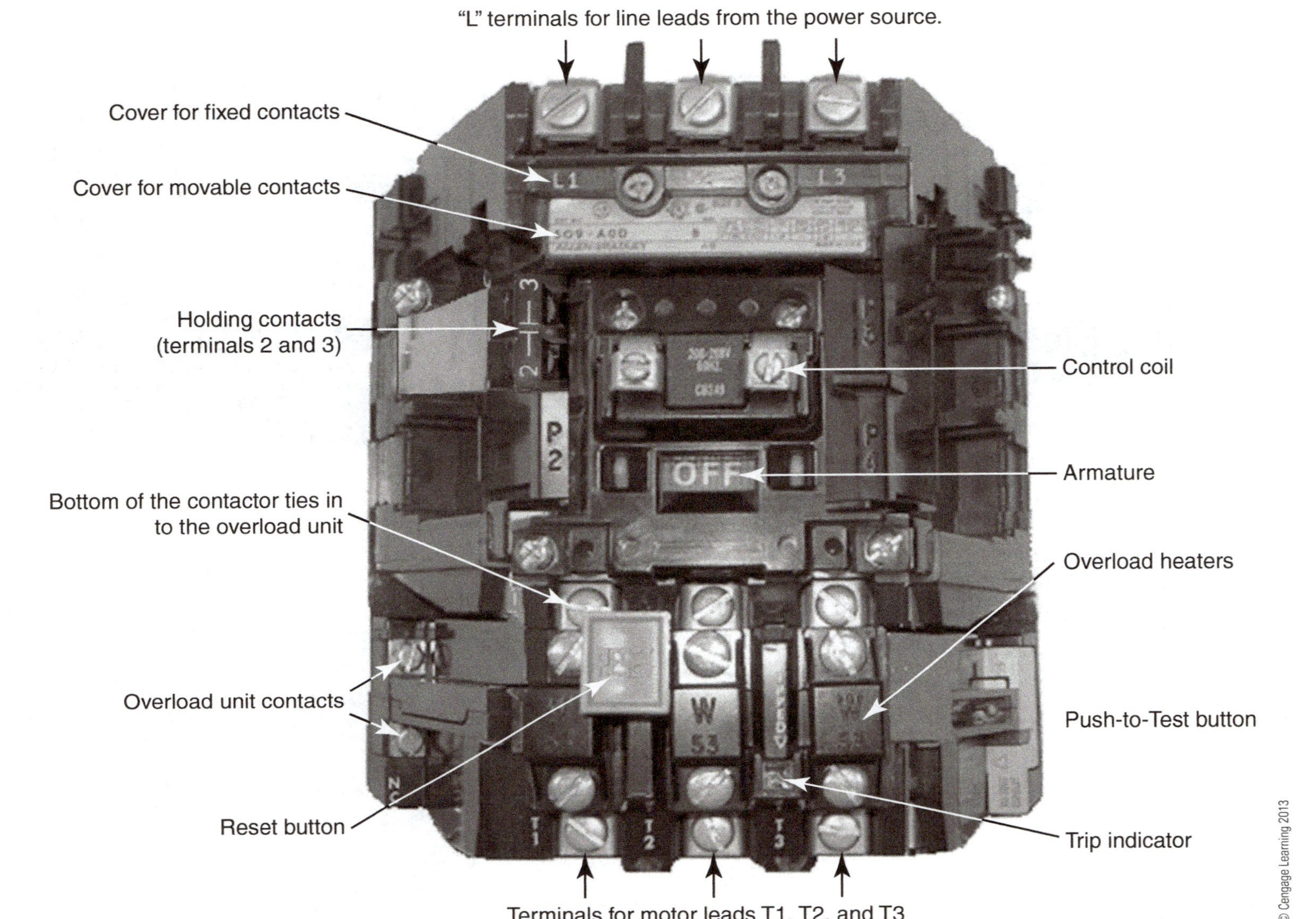

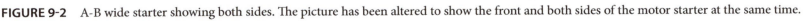

**FIGURE 9-2** A-B wide starter showing both sides. The picture has been altered to show the front and both sides of the motor starter at the same time.

## Manual Motor Starters

Two methods of across-the-line starting are manual motor starters and magnetic motor starters. A manual motor starter is a device where the power contacts that actually energize the motor load are operated manually, and can be similar to a simple toggle switch used to operate a garbage disposal. A manual motor starter unit, unlike a simple toggle switch, consists of a horsepower-rated switch that carries the motor load current, with a thermal overload device that carries the full motor current in each phase to provide overload protection, as shown in Figure 9-3. These toggle-switch-type manual motor starters also may be three-pole, for three-phase motors, as shown in Figure 9-4, but there are no overload heaters for this type of switch controller. If three-phase overload-heater-type protection is needed for a motor application, a larger three-phase manual motor starter would be needed, like the one shown in Figure 9-5. Manual motor starters do not have

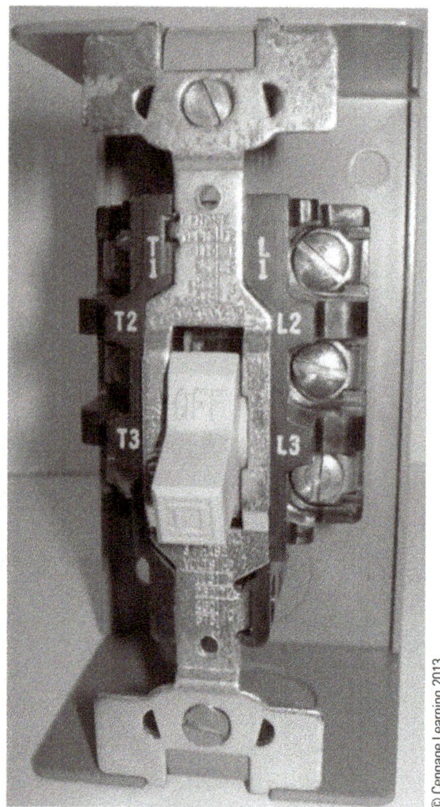



FIGURE 9-4 Three-pole manual motor starter switch

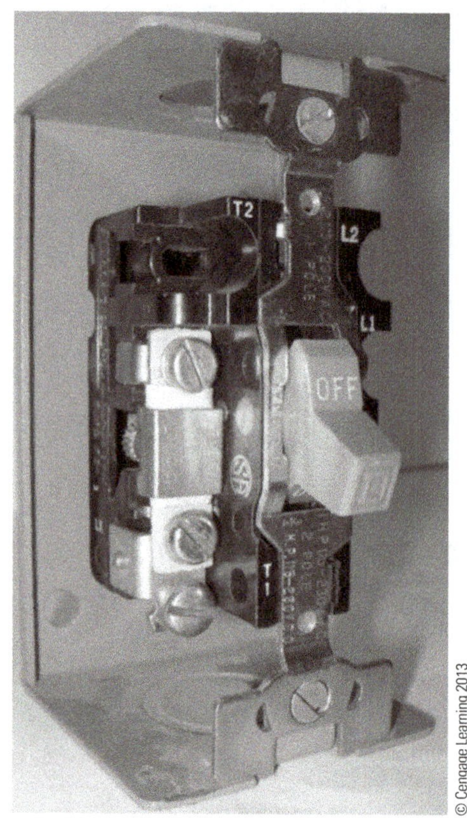



FIGURE 9-3 Single-pole manual motor starter switch

FIGURE 9-5 Manual motor starter

magnetic control coils or use electrical control circuits for their operation, so they are considered the most simple, least expensive method of electric motor control. Manual motor starters do provide the same overload protection for the motor as any other motor starter, but there are four main limitations that can make them unsuitable in some applications:

- Manual motor starters are not capable of remote control operation because the contacts are linked mechanically to the start and stop buttons, so the operator must control the motor from the starter location.
- Auxiliary control contacts are not provided to control auxiliary equipment.
- Manual motor starters are limited to small, usually fractional horsepower motors, and it would be rare to find them controlling motors greater than 10 horsepower.
- The switch contacts are a mechanical linkage function, and they will remain closed in the case of a power failure. When the power is restored, the motor will start up automatically, which may catch someone off guard and cause a safety hazard. Manual motor starters with low voltage protection (LVP) prevent automatic startup of motors after a power loss, but they are not common.

Even with these limitations, the manual motor starter still would be the appropriate choice for smaller motors that must run continuously, such as an exhaust fan.

## Magnetic Motor Starters

**Motor Starter Platforms.** The most common platform for electric motor starters in large commercial or industrial applications is the motor control center (MCC), similar to the one drawn in Figure 9-6, where all of the motor starters are located in a single location called "centralized control." The MCC provides the structure to hold and wire many motor starters in one location. A large power feeder is brought to the MCC, and from that point premounted buss bars distribute the power to the rest of the cabinet. The MCC also

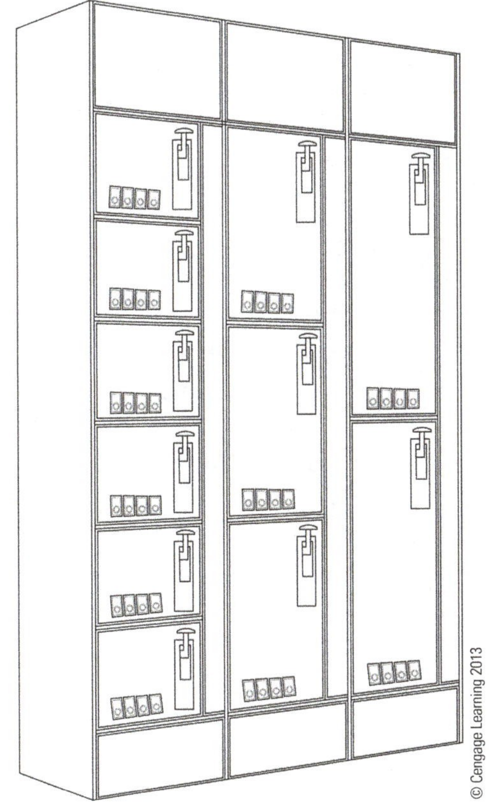


**FIGURE 9-6** MCC drawing

facilitates easy wiring, because from the wiring boxes on the top and bottom of the unit, any single compartment may be accessed through one of the vertical wireways that run along the sides of the compartments.

MCCs significantly simplify and speed up the installation time of motor starters, because the motor starters are prewired in buckets, like the one shown in Figure 9-7. These prewired buckets have terminals on the back, as shown in Figure 9-8, which plug onto power buss bars through slots inside the MCC cabinet to access power. These buckets are self-contained units that include the disconnect, overcurrent protection, motor starter, control transformer (if necessary), and a terminal strip for field wiring. The only field wiring required by the electrician is to land the T leads of the motor, and the control circuit conductors, to a terminal strip in the bucket. The buckets are manufactured in many sizes, and the compartments of the MCC may be arranged in any configuration. The drawing in Figure 9-6 shows the first column

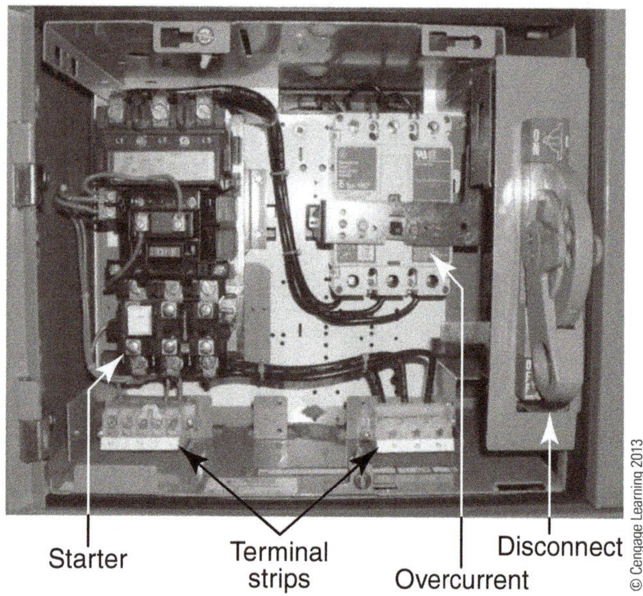


FIGURE 9-7 MCC bucket with labels

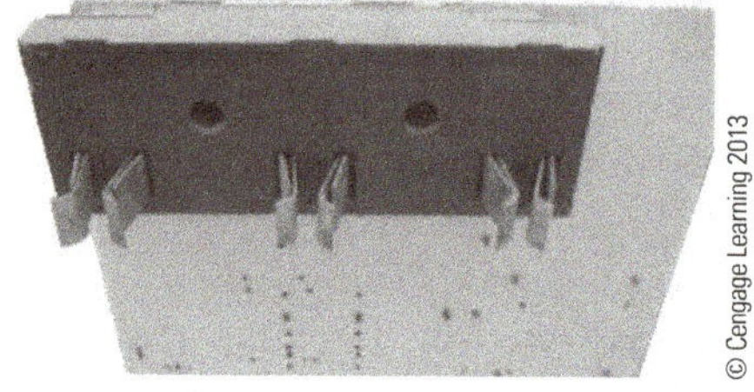


FIGURE 9-8 MCC bucket back stabs

as single-space buckets, but half-space buckets also are made. As more space is needed in the bucket, possibly because the starter is very large, the bucket may take up more than one single space, as shown in columns two and three.

Another common platform for electric motor starters in commercial and industrial applications is the combination motor starter, which combines all of the necessary control and protection components in one cabinet, as shown in Figure 9-9. Combination motor starters also significantly simplify and speed up installation time, because the motor starter is prewired in a single cabinet. Combination motor starters include the disconnect, overcurrent protection, motor starter, and a control transformer (if necessary). The only field wiring required by the electrician is to bring the power feed to the top of the disconnect, land the T leads of the motor,

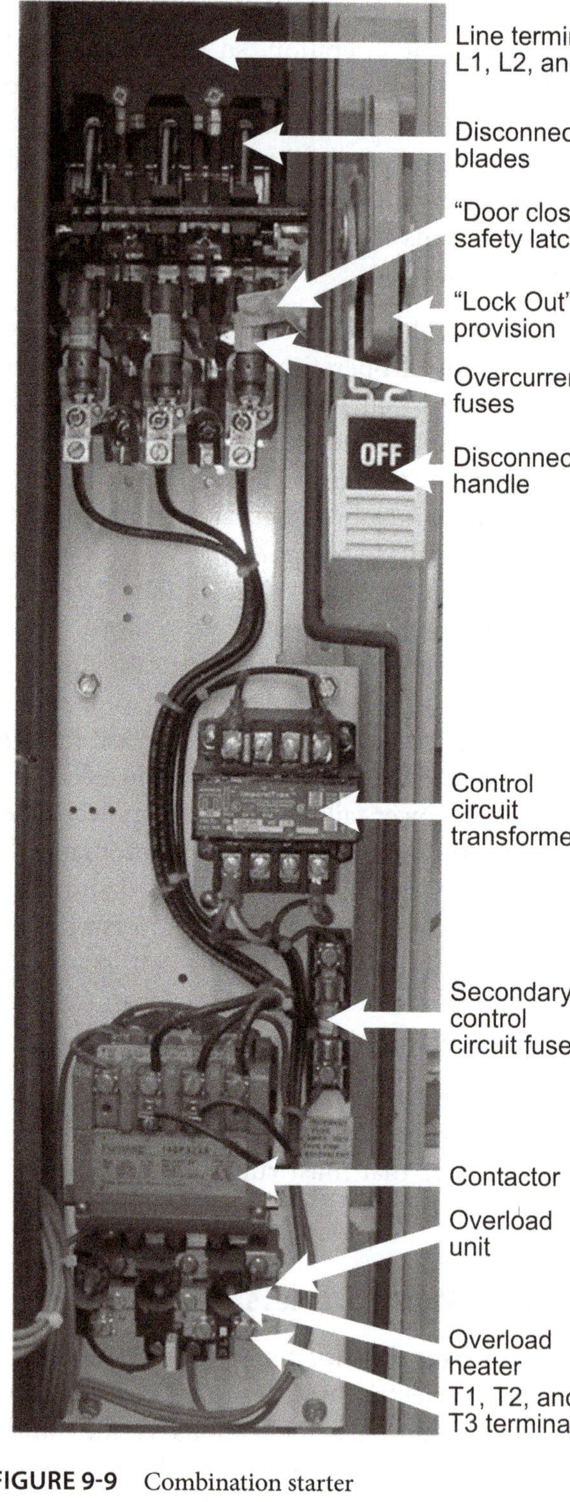


FIGURE 9-9 Combination starter

and land the control circuit conductors. Combination motor starters do not have terminal strips like MCCs. The control circuit conductors land directly on the motor starter component terminals.

Motor starters also are found as discrete components. Discrete motor starters do not have their own overcurrent protection or disconnect like the MCC motor starter bucket and the combination motor starter. All ancillary components must be supplied and field-wired separately, which increases the installation time. All power, motor, and control circuit conductors are landed directly on the motor starter component terminals. Some of the control circuit, as shown in Figure 9-10, is prewired at the factory to reduce installation time. Pictures of a discrete motor starter are shown in Figures 9-11 and 9-12, to demonstrate the factory prewiring when the starters are initially unpacked.

The factory control circuit prewiring is useful only if the motor starter contactor control coil is rated for the same line voltage as the motor. In this case the electrician would take the control circuit power from the top of the contactor unit on L1, and bring the control circuit back to terminals 2 and 3. Normally the control circuit conductors are not landed directly on the control coil, because there is always a chance of stripping the terminal screw, and it is much less expensive to replace the holding contacts than a control coil. It is not wrong to land directly on the control coil; it's just smarter to land on the holding contacts. When the control circuit and motor line voltage are different, the factory control circuit prewiring simply is removed.

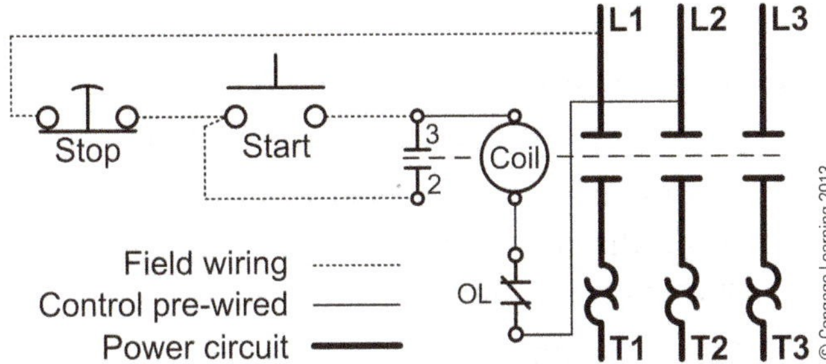


FIGURE 9-10 Starter factory prewiring

FIGURE 9-11 Discrete starter 1 with labels

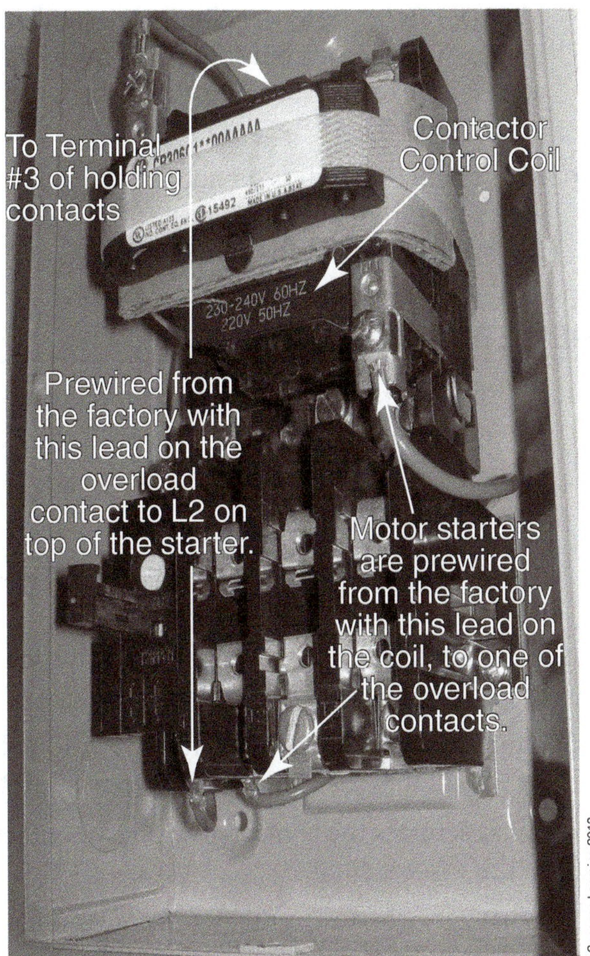


FIGURE 9-12 Discrete starter 2 with labels

## CHAPTER SUMMARY

- Motor starters are the combination of a contactor unit and an overload unit, connected to function as a single unit.
- The contactor unit is the component that directly energizes and de-energizes the motor load.
- The control circuit is present on both the contactor unit and the overload unit.
- The power circuit is present on both the contactor unit and the overload unit.
- Manual motor starters are the most simple and least expensive method of controlling a motor.
- Manual motor starters are not capable of remote control operation.
- Manual motor starters are limited to smaller horsepower motors, but still would be an appropriate choice for motors that run continuously.
- Manual motor starters do not provide protection from automatic restart when power is restored after a power outage.
- The three most common platforms to find magnetic motor starters are: motor control centers (MCCs), combination motor starters, and as discrete components.
- Motor control centers (MCCs) significantly simplify installation, because the disconnect, overcurrent protection, motor starter, and control transformer (if necessary) are prewired in "buckets" that plug into the MCC cabinet.
- Motor control centers are found in large commercial and industrial installations.
- The use of a motor control center is called "centralized control," because all the motor starters are contained in one cabinet.
- Combination motor starters significantly simplify installation, because the overcurrent protection, motor starter, and control transformer (if necessary) are prewired in a cabinet, and sold as a single unit.
- Combination motor starters are found in smaller commercial installations.

## REVIEW QUESTIONS

1. Motor starters are made by combining what two electrical components?
2. What are the names of the two electrical circuits of motor starters?
3. Which of the two units, contactor or overload unit, does the power circuit pass through?
4. Which of the two units, contactor or overload unit, does the control circuit pass through?
5. The motor load current is actually controlled by which unit, the contactor or overload unit?
6. What actuates the contactor auxiliary contacts in the control circuit?
7. What actuates the overload unit contacts in the control circuit?
8. What is the advantage of using the control circuit to energize the control coil to actuate the motor starter contactor?
9. What aspect of the magnetic motor starter circuit allows the control circuit and the power circuit to be of different voltage levels?
10. Very small motors may use what for a motor controller?
11. Do manual motor starters provide overcurrent protection?
12. What are four drawbacks of using manual motor starters rather than magnetic motor starters?
13. What are three common platforms for magnetic motor starter control?

CHAPTER

10

# Motor Starter Circuits

**PURPOSE**

To familiarize the learner with the entire magnetic motor starter circuit, documented with both ladder diagrams and wiring diagrams.

**OBJECTIVES**

After studying this chapter on motor starter circuits, the learner will be able to:

- Explain the difference between overcurrent and overload
- Explain the electrical interconnection of the power circuit between the contactor unit and the overload unit
- Explain the electrical interconnection and function of the control circuit between the contactor unit and the overload unit
- Identify magnetic motor starter components
- Explain the difference between ladder diagrams and wiring diagrams

## MOTOR CIRCUIT CONDUCTOR PROTECTION

Every electrical conductor that electricians install in electric motor circuits must be protected for both overcurrent and overload conditions. Overcurrent protection is provided at the point of supply for the circuit conductors, and is sized large enough to allow for the high locked rotor starting currents of the motor, while still protecting the conductors for short-circuit and ground-fault conditions. Overload protection is provided at the end of the motor circuit conductors with the motor starter overload unit, which is designed to mimic the thermal characteristics of the motor and open the overload control circuit contacts before excessive heat can cause damage to the motor. The reason the NEC allows the overload protection of the motor circuit conductors to be at the motor starter, rather than the point where they receive their supply, is that there is only one load in the entire circuit that could cause an overload condition: the motor.

### Overcurrent

Overcurrent is the condition where the electrical current in the circuit rises from zero to (theoretically) infinity, instantaneously, because of short circuits and ground faults. A short circuit is the unintentional condition when one or more ungrounded phase conductors of a power supply system come in contact with one or more of the other ungrounded phase conductors of that same power supply system. This is sometimes called a phase-to-phase short circuit. A ground fault is the unintentional condition where one or more ungrounded phase conductors of a power supply system come in contact with a system ground of any kind (the grounded conductor, the grounding conductor, a grounded surface, etc.). Both of these fault conditions present a serious safety hazard to working personnel, because if the fault energy is high enough it could cause an electric arc or a flash fire potential that could injure or kill people.

### Overload

Motor overload is a condition where the motor is overloaded mechanically beyond its horsepower rating, which causes the motor to draw more than its rated current. "Overload" is the term used when the circuit current is not as high as the overcurrent condition, which would trip the circuit immediately, but still is higher than the rating of the overcurrent device that is protecting the circuit conductors and motor. The overload condition may not present immediately obvious damaging affects to the circuit. If allowed to continue for an extended period of time, however, the additional current caused by overload conditions still can damage the circuit conductors and motor. The damage just would not be as violent as the damage that overcurrent conditions can cause.

## MAGNETIC MOTOR STARTER CIRCUIT

The drawing in Figure 10-1 shows the physical location of each part of the magnetic motor starter circuit, starting with the power supply, where lines 1, 2, and 3 of the power mains are landed on the top lugs of the disconnect. The disconnect is where the overcurrent protection for the motor starter circuit is located. The power mains from the bottom of the disconnect land on the top of the terminals of the contactor. Notice that the contactor contains both the main power circuit contacts, which carry the actual motor load current and turn the motor on and off, and the control circuit holding contacts, which are labeled with

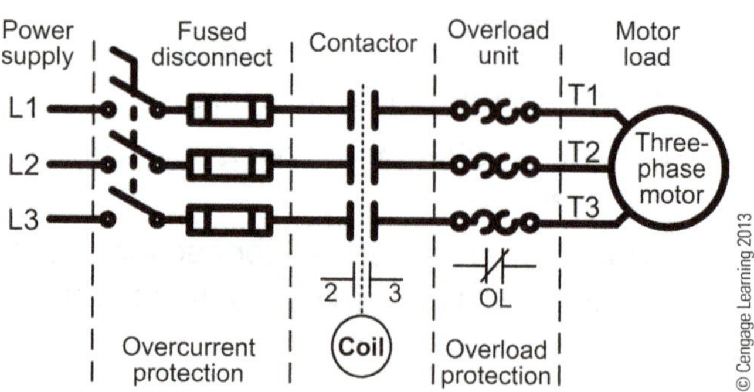



**FIGURE 10-1** Magnetic motor starter circuit

terminal numbers 2 and 3. The dotted line from the control coil, through both the power circuit contacts and the control circuit contacts, designates that both are actuated mechanically by the armature of the contactor when the control coil is energized and de-energized. The overload unit shows that the motor load current passes through the overload unit heaters (the double opposing question marks), and on to the motor. The overload unit opens the overload unit contacts when excessive motor overload current causes excessive heat in the overload unit.

## Electrical Connections Between the Contactor and Overload Unit

The drawing in Figure 10-2 shows an actual control circuit drawn in with the contactor and overload unit to help clarify where each component of the control and power circuits is located logically. The contactor control coil and contacts are encapsulated in one grey area to designate them as a single physical unit, and the overload unit heaters and contacts are encapsulated in another grey area. The contactor main contacts and the overload unit heaters are in the power circuit, which actually starts and stops the motor load. The contactor control coil and holding contacts, as well as the overload unit contacts, are in the control circuit. In this way, when the overload unit contacts open because of excessive overload motor load current, the contactor control coil will become de-energized and drop out the contactor main contacts controlling the motor.

## Different Control and Power Circuit Voltages

In some motor control situations it may be desirable to use a lower control circuit voltage than the higher voltage of the power circuit for the motor. The drawing in Figure 10-3 shows a control transformer being used to change the 480-volt power circuit voltage running the motor, into a 24-volt control circuit voltage. The motor starter contactor control coil would have to be rated for 24 volts, and then the 24-volt control circuit would operate the 480-volt motor load. Because the control circuit and power circuit are electrically insulated from one another, they may be at completely different voltage levels.

## Identifying Motor Circuit Components

The graphic in Figure 10-4 shows a picture of the component associated with each stage of the motor starter circuit. The overload unit picture shows an additional picture of the overload unit heater to demonstrate how it attaches to the overload unit with two screws. The overload unit heaters provide

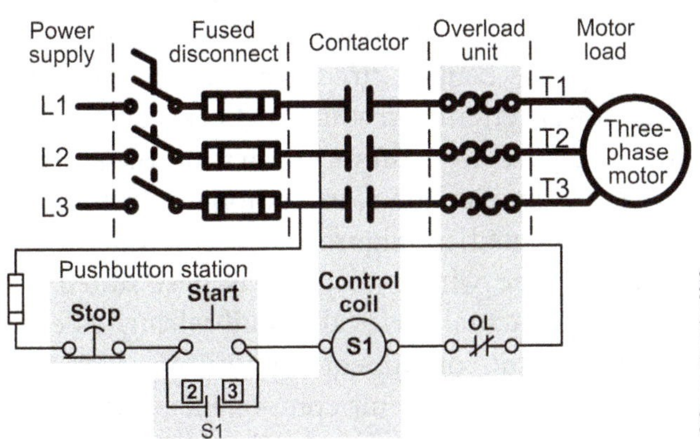



FIGURE 10-2 Electrical connections between contactor and overload units

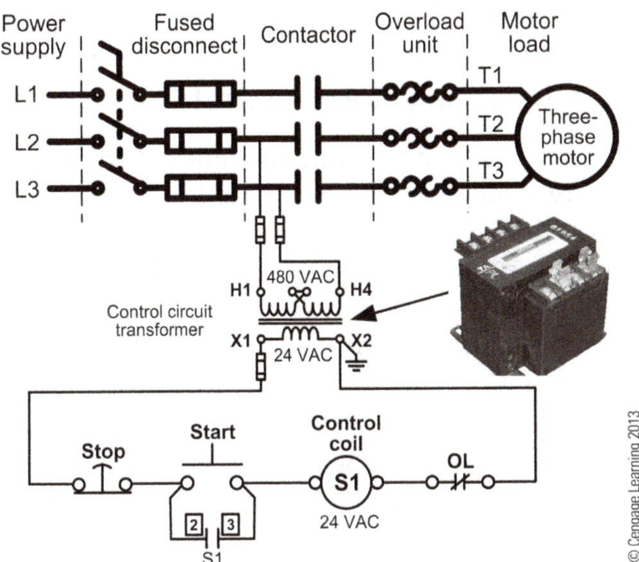



FIGURE 10-3 Lower control voltage circuit

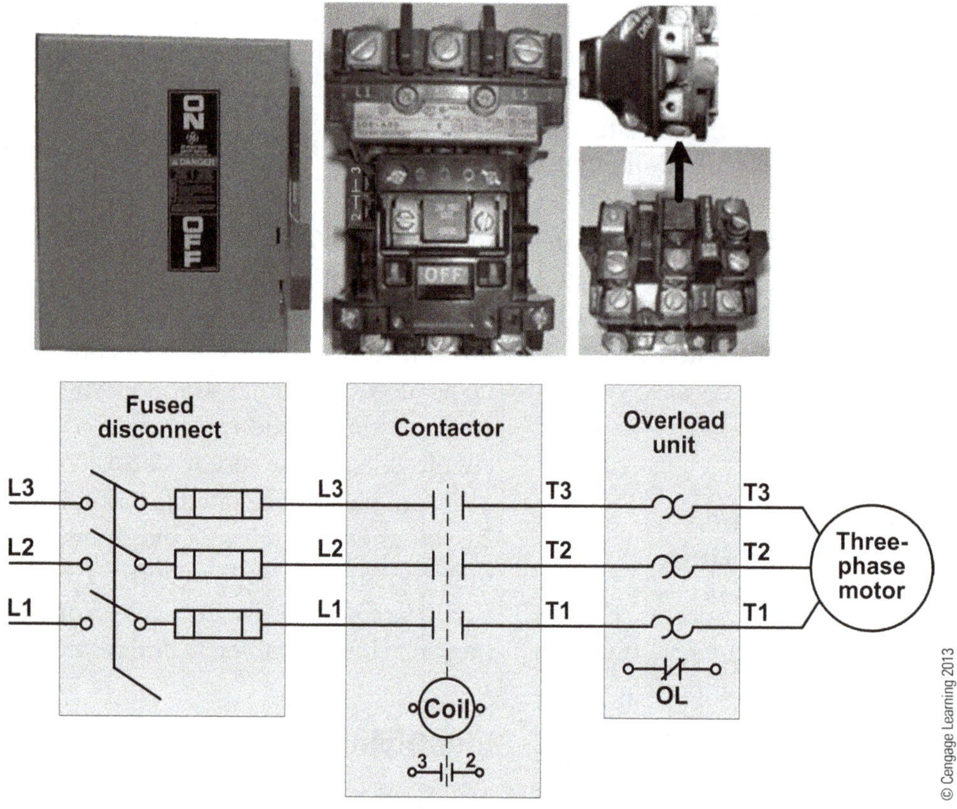


FIGURE 10-4 Picture and component association of circuit parts

the electrical jumper between the line and load sides of the overload unit, so the motor load current must pass through the heaters to get through the overload unit.

## Documenting Motor Control Circuits

Motor control circuits are commonly documented in two different forms: the ladder diagram and the wiring diagram. Both methods have their advantages and disadvantages. The ladder diagram shows the logical connections and operation of the control circuit, where the wiring diagram shows the physical connections of the circuit.

**Ladder Diagram.** The ladder diagram shows only the control circuit components and wiring, but no power circuit wiring. Ladder diagrams are named for their resemblance to an extension ladder, which consists of two vertical "rails," and any number of horizontal "rungs" between the rails. The drawing in Figure 10-5 shows a simple stop-start station control circuit documented with a ladder diagram.

**Wiring Diagram.** The wiring diagram shows both the control and power circuits on the same drawing, the control circuit in thinner lines and the power circuit in thicker lines. The drawing in Figure 10-6 shows the wiring diagram of the same simple stop-start station control circuit as documented in Figure 10-5 with the ladder diagram. The wiring diagram is best suited for identifying actual physical circuit wiring, as each control circuit component is physically related to the other control circuit components. The ladder diagram is best suited for following the logical

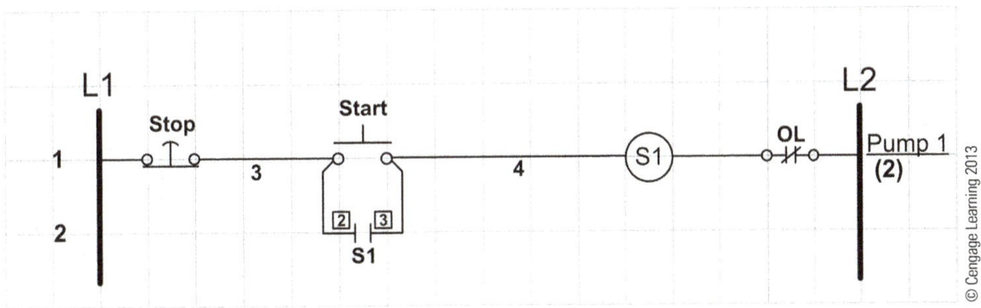


FIGURE 10-5 Ladder diagram

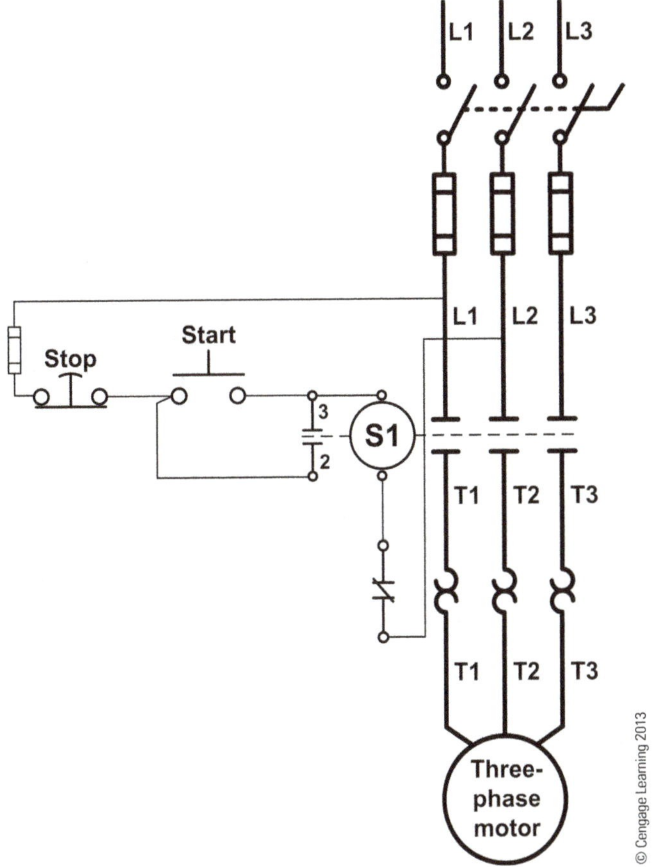


FIGURE 10-6 Wiring diagram stop-start circuit

operation of the control circuit, as each control circuit component is logically related to the other control circuit components.

As the control circuit becomes more complex, with multiple motors, contactors, and control devices, the wiring diagram quickly can become more congested with lines, making it difficult to follow for troubleshooting circuit operation problems. A reversing motor control circuit is documented first as a ladder diagram in Figure 10-7, and then as a wiring diagram in Figure 10-8, to demonstrate the differences.

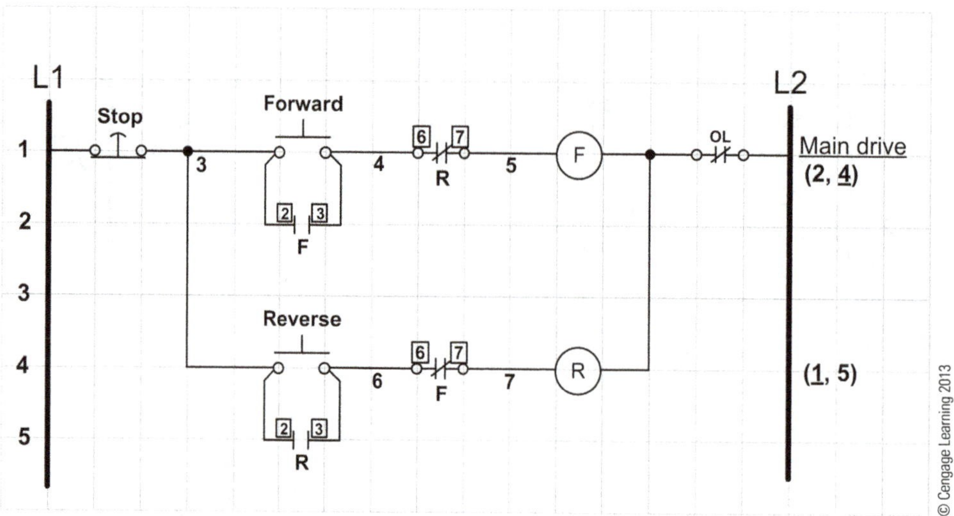


FIGURE 10-7 Ladder diagram forward-reverse circuit

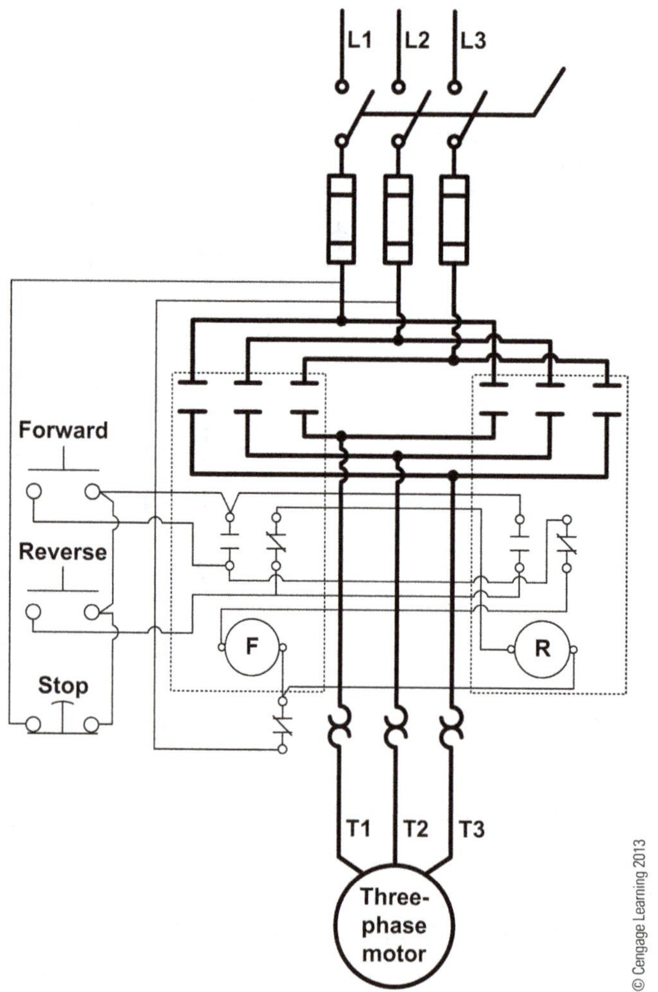


FIGURE 10-8 Wiring diagram forward-reverse circuit

**Construction Blueprints.** Construction blueprints that an electrician would encounter on the construction site of a new building normally utilize ladder-diagram-type documentation, as shown in Figure 10-9. The control circuit scheme for a single motor in a building application normally will not be extensive enough to require more than a single rung, so it probably will not look much like a ladder diagram. These prints sometimes also are referred to as one-line diagrams, with the power circuit documented once on the side to reduce clutter.

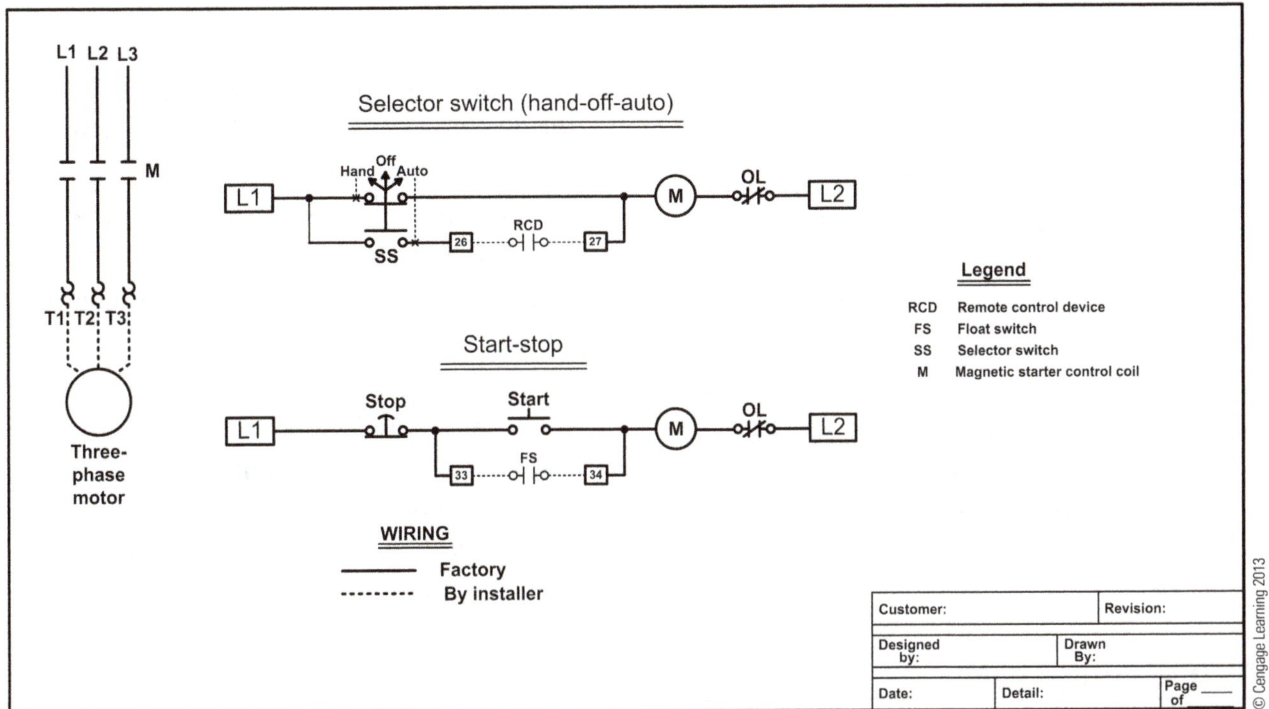


FIGURE 10-9 Construction print example

## CHAPTER SUMMARY

- All motor circuits are protected for both overcurrent and overload conditions.
- Overcurrent is an extremely high-intensity fault current condition caused by accidental short circuits and ground faults.
- Overcurrent protection is provided in the motor circuit where the conductors receive their supply.
- Overload is a lower intensity current condition caused by a mechanical overload on the motor.
- Overload protection is provided after the contactor unit, but before the motor being protected.
- The contactor control coil actuates at the same time the contactor main contacts connected in the power circuit and the holding contacts (numbers 2 and 3) connected in the control circuit.
- The overload unit controls the normally closed set of overload contacts, which are connected in the control circuit.
- The motor load current is conducted through the contactor unit through the main power contacts.
- The motor load current is conducted through the overload unit through the overload unit heaters.
- Motor control circuits are documented with ladder diagrams and wiring diagrams.
- Ladder diagrams document only the control circuit wiring.
- Wiring diagrams document both the power circuit wiring and the control circuit wiring.
- Ladder diagrams document the logical operation of the control circuit.
- Wiring diagrams document the actual physical connections on circuit components.

## REVIEW QUESTIONS

1. Every electrical conductor that electricians install in electric motor circuits must be protected for what two conditions?

   1. ______________________________

   2. ______________________________

2. Where in the motor control circuit is overcurrent protection located?
3. Where in the motor control circuit is overload protection located?
4. Why does overload protection not need to be at the point of supply for motor control circuits?
5. What are the two circuit problems that can cause an overcurrent condition?
6. Motor overload conditions are caused by what?
7. Which component, the contactor or overload unit, actually carries the motor load current?
8. Which component of the motor control circuit has the main power contacts that actually make or break the motor load current?
9. In which circuit, control or power, are the overload unit contacts wired?
10. In which circuit, control or power, are the contactor holding contacts (terminals 2 and 3) wired?
11. On what motor control circuit component does the motor load current pass through the overload unit?
12. What are two common forms for documenting motor control circuits?
13. What motor circuit wiring is not documented on a ladder diagram?
14. What is the advantage of ladder diagram documentation?
15. What is the advantage of wiring diagram documentation?

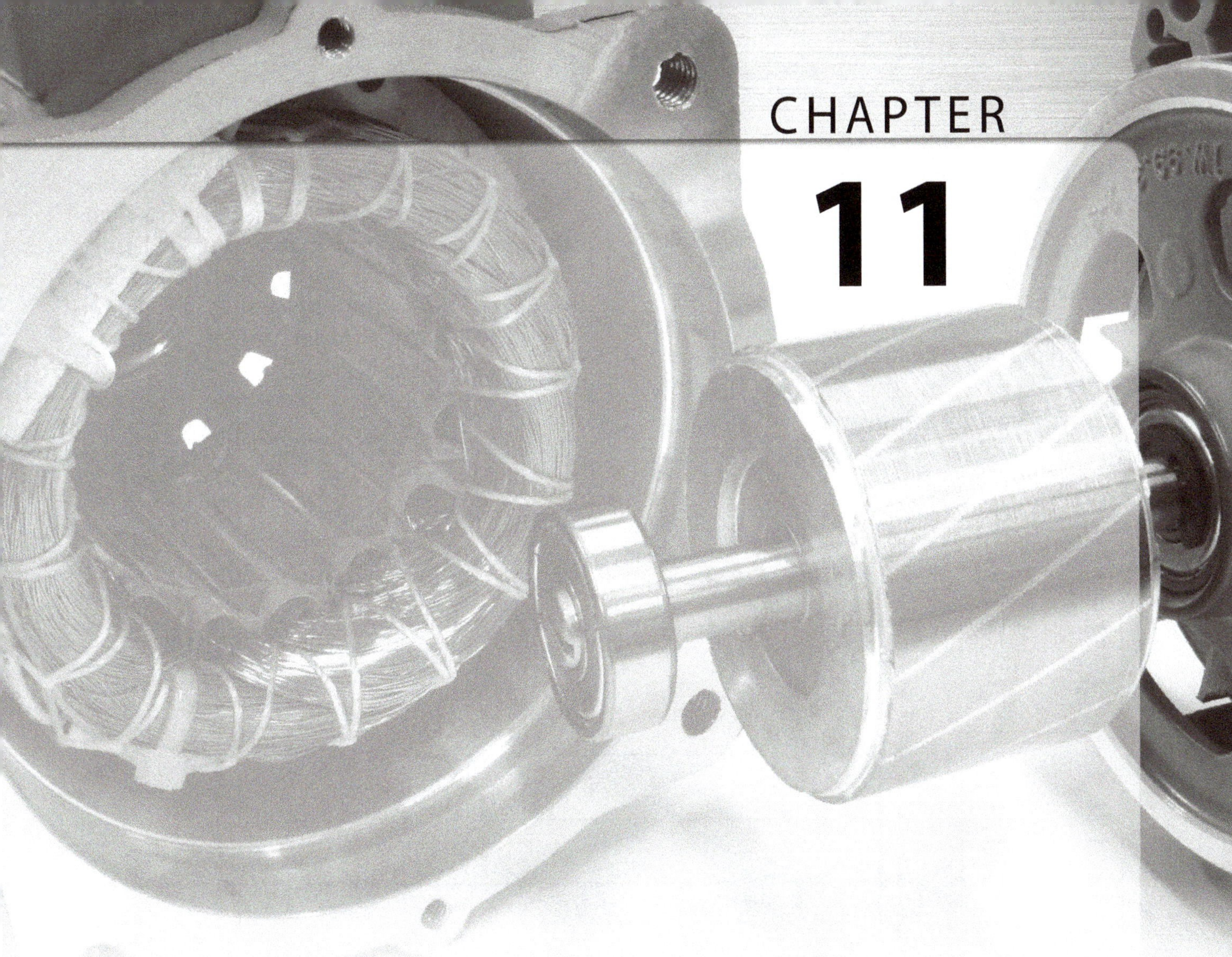

# CHAPTER 11

# Motor Control Circuit Ladder Diagram Documentation

**PURPOSE**

To study the reference documentation used to understand and analyze the logical operation of a motor control circuit ladder diagram.

**OBJECTIVES**

After studying this chapter the learner will be able to:

- Explain the placement and purpose for each point of reference documentation used on basic ladder diagrams
- Draw and document basic motor control circuit ladder diagrams
- Distinguish between motor control circuit input or control, and output or load components
- Identify basic symbol contact-switching methodologies and configurations
- Explain the Boolean logic AND and OR contact configurations for motor control circuits

## REFERENCE DOCUMENTATION

Reference documentation may not seem very important for simple ladder diagram drawings, but consistent reference documentation will reduce confusion and save countless hours of work for anyone working on the control circuit. Like each of the separate tools in a toolbox, each piece of reference documentation illustrated in Figure 11-1 has a purpose, and is essential for analyzing all ladder diagram control circuits. If you do not understand this documentation, and the purpose of each point, you certainly will be at a disadvantage to those who do.

### Example Ladder Diagram

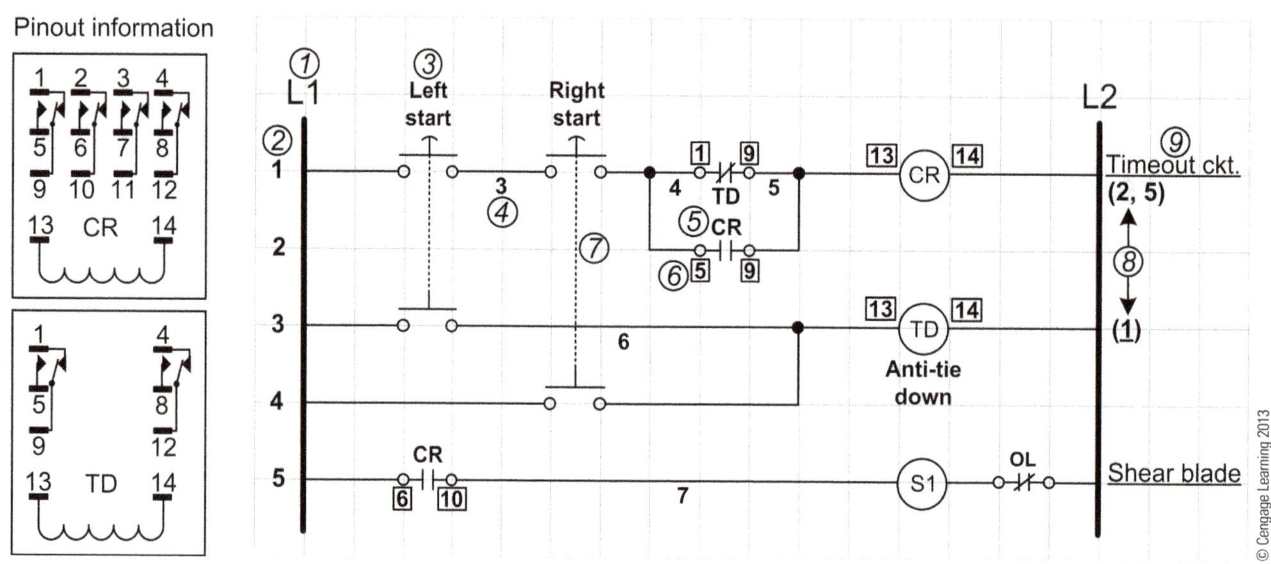


**FIGURE 11-1** Example ladder diagram with reference labels

*The nine parts of Figure 11-1 which are indicated by italicized and circled numbers are explained by corresponding number in the following section.*

### References

1. **Rail Label.** The rails of the ladder diagram distribute control circuit power to each of the rungs drawn between them, and provide an orderly manner of documenting motor control circuits. The rails are labeled L1 and L2, for line 1 and line 2. The control circuit power labels may be confusing in some situations, because they are typically labeled L1 and L2, irrespective of the actual control circuit voltage. L1 is the ungrounded conductor for the control circuit, and L2 may be either line 2 (the ungrounded conductor), or neutral (the grounded conductor), depending on the control circuit voltage. As an example, if the control circuit operating voltage is 208VAC, L2 would be the ungrounded supply circuit conductor, line 2. If the control circuit operating voltage is 120VAC, however, L2 would be the grounded supply circuit conductor, neutral. The control circuit operating voltage is determined by the voltage rating of the motor starter contactor control coil.
2. **Rung Number.** Each rung in a ladder diagram is numbered, starting with the number "1," on the drawing on the same level as the first rung, off to the left-hand side of the L1 rail. Each rung of the ladder diagram is numbered sequentially as part of the reference documentation. Once the rung interval spacing is established, the rungs still will be numbered even if no rung is drawn in a given rung area, which helps maintain a scale by having the same number of rungs per page. Counting down the rungs may not seem like a huge problem when there are only a few rungs in a drawing, but if there

are many rungs spanning many pages of a ladder diagram, numbering the rungs would save having to count down through all of the rungs each time you need to reference a specific rung.

3. **Component Label.** Even if the ladder diagram is relatively simple and the component functions seem obvious, it still is important to label each circuit component function. What seems so simple and obvious to you while you are working on the circuit may not seem obvious to another electrician that works on the circuit after you. Furthermore, what seems so simple and obvious to you when the circuit is fresh in your mind may not seem so simple or obvious to you when you work on the circuit a month, or a year, later.
4. **Wire Number Label.** Using wire number reference documentation is indispensable when wiring the control circuit, and again when troubleshooting. By comparing the wire numbers of the hard-wired circuit in the field with the wire numbers on the motor control circuit ladder diagram, the troubleshooter can quickly identify any circuit component and its purpose. Wire numbers also can help the troubleshooter follow the logical operation of the motor control circuit without trying to trace control circuit wires through raceways and junction boxes. Wire numbers are an important piece of reference documentation, and worth an extended explanation.

There is no universally required method or sequence that must be followed for numbering the wires on the ladder diagram of a motor control circuit. There are, however, a few concepts that can be applied to help avoid confusion:

- No two wires can have the same number.
- No single wire can have more than one number.
- Wires that junction, but do not pass through a control circuit component, remain the same number.
- Wires before the first component in a rung are not numbered because they are L1, and wires after the load normally are not numbered because they are L2 (even when they still pass through the overload unit contacts).
- Wire numbers generally are assigned from the lower numbers to the higher numbers, from the top-left-hand side of the ladder diagram to the bottom-right-hand side.

In the ladder diagram shown in Figure 11-2, L2 would be found on one side of the warning lamp, but L1 would not. L1 ends at the float switch, and a new wire, wire number 3, goes on to the warning lamp. When the electrician connects this control circuit, the high-level float would connect to wire numbers L1 and 3, and the warning lamp would connect to wire numbers 3 and L2. Wires do not change numbers unless they go through a component that breaks them from the previously numbered wire.

***The Argument for Starting Wire Numbering with the Number 3.*** It is not uncommon to find motor control circuit wiring where the "L" has been dropped from the L1 and L2 labels, leaving only the numbers 1 and 2 to remain for the control circuit power rails. This practice probably started when an installer did not have the proper L1 and L2 labels, and substituted the numbers 1 and 2 out of practicality and convenience, as shown in the drawing of Figure 11-3. Every wire number book has the numbers 1 and 2 in them, but L1 and L2 may not always be available. If the new circuit wires then were labeled starting with the number 1, the result would be having two different circuit wires with the number 1 and number 2 labels. This situation would cause a great deal of confusion. Starting to number the new control

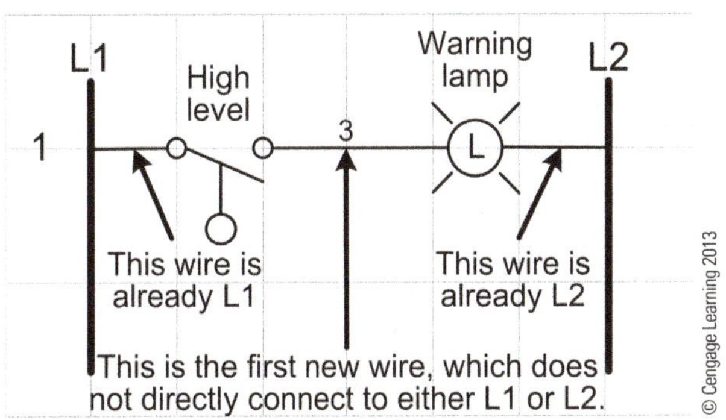



**FIGURE 11-2** Wire numbering example

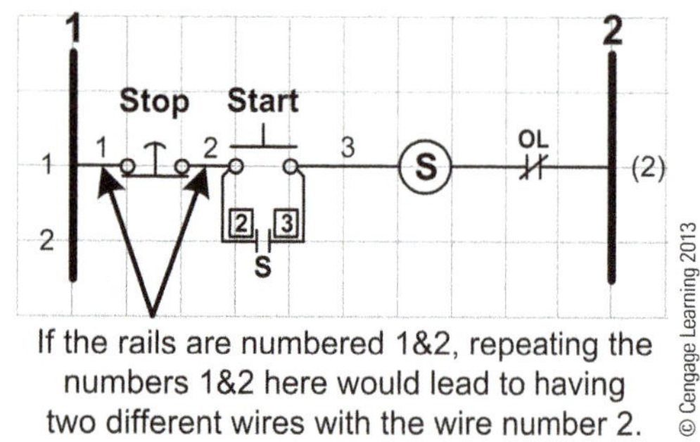


**FIGURE 11-3** Rationale for starting with number 3

circuit wires with the number 3 label automatically lessens the possibility of this confusion.

5. **Contact Label.** All contacts are labeled the same as the control coil by which they are actuated. In Figure 11-3 the control coil is labeled "S," and the holding contacts that are actuated by the control coil are also labeled with an "S." This label is especially helpful when there are multiple contacts from multiple control coils, spread across multiple rungs of a ladder diagram. Holding contacts, contactor terminal numbers 2 and 3, are such a common application that sometimes this documentation is skipped, but getting in the habit of labeling all contacts is the best practice, and is indispensable when following or troubleshooting the logical operation of the control circuit.
6. **Terminal Number Label.** Relays, contactors, and many other electrical components have numbered or labeled terminals. These are numbers or labels that the manufacturer has put on the component terminals to facilitate connecting them in the circuit correctly. With all of the different points of ladder diagram documentation, the drawing starts to get congested with numbers, and some type of marking must be used to differentiate between them. In the example, a box is drawn around the terminal numbers to distinguish them from the wire numbers. It is not necessary to use a box. It may be any marking to prevent them from being confused with other documentation numbers in the same area of the drawing. The terminal numbers 2 and 3 on the holding contacts of a motor starter contactor are a common example of terminal markings. Also, providing the contact number documentation on the ladder diagram helps when troubleshooting, because an incorrect termination is identified easily. Not all component terminals in a ladder diagram will have terminal numbers.

   In rung number 2 of the example ladder diagram in Figure 11-1, the control relay (CR) contacts are labeled with terminal numbers 5 and 9. Reference the CR pin-out information to the left of the ladder diagram, and notice that terminals 5 and 9 are the first set of normally open contacts available on the relay. Now look down in rung number 5 and locate the normally open CR contacts, which are labeled with terminal numbers 6 and 10. The same normally open contact of the CR relay cannot be used again in another rung, so the next normally open contact available in the relay has to be used, terminals 6 and 10.
7. **Mechanical Linkage.** A continuous dotted line, like the ones drawn on the example ladder diagram between the left start switch contacts and the right start switch contacts, indicates a mechanical linkage between the contacts that the dotted line goes between. A mechanical linkage means that when one of the contacts actuates and changes states, all of the other contacts associated with the mechanical linkage will actuate and change states. When a mechanical link is drawn that must pass through other control circuit components that are not mechanically linked, continuance arrows are used, as shown in the ladder diagram in Figure 11-4.

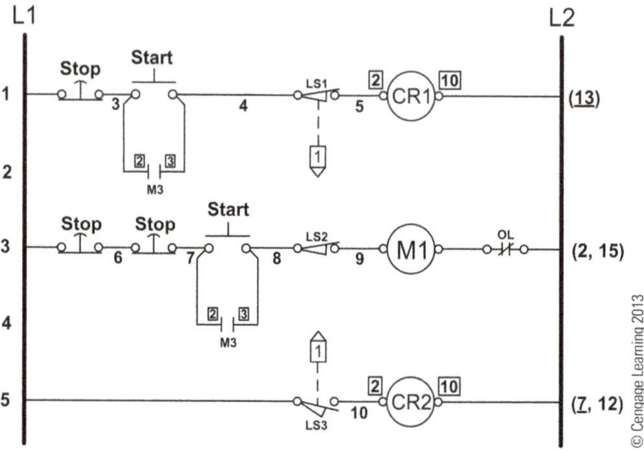


**FIGURE 11-4** Arrow mechanical linkage

8. **Control Coil Contact Reference Label.** Control coil contact reference means that control coils typically have contacts associated with them that change state when the control coil is energized and de-energized. In the ladder drawing, all the contacts associated with a particular control coil can be located almost instantly using this cross-reference documentation. For every rung with a control coil, numbers in parentheses on the right side of the L2 rail will reference all of the rung numbers of the ladder diagram that contain contacts actuated by that coil. To clarify which control contacts are actuated by which control coils, the contacts are labeled the same as the control coil that actuates them. For example, if a control coil is labeled M1, all contacts, including the holding contacts, actuated by that control coil also will be labeled M1. Notice in the example ladder diagram, the time delay (TD) control coil contact reference in rung number 3 is an underlined number 1. The underline documents that the contact in rung number 1 of the ladder diagram that is associated with the TD control coil is a normally closed contact configuration. If you look in rung number 1 of the ladder diagram for the contact labeled TD, you will find that it is a normally closed contact.
9. **Ladder Rung Note.** Ladder rung notes are documented to the right of the L2 rail, and describe the purpose or intention of the rung. These notes are a short, underlined phrase intended to explain the rung at a very basic level. The simple motor control circuit examples used in this book may not make rung notes seem like a huge problem when the purpose is so apparent, but what if there were a hundred rungs in a ladder diagram control circuit controlling many different functions? It is much easier to document the purpose or function of each rung initially, rather than trying to determine the purpose or function of each rung every time you use the ladder diagram after that.

## Symbols

Symbols are drawings that represent or stand for some material object to derive its meaning or purpose. In this case they represent electrical components in electrical control circuits. There are thousands of electrical motor control symbols four of which are shown in Figure 11-5 pressure switches, float switches, temperature switches. Given the vast quantity of control circuit component symbols, and the lack of standardization for them, it would be a waste of time to try to memorize them. Any unusual symbol used in a control circuit print for which the identity is not readily obvious should be identified in the symbol legend.

It is more important to notice that each of the different types of control devices utilizes only a handful of different switching configurations, as shown in Figure 11-6, and it is the actual switching configuration that determines the logical operation

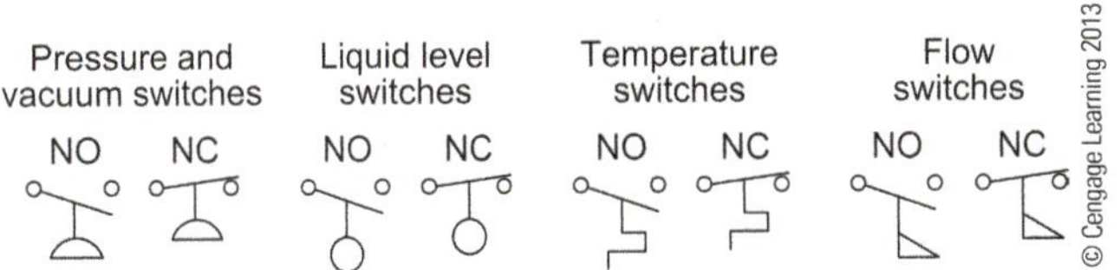

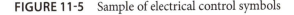

**FIGURE 11-5** Sample of electrical control symbols

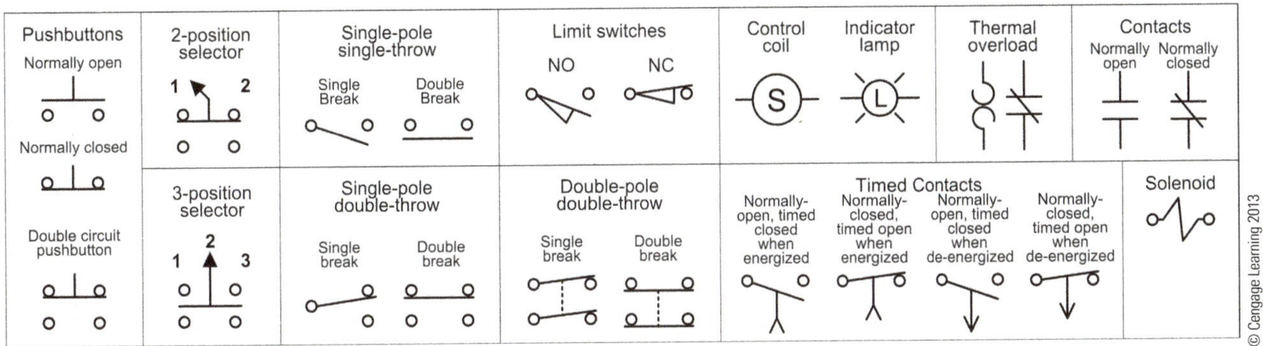


**FIGURE 11-6** Switching configurations of common control components

of the control circuit. For that reason, the symbols included here focus on switching configurations and a few common control circuit components.

**Pushbuttons.** Signified by the straight vertical stem with no arrow on it, pushbuttons are momentary contact devices, called out as either normally open or normally closed. When the pushbutton is pressed, the contact will change states (for example, a normally open contact will close), but when the pushbutton is released the contact will revert to its original call-out contact state.

**Selector Switches.** Signified by the vertical stem, either straight or bent and capped with an arrow, selector switches are manually operated, multi-position switches manufactured in many different configurations. Selector switches may be manufactured so that when manually switched into a position, they will hold that same position until they are manually switched out of it, which is called the maintained contact. Selector switches also may be manufactured like the momentary contacts of pushbuttons, where the selector switch returns to its initial state when it is released, which is called the return type.

**Switch Poles and Throws.** The term "pole," in reference to an electrical switch, determines how many different electrical circuits may be conducted through and controlled by the switch. A single-pole switch can control only one electrical circuit, and a two-pole switch can control two different circuits, one circuit on each pole. The term "throw," in reference to an electrical switch, determines how many different switching positions the switch can be in. A single-throw switch only can turn on and off, and a double-throw switch can toggle a common connection between two different electrical circuits.

**Single-Break and Double-Break Contacts.** Electrically, there is no difference in the operation of the electrical circuit if the control device uses single-break or double-break contact design. The single-break design normally is sufficient for relay contacts, toggle switches, and other lower-energy control devices of a motor control circuit, because the current levels are relatively low. In other motor control situations where the control device may actually control the motor load directly, such as a small sump pump float switch, or air compressor pressure switch, the higher circuit currents may require the use of the double-break design. The double-break design may use flat contact faces with greater surface contact area to dissipate arc heat, and because the two opening spaces are additive, each contact only has to move one-half the total distance required to open the motor load current.

**Limit Switches.** Limit switches are manufactured in a wide variety of actuator types, with the purpose of physically sensing the presence of different mediums. The word "limit" may be deceiving, because even though it may be used to define travel limits or boundaries in a control process, "limit switch" is a general term used for any switch used to sense physically the presence of something, such as a box, a closed gate, or a correctly positioned part.

**Control Coils.** Control coils are wired in the control circuit, and when energized create the magnetic flux to actuate the armature of a magnetic relay or contactor. Not all control coils are magnetic. They also may be electronic, as is the case with solid-state relays and contactors.

**Indicator Lamps.** Indicator lamps are used to provide a visual indication for an operator, to verify that a certain operation is either energized or de-energized. One such indication is paralleling an indicator lamp with a motor starter contactor control coil, so that when the contactor control coil is energized to run the motor, the indicator lamp also is energized to signal the operator that the motor is running.

**Timing Relays.** Timing relays, sometimes called time-delay relays, provide a delayed action, changing operating states of the contacts associated with their control coil. The timing relay contacts may be either normally open or normally closed, and change states after the specified time delay when the relay control coil is energized, which is called "on delay."

Or, the timing relay contacts may be either normally open or normally closed, and change states after the specified time delay when the relay control coil is de-energized, which is called "off delay."

**Thermal Overloads.** Thermal overloads are designated with what is sometimes called a double question mark design symbol, next to the symbol for a normally closed set of contacts. The double question mark is the symbol for the actual thermal overload heaters that carry the motor load current through the overload unit. There is one overload heater, and one normally closed overload contact for each phase of the motor starter, but only the wiring diagram actually shows each individual heater and each individual contact. The ladder diagram also documents the normally closed overload unit contacts after the motor starter contactor control coil, but only one overload contact is usually drawn to represent all three. The double question mark thermal overload heaters are not drawn in the ladder diagram. Overload heaters are drawn only in the wiring diagram, because they are part of the power circuit.

**Contacts.** Normally open electrical contacts are drawn with two straight, parallel lines, similar to the symbol sometimes used to designate a capacitor. Normally closed electrical contacts are drawn the same way, but a third line is drawn at an angle to pass through both of the other lines. The normally open or normally closed configuration states of electrical contacts are called out when the control coil is de-energized and the device is in the relaxed state, as it was before being connected into a circuit. Regardless of the function an electrical contact performs in a control circuit, possibly being held open or closed, its configuration is always called out in the device's de-energized or relaxed state.

**Solenoid.** "Solenoid" is a general, all-encompassing term that includes any device that produces mechanical motion from the forces of the magnetic flux created by an electromagnet. Solenoid devices found in motor control circuits include electrical solenoid valves for compressed air, water, vacuum suction, hydraulic pressure, and a host of electrical-mechanical actuators.

## Ladder Rung Components

Basically, there are two types of rung components found in motor control circuit ladder diagrams: load components, which are circuit output components, and control components, which are circuit switching components.

**Examples of Control Circuit Load, or Output Components.** Indicator lamps, audible indicator devices, and control coils of relays and contactors all are examples of output load components. Every rung must contain at least one load component with sufficient resistance (or impedance) to drop the applied voltage and limit the circuit current. Load, or output, components are easy to identify, because they dissipate power. When energized by the control circuit voltage, the resulting current through the component resistance will cause power to be dissipated ($P=I^2R$). Load, or output, components are inserted on the right-hand side of the ladder rung, the last component before the L2 rail (disregarding the overload contacts). When talking about loads in the control circuit, it is important to remember that we are never talking about the actual motor load, which is carried by the main contactor contacts, and documented in the power circuit.

It is important to remember that whenever multiple output components are utilized in a single rung, which is perfectly acceptable and done all the time, they must be electrically connected in parallel. If two or more output components are placed on a single control circuit rung in series, the applied voltage will distribute itself across each of the components, proportional to the resistance, or impedance, of each. This means that each of the components connected in series will have some voltage value, less than the supply voltage that they are rated for, dropped across them. If two output components are accidentally wired in series in the control circuit, neither will receive sufficient voltage to operate properly, because the supply voltage will be divided between them.

**Examples of Control Circuit Input Components.** To avoid confusion between the power circuit contacts and the control circuit contacts, we will refer to the power circuit contacts as the main contacts, and the control circuit contacts as input components. As mentioned earlier, input components do not dissipate a sufficient level of power to be considered a power-dissipating device. Switches and relay contacts are manufactured with the least possible resistance (as electricians we assume zero Ohms), so when current flows through them, no significant power (heat) will be dissipated. $P=I^2*R$, and zero Ohms times any current squared is still zero watts. Input components are inserted on the left-hand side of the ladder rung. Input components are the devices that determine the control circuit operating logic, which may require them to be connected in series, parallel, or a combination of series and parallel to meet the operating logic needs of the control circuit. There is no practical limit as to how many input components may be connected in a single rung, as they are basically switches.

## Boolean Logic

Switch and relay input contacts in the ladder diagram electrical control circuit are often configured to implement Boolean logic expressions that determine the logical operation of the control circuit. Although there are sophisticated logical expressions that are used in advanced control circuit applications, this study will concern itself only with the two most basic of those logic configurations: the AND and OR functions. When any number of control input devices are connected in series, together they perform an AND function because all of the devices would have to be closed for the control power current to pass through. When any number of control input devices are connected in parallel, together they perform an OR function because any device closing would pass the control power current through the rung.

The Boolean logic AND function circuit is demonstrated in Figure 11-7.

The Boolean logic OR function circuit is demonstrated in Figure 11-8.

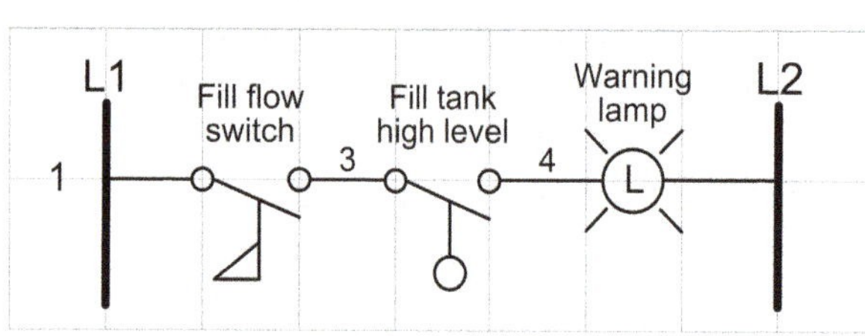

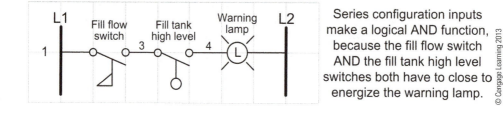

Series configuration inputs make a logical AND function, because the fill flow switch AND the fill tank high level switches both have to close to energize the warning lamp.



**FIGURE 11-7** Boolean AND logic function

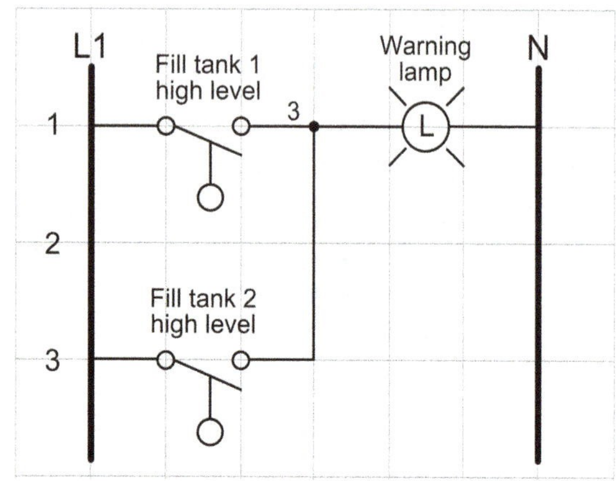

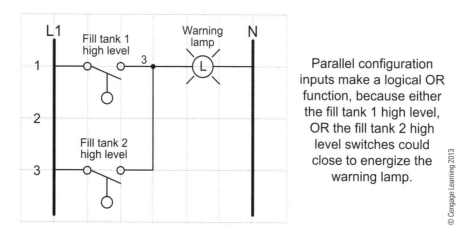

Parallel configuration inputs make a logical OR function, because either the fill tank 1 high level, OR the fill tank 2 high level switches could close to energize the warning lamp.



**FIGURE 11-8** Boolean OR logic function

## CHAPTER SUMMARY

The diagram in Figure 11-9 reviews each point of reference documentation for a quick reference.

- Symbols are drawings that represent or stand for some material object to derive its meaning or purpose.
- Pushbuttons are momentary contact devices; called out as either normally open or normally closed.
- Selector switches are manually operated multi-position switches that can be manufactured as either maintained contact or return type.
- The term "pole" refers to how many different electrical circuits may be conducted through and controlled by a switch.
- The term "throw" refers to how many different switching positions a switch can be in.
- The term "limit switch" is a general term used for any switch being used to sense the physical presence of something.
- Timing relays provide a delayed action changing operating states of the contacts associated with their control coil.
- The term "solenoid" is a general, all-encompassing term that includes any device that produces mechanical motion from the forces of the magnetic flux created by an electromagnet.
- Two types of rung components are found in motor control circuit ladder diagrams: load components and control components.
- Examples of control components include indicator lamps, audible indicator devices, and control coils.
- Whenever multiple output components are connected in a single rung, they must be connected electrically in parallel.
- Input components are the devices that determine the control circuit operating logic, which may require them to be connected in series, parallel, or a combination of series and parallel to meet the operating logic needs of the control circuit.
- When any number of control input devices are connected in series, together they perform an AND function.
- When any number of control input devices are connected in parallel, together they perform an OR function.

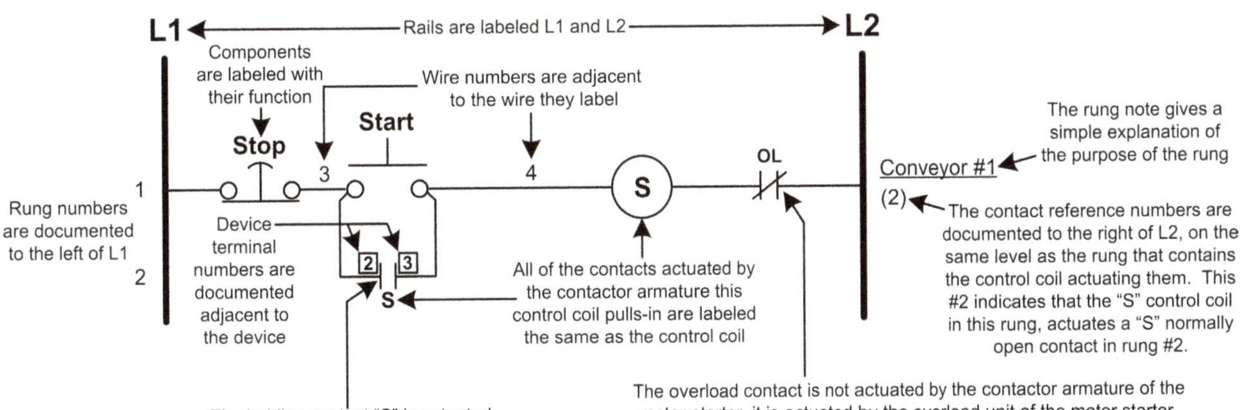

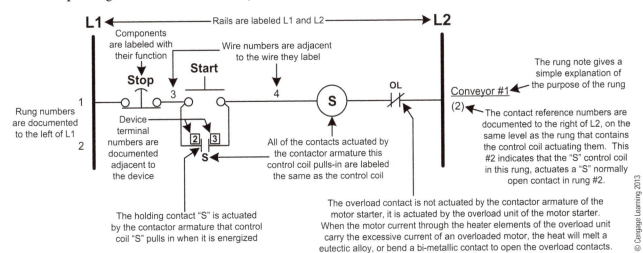


**FIGURE 11-9** Review ladder diagram

## REVIEW QUESTIONS

1. What are four basic concepts for wire numbering that can be applied to help avoid confusion?

   1. ______________________________
   2. ______________________________
   3. ______________________________
   4. ______________________________

2. How are contacts in the ladder diagram rungs labeled?
3. Who provides terminal number labels to facilitate connecting the component in the circuit correctly?
4. What are two methods used on ladder diagrams to indicate a mechanical linkage between two or more control circuit components drawn in a ladder diagram?
5. What is the purpose of the numbers documented in parentheses, to the right of rail L2, on the same rung level as a control coil?
6. How is the control coil contact rung reference label documented for a normally closed contact configuration?
7. What piece of reference documentation is used to document the purpose of each ladder rung?
8. What is the purpose of using symbols in motor control circuits?
9. What does it mean that pushbuttons are momentary contact devices?
10. Selector switches are manufactured in what two types of switching?
11. What does the term "pole" refer to for electrical switches and contacts?
12. What does the term "throw" refer to for electrical switches and contacts?
13. What two parts of the documentation for timing relay contacts is necessary to determine the function of a timing contact in a control circuit?
14. What are the two types of rung components found in motor control circuit ladder diagrams?
15. What are two common types of motor control circuit load components?
16. How are load components differentiated from non-load components?
17. What purpose do non-load, or input, components serve in motor control circuits?
18. Why must there be at least one load component in every rung of the motor control circuit ladder diagram?
19. When there is more than one load component in a single rung of a motor control circuit ladder diagram, how must they be connected?
20. How are motor control circuit input components connected to perform the Boolean logic AND function?
21. How are motor control circuit input components connected to perform the Boolean logic OR function?

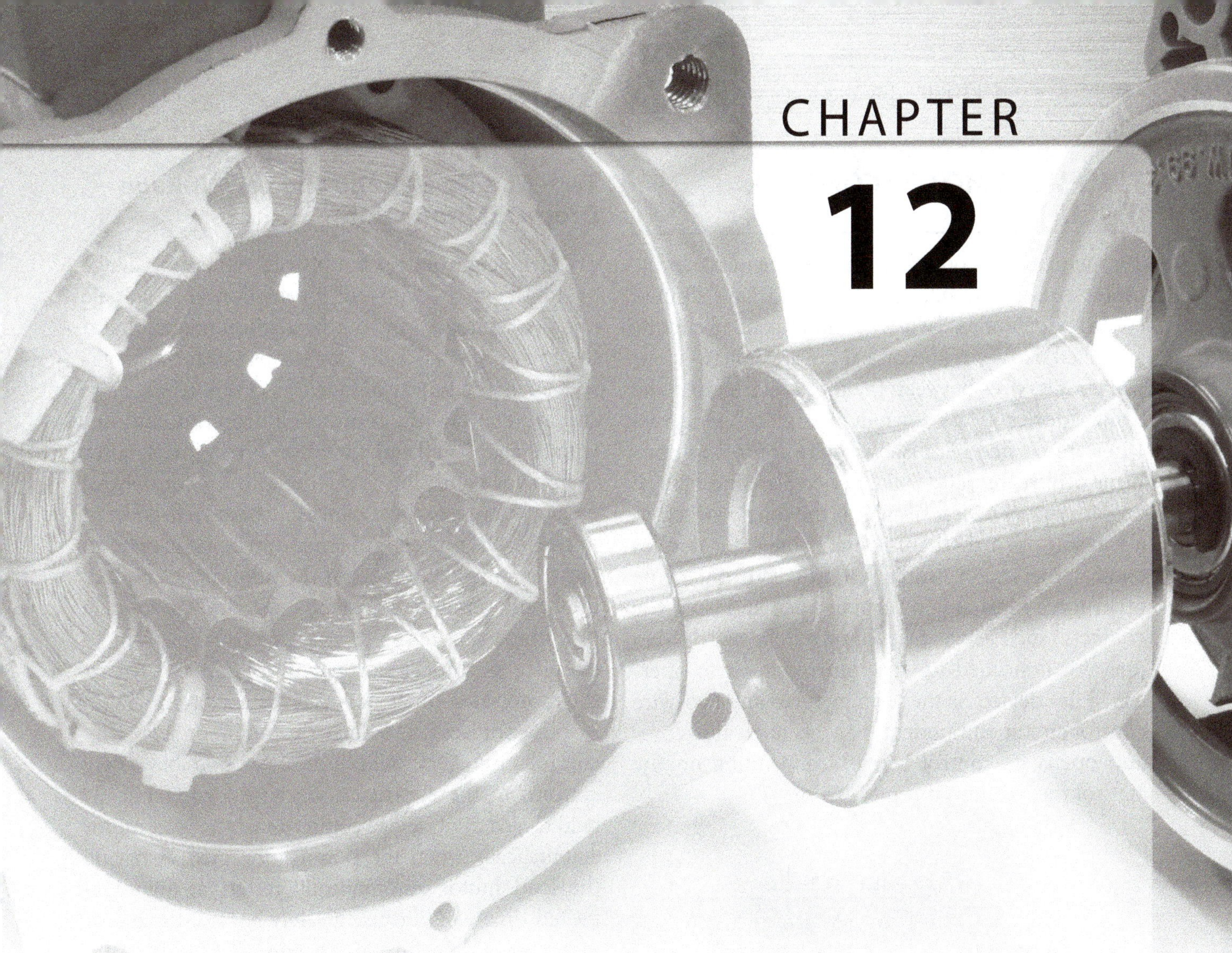

CHAPTER

12

# Two-Wire Motor Control

**PURPOSE**

To study the motor control schemes and use of two-wire motor control.

**OBJECTIVES**

After studying this chapter on motor starters the learner will be able to:

- Explain what two-wire motor control is
- Explain no-voltage and low-voltage drop-out protection
- Explain differential pilot devices
- Explain the operation of a hand-off-automatic (HOA) circuit

## TWO-WIRE MOTOR CONTROL CIRCUITS

Two-wire motor control circuits are the most simple motor control circuits, but they also have limited control capabilities. Two-wire motor control circuits are called "two-wire" because there are only two control circuit wires between the motor starter and the control device(s), as shown in Figure 12-1. Only two wires are needed because two-wire control components, which are called "pilot devices," are mechanical on-off-type switches, similar to the toggle switch in a room to turn the lights on and off. Pilot devices control the motor starter control coil, but some other force must act on them to open and close the contacts (such as water level for a float switch). No operator is needed for the control circuit to operate. "Pilot device" is the term given to any two-wire control circuit component that mechanically closes and opens its control contacts depending on the state of the medium it is measuring.

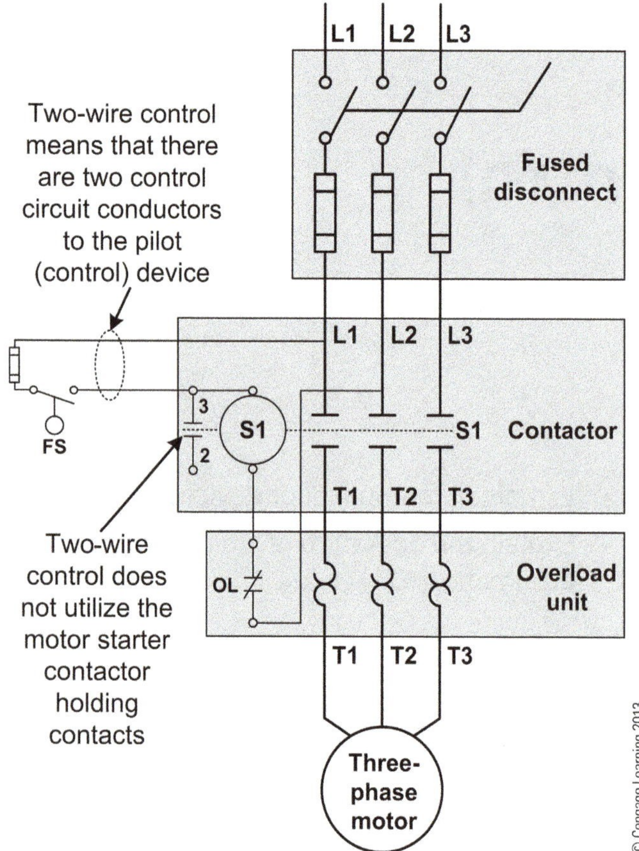




**FIGURE 12-1** Two-wire control circuit diagram

### Two-Wire No-Voltage and Low-Voltage Drop-Out Protection

Two-wire control does not provide either no-voltage or low-voltage drop-out protection. No-voltage and low-voltage drop-out protection means that if the motor starter contactor drops out for any reason, such as a power outage or a severe control circuit voltage sag, the motor will drop out, and not restart again without operator intervention after the power condition is restored. The pilot control device is a mechanical control, which is unaffected by the loss of power. If the medium being measured has the control contacts of the pilot device closed, they will remain closed during a power outage or voltage sag, ready to pass electrical current to energize the motor starter control coil and restart the motor as soon as power is restored to normal.

The phrases "no-voltage drop-out," or "two-wire control circuit," should indicate a possible safety concern to the electrician, because a mechanically held closed pilot device could restart the load motor automatically when the power is restored. If having the load motor automatically restart at an unexpected time would cause a safety hazard, the motor power should be turned off manually at the motor disconnect in the event of a power outage, and then restarted again under operator-controlled conditions. In some motor control applications, the automatic restart operation of two-wire control actually may be preferred, because operator intervention would not be necessary for the application to resume normal operation, such as a sump pump application.

### Differential Control Devices

A differential control device is a two-position (on-off) switch that actuates when the measured medium reaches either of two predetermined extreme high- or low-range values, with a dead zone in between where the switch does not change states. A pressure switch pilot control device without differential control would cycle the load, such as an air compressor motor, excessively as it tried to keep the air tank pressure at exactly the set point. It is, therefore, common to find differential switches for pilot devices in these motor control circuits.

Consider the following example to illustrate differential control. A differential switch could be adjusted to provide a mechanical means that would open the circuit and stop the compressor motor at 120 PSI for the high-pressure shut-off, but not close the pressure switch contact to start the compressor motor again until the pressure tank dropped down to 80 PSI for the low-pressure compressor restart. This would give the air tank a 40 PSI differential where the compressor motor would not run as compressed air is being used. This would cycle the air compressor motor fewer times than a pressure switch that stopped the motor at 120 PSI but then restarted the motor again the instant the air tank pressure dropped below 120 PSI. Differential controls provide a mechanical means independent of the electrical control circuit that acts like a mechanical holding circuit.

Figure 12-2 shows a float differential switch for a sump pump application. The plates on either side of the actuator arm may be adjusted by loosening the screw and moving them. When the actuator arm is between the two plates, the assembly can turn as the float moves up and down, but no switching will take place until one of the plates pushes against the actuator arm. Figure 12-3 shows a pressure differential switch for an air compressor tank application. The air pressure in the tank pushes against the back actuator plate of the pressure switch, using a diaphragm to transfer the pressure without allowing air to leak out of the tank. The differential adjustment is made by adjusting the spring tension pushing against the back actuator plate, against the pressure of the air in the tank.

## Two-Wire Motor Control Circuit

The ladder diagram shown in Figure 12-4 is a two-wire motor control circuit using a differential control pressure switch in the pressure tank of an air compressor. The pressure tank switch then is the automatic pilot device that starts and stops the air compressor to keep the tank air pressure between the low and high set points of the differential.

The load component in the drawing of Figure 12-4 is a control coil on the contactor of a motor starter, and is labeled "S1," for Starter 1. The control coil does not have to be labeled with an "S." It may be labeled any way that makes sense to the application. Many times the control coils of motor starters on ladder diagrams will be labeled as "M1," "M2," etc. (for Motor 1…), but new learners may confuse the M1 on the contactor control coil as the actual motor, and they try to wire the motor load in the control circuit. Even though the "M" coil designator is common in the industry, other coil designators will be used in this text to avoid that possible confusion.

**FIGURE 12-2** Float differential switch adjustment

**FIGURE 12-3** Pressure differential switch adjustment

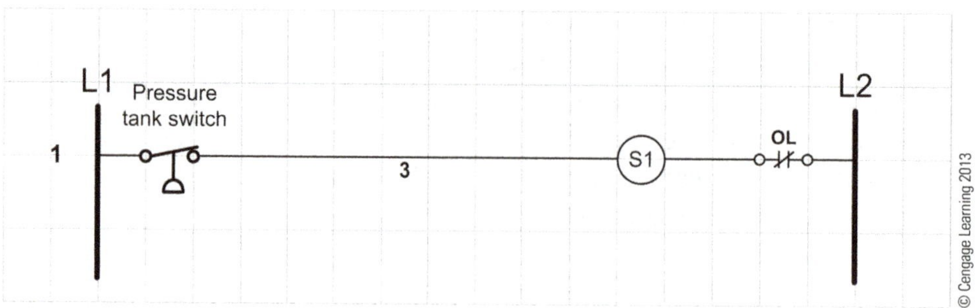


FIGURE 12-4 Two-wire control ladder diagram

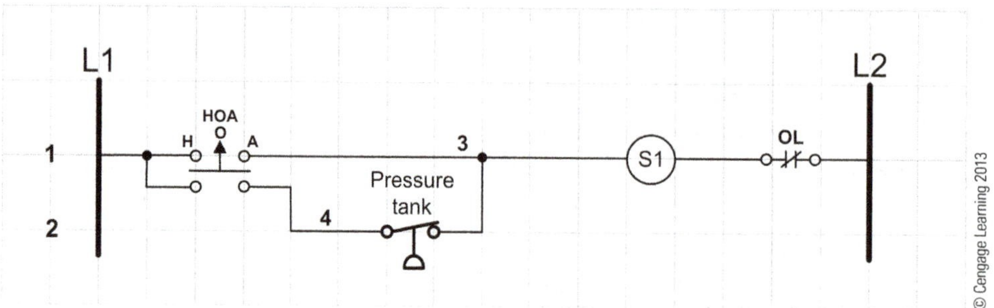


FIGURE 12-5 Hand-off-automatic circuit ladder diagram

## Hand-Off-Automatic (HOA) Control

The purpose of a hand-off-automatic circuit, shown in Figure 12-5, is to give the operator three operating modes: a manual hand (on) position, an off mode, and an automatic position where the pilot device automatically controls the control coil. When the selector switch is actuated so that it does not make contact with the top or bottom contacts, the selector switch is in the off position (shown) and the motor cannot start, regardless of the pilot device state. When the selector switch is actuated so that it makes contact across the two bottom contacts, it makes the automatic circuit, and the pilot device will control the circuit automatically without operator intervention. When the selector switch is actuated so that it makes contact across the two top contacts, it makes the hand circuit, which causes the control coil to energize, manually starting the motor regardless of the pilot device state. The following ladder diagram shows the two-wire hand-off-automatic (HOA) motor control circuit ladder diagram with a three-position selector switch, and a pressure switch for the pilot device in the automatic circuit.

## CHAPTER SUMMARY

- Two-wire motor control circuits are called "two-wire" because there are two control circuit wires between the motor starter and the control device(s).
- Two-wire motor control still uses the magnetic contactor motor starter to control the motor load current, but it does not use the holding contacts.
- The control devices used in two-wire controls are called "pilot devices," which mechanically open and close the control contacts by the action of the medium being measured.
- Two-wire control does not provide either no-voltage or low-voltage drop-out protection.
- Two-wire motor control circuits may be a safety hazard in the event of a power failure, because the motor will restart automatically when the power is restored if the pilot device control contacts are closed.

## REVIEW QUESTIONS

1. What is the name given to two-wire control circuit components that directly energize and de-energize the motor starter contactor control coil?
2. What is the term given to motor control circuits that will not restart the motor automatically when power is restored after a power failure?
3. Does two-wire motor control provide no-voltage and low-voltage drop-out protection?
4. What precaution may be taken in the case of a power failure with two-wire motor control circuits to prevent the motor from restarting automatically when the power is restored?
5. Are the contactor holding contacts on the motor starter used for two-wire motor control?
6. What is a differential control device?
7. What is the purpose of using a differential control pilot device in two-wire motor control circuits?

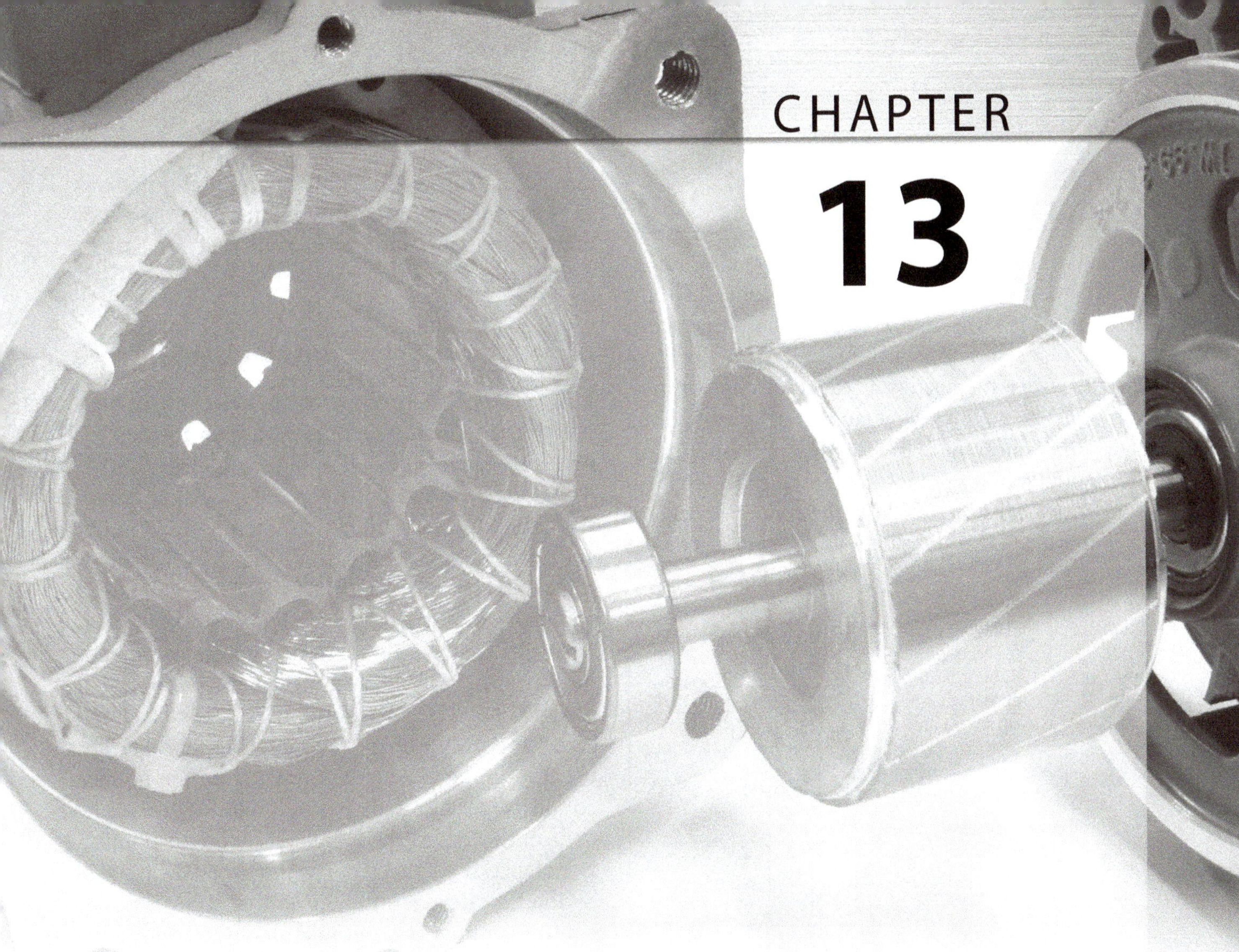

CHAPTER

13

# Three-Wire Motor Control

**PURPOSE**

To study the use of three-wire motor control.

**OBJECTIVES**

After studying this chapter on motor starters the learner will be able to:

- Explain what three-wire motor control is
- Explain the function of momentary contact control devices
- Explain the holding circuit
- Explain the operation of a three-wire motor control circuit
- Explain the terms "inhibit" and "enable" in reference to control circuit contacts
- Explain no-voltage and low-voltage drop-out protection

## THREE-WIRE MOTOR CONTROL CIRCUITS

Three-wire motor control circuits are called "three-wire" because there are three control circuit wires between the motor starter and the control device(s), as shown in Figure 13-1. Three control circuit wires are needed because of the way the momentary contact stop and start pushbuttons work in conjunction with the holding contacts on the motor starter contactor to energize the contactor control coil and then keep the contactor control coil energized through the holding contacts of the contactor. Three-wire control circuit components may be either maintained contact or momentary contact, but the holding contact function is specific to three-wire control.

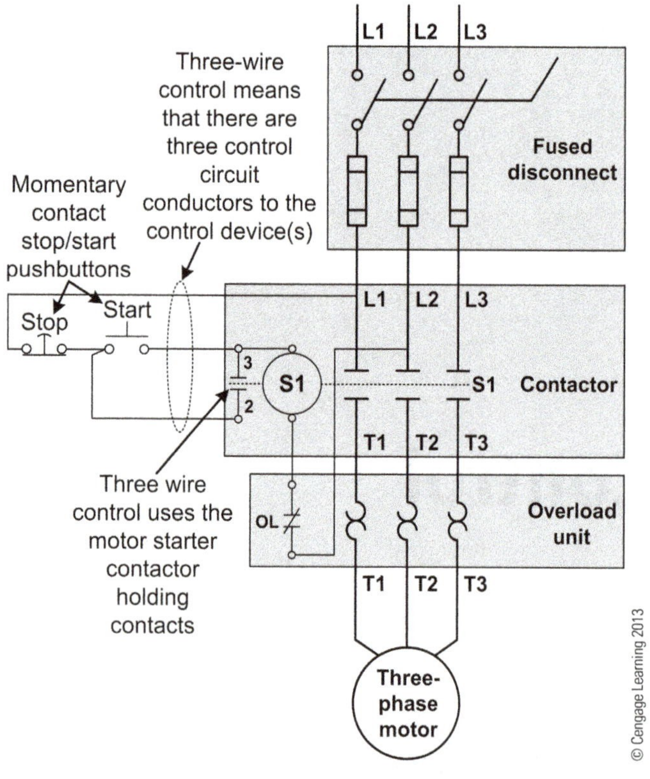

**FIGURE 13-1** Three-wire control circuit diagram

### Momentary Contact Pushbutton Switches

Momentary contact control devices change states when acted upon, but then return to their original state when released, as shown in Figure 13-2. Three-wire motor control circuits utilize momentary contact pushbuttons to energize the motor starter contactor control coil, and utilize holding contacts to keep the control coil energized when the momentary contact start pushbutton is released. The first device in the motor control ladder diagram rung, shown in Figure 13-3 from left to right, is the stop pushbutton. A normally closed contact pushbutton is used for the stop device, because under normal circumstances the control circuit power has to be passed on to the rest of the control devices in the rung if the stop pushbutton is not pressed. The stop pushbutton always will be the first device off of L1 so that it has "master control" of the rung, because all of the control circuit power to that rung first must pass through the stop pushbutton. No matter what else in the rung is closed or energized, pushing the stop pushbutton will de-energize the entire rung. The stop pushbutton only breaks contact between the two stationary contacts when the pushbutton is pressed. As soon as the pushbutton is released the contacts will close again.

The second device in the motor control ladder diagram rung shown above is the start pushbutton, which is shown in Figure 13-4. A normally open

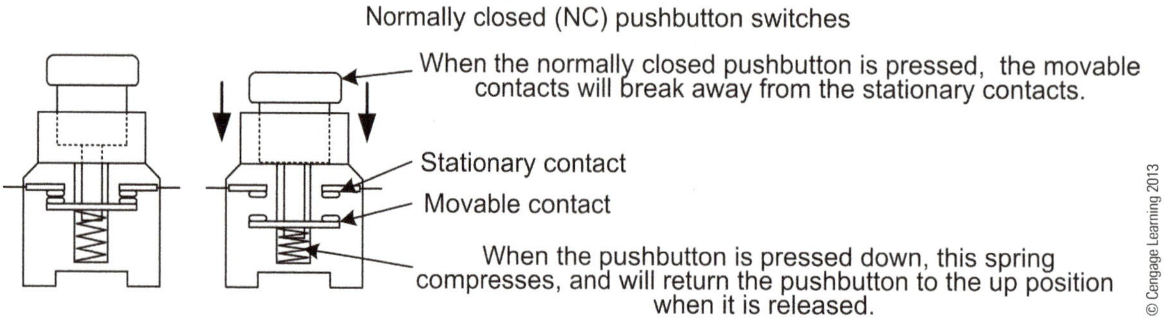

**FIGURE 13-2** Normally closed pushbutton switches

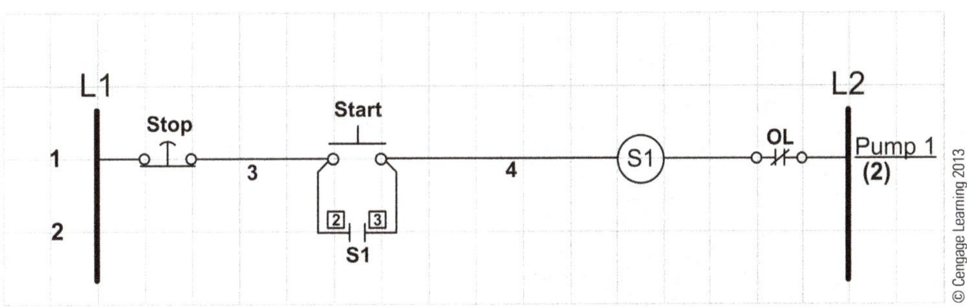
L1
L2
1
2
Stop
Start
3
4
S1
OL
Pump 1
(2)
2
3
S1
© Cengage Learning 2013

FIGURE 13-3 Ladder diagram

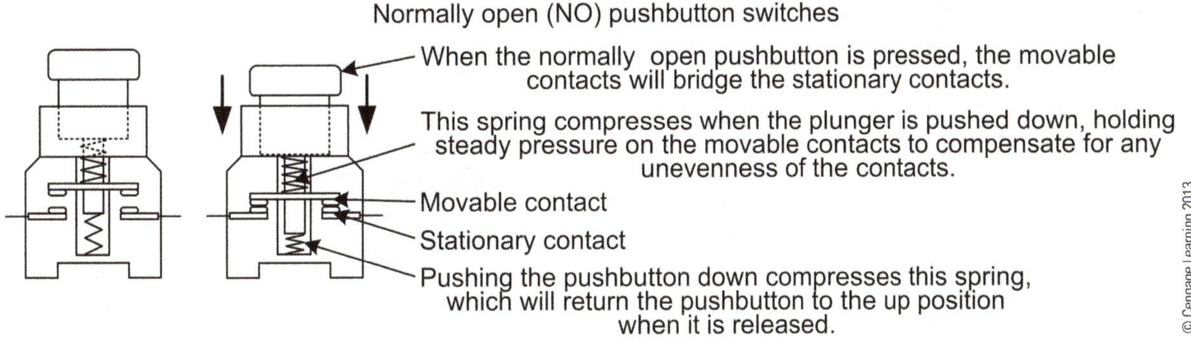
Normally open (NO) pushbutton switches
When the normally open pushbutton is pressed, the movable contacts will bridge the stationary contacts.
This spring compresses when the plunger is pushed down, holding steady pressure on the movable contacts to compensate for any unevenness of the contacts.
Movable contact
Stationary contact
Pushing the pushbutton down compresses this spring, which will return the pushbutton to the up position when it is released.
© Cengage Learning 2013

FIGURE 13-4 Normally open pushbutton switches

contact pushbutton is used for the start device, because under normal circumstances the control circuit power should be blocked from energizing the motor starter contactor control coil in the rung until the start pushbutton is pressed. When the start pushbutton is pressed, the control circuit power will pass through to the motor starter contactor control coil S1. When the S1 control coil is energized, it will pull in the contactor armature magnetically, closing the three main power circuit contacts and the control circuit holding contacts, each of which is actuated mechanically by the contactor armature. The start pushbutton only makes contact between the two stationary contacts when the pushbutton is pressed. As soon as the pushbutton is released, the contacts will open again. That is why the holding contacts on the contactor unit are needed to maintain the energized control circuit.

## Holding Contacts

Holding contacts are a normally open set of contacts mounted on the contactor portion of the motor starter, and are actuated by the armature of the contactor unit. Holding contacts actually have a couple of alternative names, "maintaining contacts" and "memory contacts," all of which are interchangeable designations, but point to the same function of the contact. Sometimes it is said that the holding contacts hold, remember, or maintain the last state of the contactor control coil. Once the start pushbutton energizes the contactor control coil, the holding contacts will make a redundant current path around the momentary contact start pushbutton to keep the control coil energized when the start pushbutton is released. Likewise, when the stop pushbutton is pressed to de-energize the contactor control coil, the contactor armature will drop out, physically opening the holding contact so that when the momentary stop pushbutton is released, the holding contacts now will be open (and so will the start pushbutton) and will prevent the contactor control coil from becoming energized.

## Integral Holding Contacts

Most motor starter manufacturers build the holding contacts into the contactor portion of the starter as an integral part. Other manufacturers provide the holding contacts as an accessory that attaches to the contactor unit, much like auxiliary

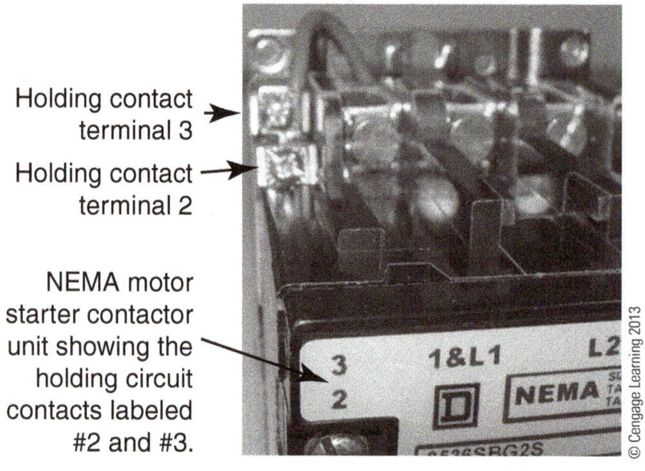
Holding contact terminal 3
Holding contact terminal 2
NEMA motor starter contactor unit showing the holding circuit contacts labeled #2 and #3.
3
2
1&L1
L2
NEMA
© Cengage Learning 2013

FIGURE 13-5 NEMA integral holding contacts on motor starter contactor

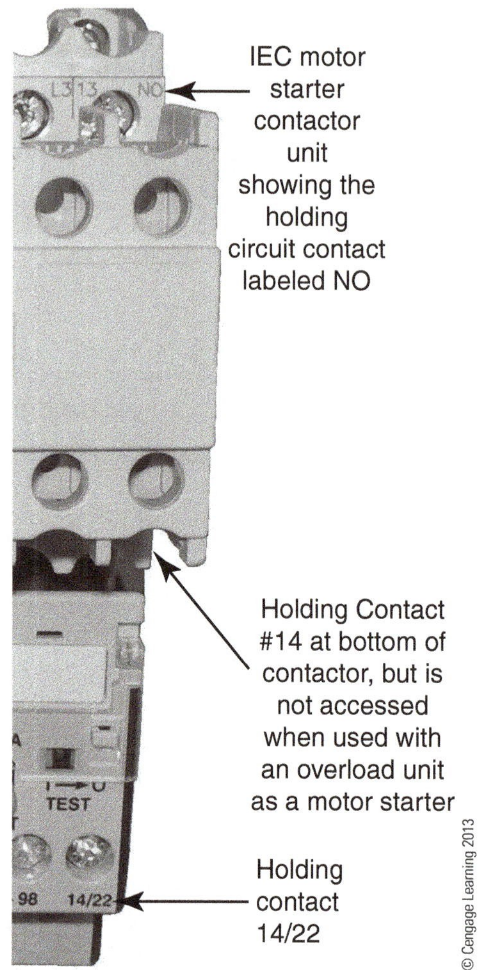
IEC motor starter contactor unit showing the holding circuit contact labeled NO
L3
13
NO
Holding Contact #14 at bottom of contactor, but is not accessed when used with an overload unit as a motor starter
TEST
98
14/22
Holding contact 14/22
© Cengage Learning 2013

FIGURE 13-6 IEC integral holding contacts on motor starter

contacts do. When looking on a motor starter for the holding contacts, they will be labeled with the numbers 2 and 3 on NEMA motor starters, as shown in Figure 13-5, and NO and 14/22 on IEC motor starters, as shown in Figure 13-6.

Remember, when documenting switch and relay contacts, the normally open or closed configuration status refers to the relaxed, de-energized, or shelf state of the device. If a device is normally open, a continuity check of the device with a digital multi-meter will give a reading of OL (overload, because the resistance is higher than the meter is capable of reading). If the device is normally closed, a resistance check will give a reading of 0.0 Ω. The normally open and normally closed state of the devices is usually not labeled on the ladder drawing. Rather, you must be able to recognize the symbol and be aware of the device's correct operation without additional explanatory material.

## Operation of Three-Wire Motor Control Circuits

If holding contacts were not used in three-wire control circuits, the control circuit operation would not be any different than a two-wire control circuit, where the motor runs only while the pilot device contact is closed and is energizing the motor starter control coil. Without a holding circuit, the operator would have to hold the start button down continuously to keep the motor starter coil energized and the load motor running. The following sequence explains the operation of the holding circuit in a three-wire motor control circuit.

When control circuit power first is applied to the three-wire motor control circuit, the rails become energized. In rung number 1 of Figure 13-7, the control power passes through the normally closed, momentary contact stop pushbutton, on to the normally open, momentary contact start pushbutton, where the open contacts prevent the control circuit power from passing any further. The control circuit power also is passed down to the holding contacts in rung number 2, because wire number 3 goes to both places, but the open holding contacts also would prevent the control circuit power from passing any further.

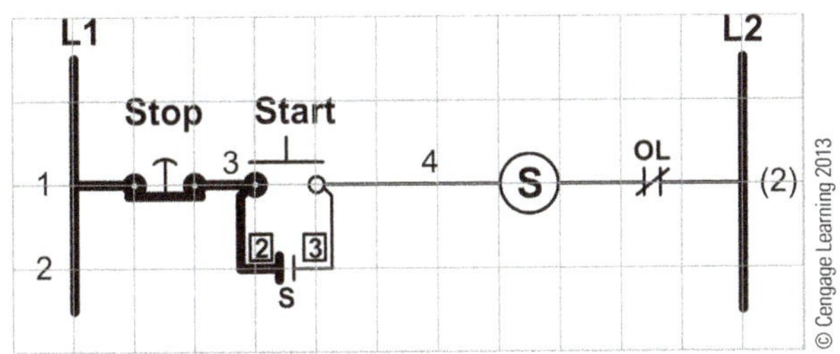


FIGURE 13-7 Three-wire circuit explanation when first energized

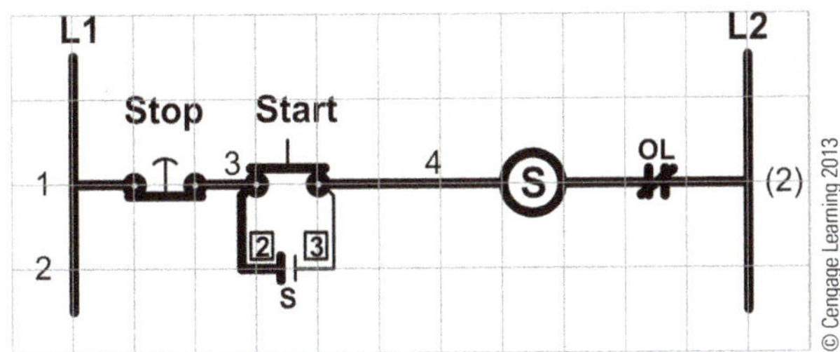


FIGURE 13-8 Three-wire circuit explanation with start pushbutton pressed

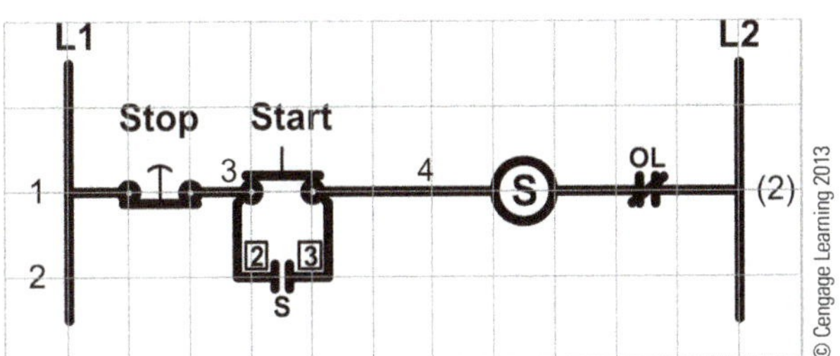


FIGURE 13-9 Three-wire circuit explanation when holding contacts closed

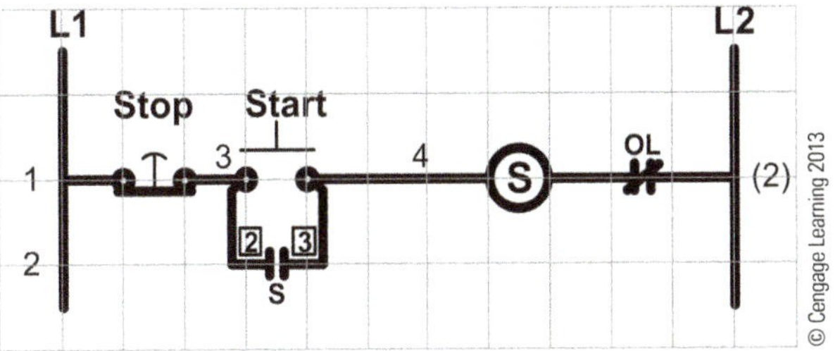


FIGURE 13-10 Three-wire circuit explanation when start pushbutton is released

The instant the start pushbutton is pressed, but before the contactor armature has the chance to move, the contactor control coil will become energized, as shown in Figure 13-8. The overload contacts between the contactor control coil and L2 are normally closed, so if the overload unit is not tripped the rung will have a complete current path to L2. Because the contactor armature has not yet had the chance to move at this instant, the holding contacts will remain open.

The holding function of a motor starter is performed when the control coil is energized and the armature pulls in, closing both main power contacts to run the load motor and holding circuit contacts. Once the contactor armature does pull in, the holding circuit contacts, which are also on the contactor unit and physically actuated by the contactor armature, close. The closed holding contacts form an electrically redundant current path around the normally open start pushbutton to keep the control coil energized after the start button is released, as shown in Figure 13-9. Remember that when control devices are connected in parallel, they form the Boolean OR function; either the start pushbutton or the holding contacts can keep the ladder rung energized.

Now when the start pushbutton is released and the original current path through the start pushbutton is opened, the holding contacts will maintain a current path around the start pushbutton to keep the contactor control coil energized, as shown in Figure 13-10. As long as the contactor control coil is energized, the contactor armature will remain pulled in, and the holding contacts will remain closed. And as long as the holding contacts remain closed, the control coil will remain energized.

If the control circuit power is interrupted for any reason (a power outage, etc., but in this case by pressing the stop pushbutton), the contactor control coil will become de-energized and the contactor armature will drop out. When the contactor armature drops out, it will open both the power contacts energizing the motor and the holding contacts keeping the contactor control coil energized, as shown in Figure 13-11.

Now when the stop pushbutton is released, neither the stop pushbutton nor the holding contacts will provide a current path to energize the ladder rung, as shown in Figure 13-12, and the contactor control coil will not become energized. When the operator presses the stop pushbutton to interrupt the electrical circuit to the motor starter

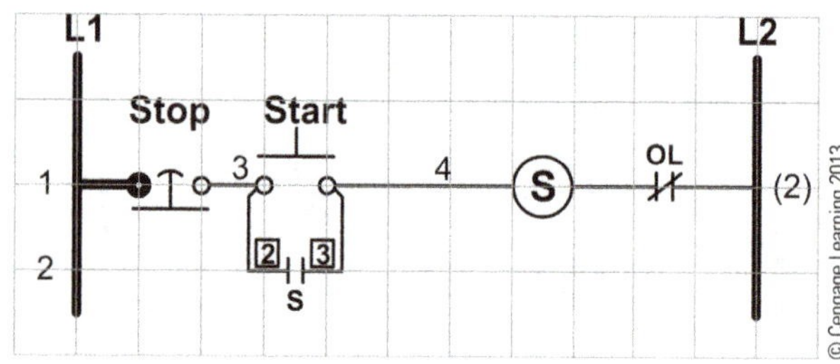


**FIGURE 13-11** Three-wire circuit explanation when stop pushbutton is pressed

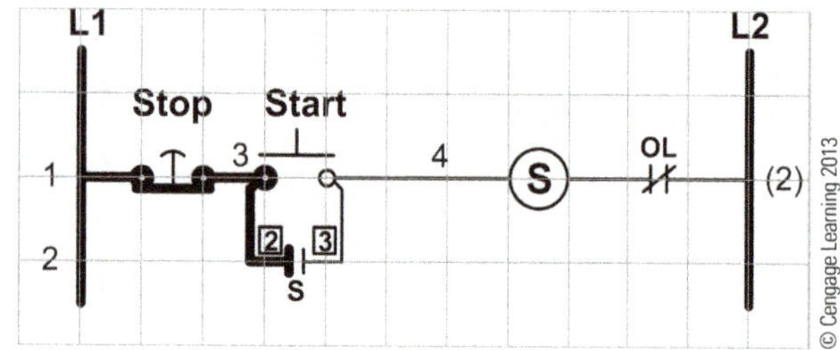


**FIGURE 13-12** Three-wire circuit explanation when back to energized

control coil to stop the load motor, releasing the stop pushbutton and restoring power to the control circuit will not restart the load motor, because both the start pushbutton contacts (normally open contacts) and the holding circuit contacts on the motor starter (normally open contacts when the armature is at rest) are open, so the starter control coil will remain de-energized.

## Inhibit and Enable

More often than not, when normally closed control contacts under the operation of a coil are used in the ladder rung logic, their purpose is to prevent two operations from happening at the same time. Because of this, normally closed contact devices on ladder rungs are sometimes called "inhibit" contacts or devices, because the action of one operation inhibits some other operation from acting simultaneously. Likewise, more often than not, when normally open control contacts under the operation of a coil are used in the ladder rung logic, their purpose is to enable some other operation to happen at the same time. Because of this, normally open contact devices on ladder rungs are sometimes called "enable" contacts, because the action of one operation enables some other operation to act simultaneously.

## Three-Wire No-Voltage and Low-Voltage Drop-Out Protection

Three-wire control does provide true no-voltage and low-voltage drop-out protection. In the event of a power failure or electrical feeder voltage dip, the starter control coil will become de-energized, the armature will drop out, and both the power and control circuits will open, no differently than if the stop button had been depressed. Now when power is restored, the motor will not restart automatically, because both the start pushbutton contacts and the holding circuit contacts are open, preventing the motor starter control coil from becoming energized until the momentary contact start pushbutton is pressed once again.

## CHAPTER SUMMARY

- Three-wire motor control circuits have three control circuit wires between the motor starter and the control device(s).
- Momentary contact control devices are called out as either normally open or normally closed.
- Momentary contact control devices change states when acted upon, but return to their call-out state when released.
- The motor starter contactor normally open holding contacts are wired in parallel with the normally open momentary contact start pushbutton, which perform the Boolean logic OR function, meaning either the start pushbutton or the holding contacts may energize the contactor control coil.
- When the three-wire motor control circuit first is energized, the circuit is broken by both the normally open start pushbutton contacts and the normally open contactor holding contacts, so the control coil cannot be energized until the start pushbutton is actuated.
- Once the motor starter contactor is energized and closes the holding contacts, the holding contacts provide a redundant electrical path around the normally open start pushbutton contacts, to keep the contactor control coil energized.
- The normally closed stop pushbutton is wired as the first control component in the ladder rung, so that it has master control over the entire rung.
- Once the motor starter contactor control coil is energized and being held energized by the holding contacts, pressing the normally closed stop pushbutton will de-energize the control coil, causing the contactor holding contacts to open.
- The holding contacts of a motor starter contactor unit normally are manufactured as an integral part, whereas auxiliary contacts that are also actuated by the contactor armature normally are add-ons.
- Normally closed contact devices on ladder rungs are sometimes called "inhibit" contacts or devices, and normally open contact devices on ladder rungs are sometimes called "enable" contacts.
- Three-wire motor control circuits provide true no-voltage and low-voltage drop-out protection.

## REVIEW QUESTIONS

1. What does the term "three-wire" motor control circuit signify?
2. What does the term "momentary contact" for the pushbutton switches signify?
3. Why are stop pushbuttons normally closed contact configurations?
4. Why are stop pushbuttons wired as the first component in the control circuit rung?
5. Why are start pushbuttons normally open contact configurations?
6. What actually actuates the motor starter contactor holding contacts?
7. How are the holding contacts of a three-wire motor control circuit wired in relation to the normally open start pushbutton?
8. How are holding contact terminals numbered on motor starter contactors for NEMA and IEC types?

   1. NEMA ______________________________

   2. IEC ________________________________

9. Why are normally closed control contacts sometimes called "inhibit" contacts?
10. Why are normally open control contacts sometimes called "enable" contacts?

# Control Circuits

### PURPOSE

To familiarize the learner with common motor control circuit schemes.

### OBJECTIVES

After studying this chapter on control circuits the learner will be able to:

- Follow common motor control circuit schemes

This chapter contains several basic motor control circuits common in the electrical industry, with a short explanation of each circuit preceding the drawing. The control circuits start very simply and build to more complex motor control schemes. Many of the control circuits are variations on the same control scheme. For instance, there are four different jog circuits.

For the control circuits that have devices, such as relays that require terminal number information to connect them correctly, the pin-out information will be given graphically next to the drawing. The control circuit diagrams in this chapter change between NEMA and IEC motor starters, different auxiliary contacts, and different types of relays to expose the learner to a variety of devices. All of the control circuits, however, may be connected with different devices if these are not available, with only minor changes to the terminal numbering

## TWO-WIRE CONTROL CIRCUITS

### Control Circuit 1. Pilot Device Control

The pressure tank switch pilot device, shown in Figure 14-1, determines when the motor is energized and de-energized by energizing and de-energizing the control coil. When the pilot device is open, the control circuit power cannot pass on to energize control coil S1 and start the motor. When the pilot device closes, control coil S1 will energize if the overload unit contacts also are closed, which then will energize the motor.

### Control Circuit 2. Hand-Off-Automatic (HOA) Control

The hand-off-automatic control circuit shown in Figure 14-2 is a variation on the simple pilot device control circuit that gives the operator manual control of the circuit. This circuit requires a three-position selector switch, shown in rung 1 of the drawing, where the center bar may be drawn up to make contact across the top two switch contacts; held in the middle, where it will not make contact across any contacts; or pushed down, where it will make contact across the bottom two switch contacts. The "hand" switch position, when the center bar is drawn up against the top contacts, will bypass the pressure tank switch pilot device and cause control coil S1 to become energized and run the motor, regardless of what state the pilot device is in. The "off" switch position, when the center bar is held in the middle, will break any possible electrical circuit path that could energize control coil S1, and thus will prevent the motor from starting. The "automatic" switch position, when the center bar is pushed down against the bottom contacts, will energize and de-energize the control coil to start and stop the motor under the control of the pilot device.

## THREE-WIRE CONTROL CIRCUITS

### Control Circuit 3. Single Motor Starter with One Pushbutton Control Station

When the control circuit shown in Figure 14-3 first is energized, the normally closed stop pushbutton will pass control circuit power to wire 3, but wire 4 will not become energized because both the normally open start pushbutton and the normally open holding contacts (terminal 2 and 3) both are open.

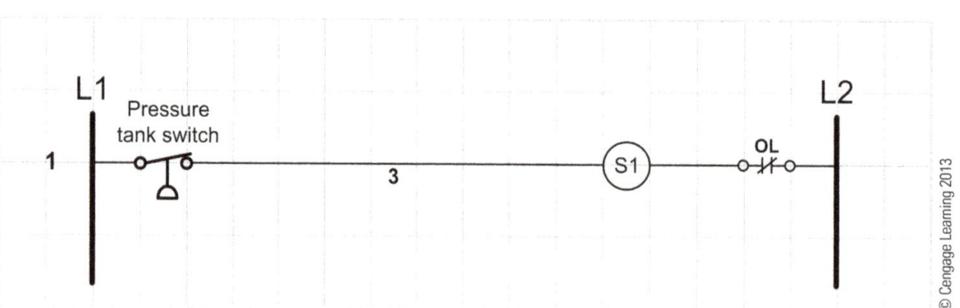


**FIGURE 14-1** Circuit 1. Two-wire control ladder diagram

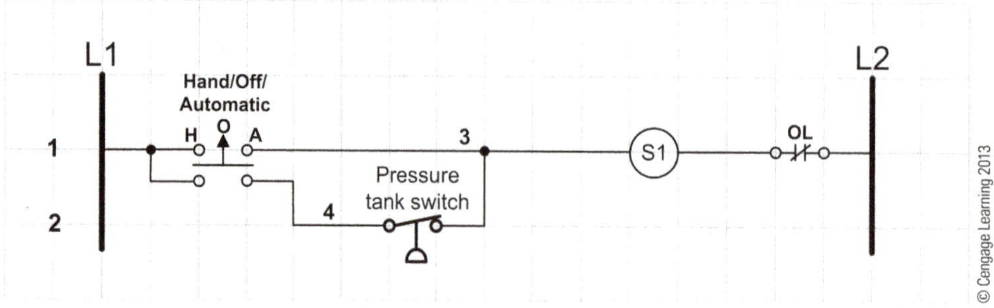


**FIGURE 14-2** Circuit 2. Hand-off-automatic circuit ladder diagram

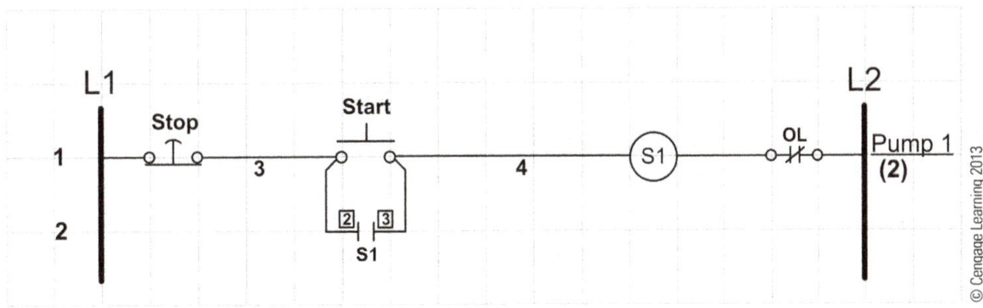


FIGURE 14-3 Circuit 3. Single motor starter with one stop-start pushbutton station

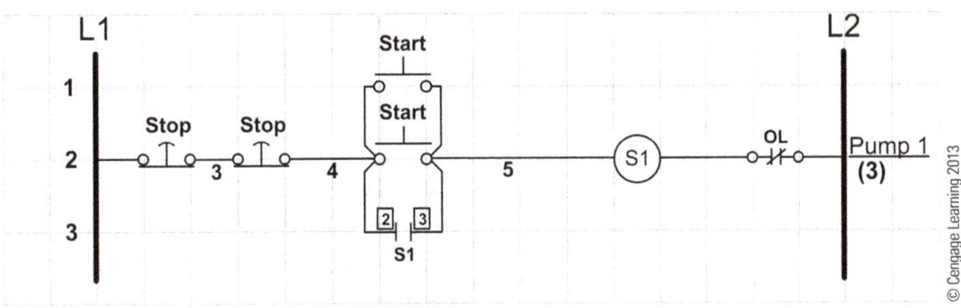


FIGURE 14-4 Circuit 4. Single motor starter with two pushbutton control stations

When the start pushbutton is pressed, wire 4 will become energized, and energize control coil S1. When control coil S1 becomes energized, the normally open S1 holding contacts in rung 2 will close to provide a holding circuit around the normally open, momentary-contact start pushbutton. When the start pushbutton is released, control coil S1 will remain energized through the S1 holding contacts in rung 2. When the normally closed stop pushbutton is pressed, the entire rung will be de-energized. When control coil S1 becomes de-energized, the S1 holding contacts in rung 2 also will open. When the stop pushbutton is released, control circuit power again will energize wire 3, but the normally open start pushbutton and the normally open S1 holding contacts will prevent wire 4 from becoming energized until the start pushbutton is pressed again.

## Control Circuit 4. Single Motor Starter with Two Pushbutton Control Stations

The operation of the control circuit shown in Figure 14-4 is not much different than a single pushbutton control station circuit, but two rules must be observed whenever two or more control stations are used. The first rule is that all of the stop pushbuttons from all of the control stations must be wired in series at the start of the rung. The reason for connecting all of them in series at the start of the rung is so that any stop pushbutton in any of the control stations will de-energize the entire rung. The second rule is that all of the start pushbuttons, and the holding contact on the contactor unit, are wired in parallel. Remember that wiring switch components in parallel will create a Boolean logic OR function, meaning that pressing any of the start pushbutton switches will energize the control coil, and the holding contacts will hold the circuit for all of them.

## Control Circuit 5. Single Motor Starter with a Single Pushbutton Control Station, with Auxiliary Contact Controlled Status Indicator Lamps

The control circuit in Figure 14-5 operates the same as the single motor starter with a single pushbutton control station in Circuit 3, but auxiliary contacts have been added to the motor starter to operate status indicator lamps. The green indicator lamp is to indicate the control circuit is energized and the motor is ready to run, and the red indicator to indicate that the motor actually is running. With the circuit de-energized and the relay at rest, the green indicator lamp should be lit because the normally closed auxiliary contacts will pass the control circuit power to the green indicator lamp; and red indicator lamp should be off because the normally open auxiliary contacts will block control circuit power to the red indicator lamp. When the start pushbutton is pressed, the S1 control coil will become energized, and as in previous circuits, energizing control coil S1 will start the motor. What also happens in this control circuit when control coil S1 is energized is that the contactor armature also will actuate the auxiliary contacts mechanically and cause them to change states. The green indicator lamp will go out because the normally closed auxiliary contacts will open; and the red indicator lamp will light to indicate that the motor is running, because the normally open auxiliary contacts will close.

This control diagram uses an IEC motor starter with normally open and normally closed auxiliary contacts. The pin-out information for the auxiliary contacts is provided to the left of the diagram. All of the terminal numbers shown for this circuit are for an IEC motor starter.

In some motor control situations it may be preferable to perform this same operation using a control relay. These cases might include a process where the operation of one part is contingent on the operation of another part, or when more switched poles are needed than are available using auxiliary contacts. Normally a general-purpose control relay, such as an ice cube relay, is used for this purpose, but in situations where safety is a primary concern, a force-guided relay would be used. Remember from Chapter 7 that a force-guided relay has mechanically interlocked poles, so that no single pole can change states without all poles changing states. The next two circuits demonstrate two different methods of achieving the same purpose, using an octagon base ice cube relay.

## Control Circuit 6. Single Motor Starter with a Single Pushbutton Control Station, with Relay-Controlled Status Indicator Lamps

The control circuit in Figure 14-6 operates the same as the single motor starter with a single pushbutton control station, using auxiliary contact-controlled status indicator lamps as in Control Circuit 5, but with a relay added in place of the auxiliary contacts to operate status indicator lamps. With the circuit de-energized and the relay at rest, the green indicator lamp should be lit because the normally closed relay contacts will pass the control circuit power; and red indicator lamp should be off because the normally open

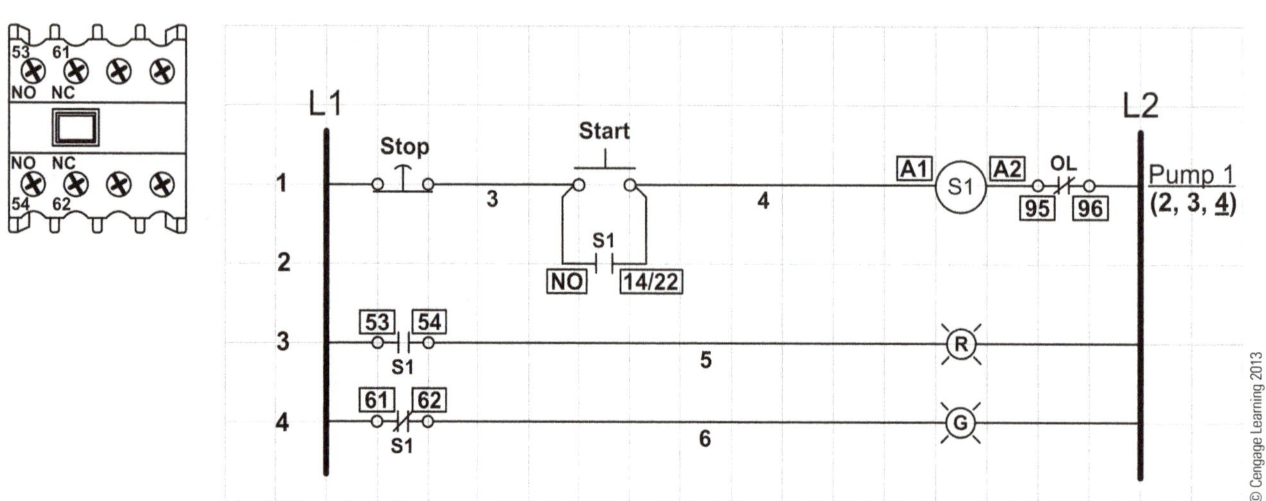


FIGURE 14-5 Circuit 5. Single motor starter with a single pushbutton control station, with auxiliary contact-controlled status indicator lamps

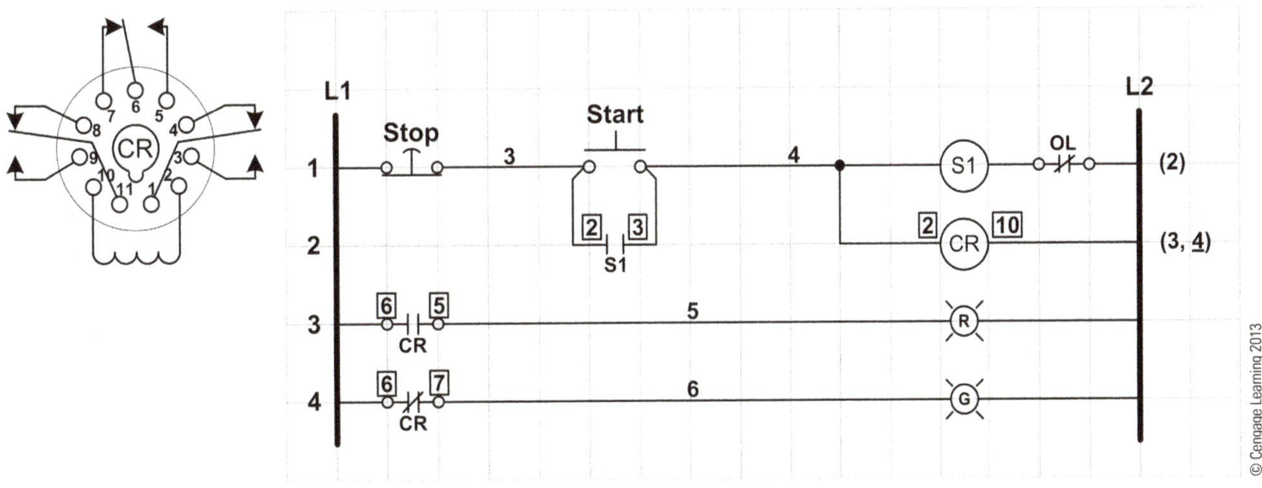


**FIGURE 14-6** Circuit 6. Single motor starter with a single pushbutton control station, with relay-controlled status indicator lamps

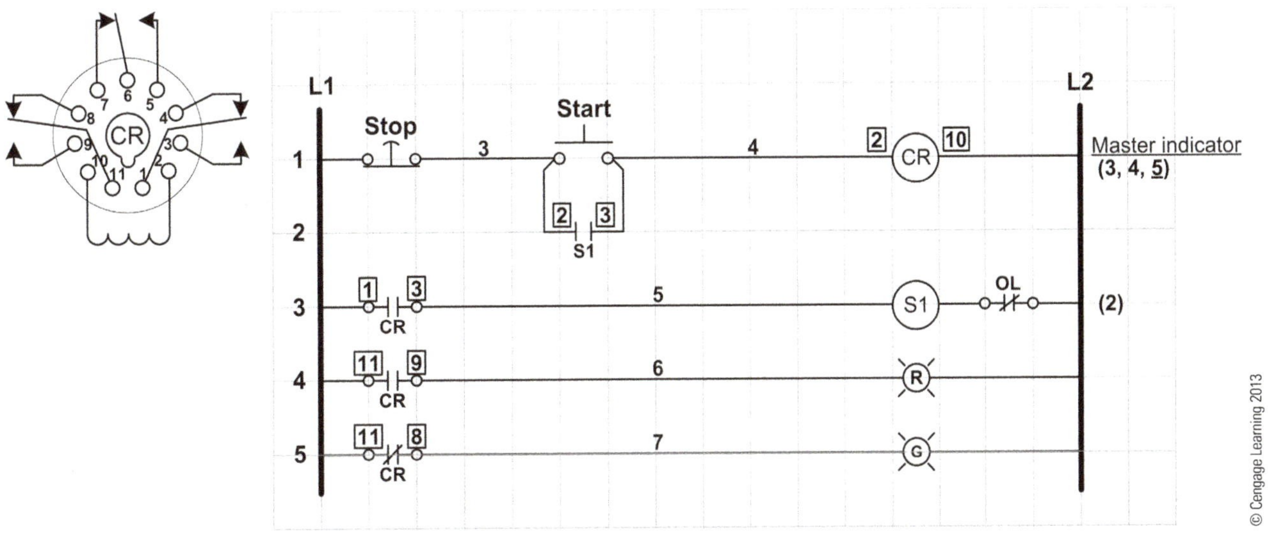


**FIGURE 14-7** Circuit 7. Single motor starter with a single pushbutton control station, with relay-controlled starter and status indicator lamps

relay contacts will not pass the control circuit power. When the start pushbutton is pressed, both the S1 control coil and the CR control coil will become energized. As in previous circuits, energizing control coil S1 will start the motor. When the control relay control coil CR is energized, both lamps will change states: the green indicator lamp will go out, and the red indicator lamp will light to indicate that the motor is running. This control diagram uses an eleven-pin, octagon base ice cube relay. The pin-out information for the relay is to the left of the diagram.

## Control Circuit 7. Single Motor Starter with a Single Pushbutton Control Station, with Relay-Controlled Motor Starter and Status Indicator Lamps

The control circuit in Figure 14-7 will operate the same as the last circuit, only the control relay is being used to energize the S1 control coil in addition to the indicator lamps. The advantage of this control circuit becomes quite obvious if control relay CR is a force-guided relay. By controlling all circuit components with a force-guided relay, there is no chance that one part of the circuit could change states without all parts of the circuit changing states. Controlling indicator lamps together with a motor starter control coil may not seem like a life-or-death safety issue, but this circuit demonstrates that several parts of a process can be controlled together with a control relay.

Also, notice that the holding circuit for control relay CR uses the motor starter contactor holding contact numbers 2 and 3 rather than a normally open set of contacts on the relay itself. The reason for using the contactor holding contacts is that if the overload unit contacts open, the control circuit will become de-energized until the start pushbutton is pressed again. If a set of normally open contacts on the control relay CR is used for the holding circuit, the motor still will stop if the overload unit contacts open, but the control circuit will remain energized.

## Control Circuit 8. Single Motor Starter with a Single Pushbutton Control Station, with a Push-to-Test Indicator Lamp Circuit

The purpose of the control circuit in Figure 14-8 is to have an indicator lamp light whenever the motor S1 control coil is energized, to indicate that the motor is running, and to provide a test pushbutton to prove the lamp is not burned out. If the motor is located a distance from the control point, an indicator lamp often is used to notify the operator of the running status. If the indicator lamp is out, the operator may not know if that is because the motor is not running or because the indicator lamp is burned out. The push-to-test function allows the operator to test the lamp by applying the control circuit supply voltage directly across the lamp to see if it lights. The S1 coil and the indicator lamp are both load components of the ladder rung, and therefore are wired in parallel so that they both get the correct operating voltage. The push-to-test switch is a two-circuit momentary-contact pushbutton. The normally closed circuit through the push-to-test pushbutton energizes the indicator lamp with the control circuit power from rung 1 whenever the S1 control coil is energized. When the push-to-test pushbutton is pressed, the normally closed contacts will open to disconnect the indicator lamp from the control circuit power in rung 1, and the normally open contacts will close to connect the indicator lamp directly across L1 and L2 to test whether the lamp is good.

## Control Circuit 9. One Motor Automatically Starting a Second Motor Using Auxiliary Contacts

The purpose of the control circuit in Figure 14-9 is to have a second motor start and run whenever the first motor is energized. One application for this circuit might be a situation where every time a certain conveyor is started, a second conveyor also must be

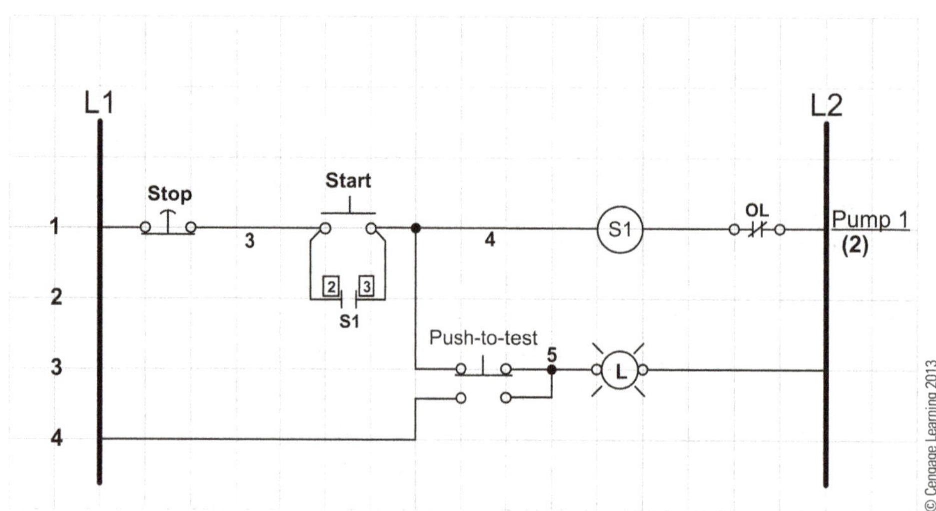



FIGURE 14-8 Circuit 8. Single motor starter with a single pushbutton control station, with a push-to-test indicator lamp circuit

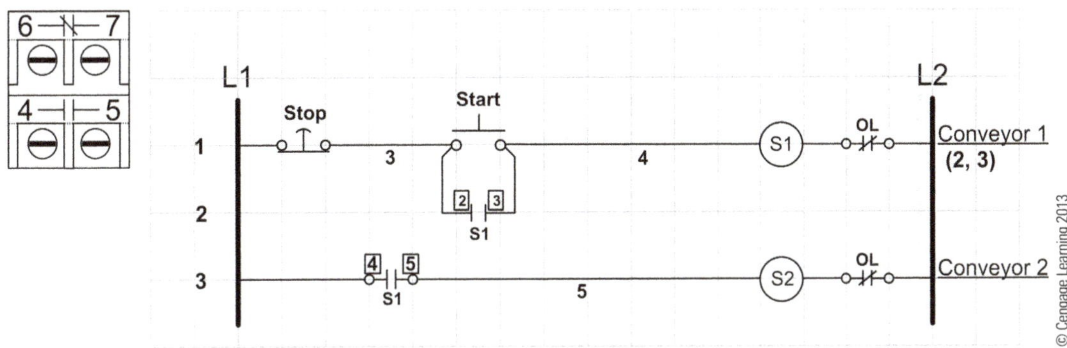


**FIGURE 14-9** Circuit 9. One motor automatically starting a second motor using auxiliary contacts

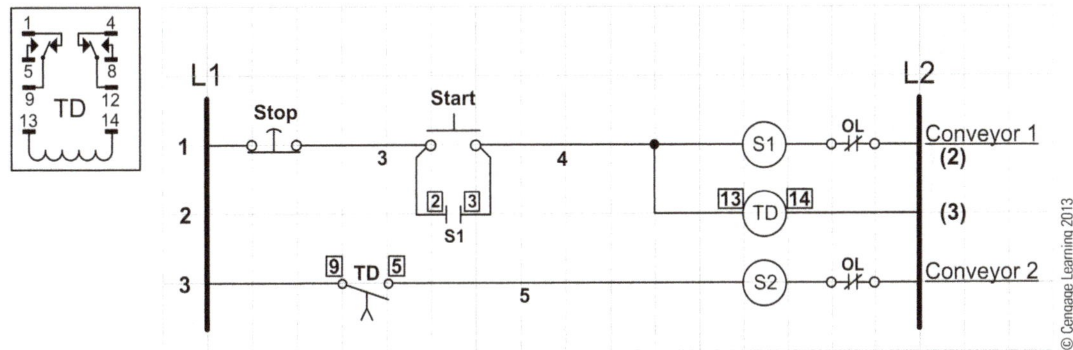


**FIGURE 14-10** Circuit 10. One motor automatically starting a second motor after a time delay, using a timing relay

started as part of the process. By using the auxiliary contacts on the first motor starter automatically to start the second motor starter, the chance is eliminated of the operator forgetting to start the second conveyor. An auxiliary contact for the motor starter is provided on the left of the drawing for terminal numbers. The auxiliary contacts for this circuit are NEMA type, and the terminal number diagram is provided on the left side of the drawing.

## Control Circuit 10. One Motor Automatically Starting a Second Motor After a Time Delay, Using a Timing Relay

The control circuit in Figure 14-10 shows a circuit where starting the first motor automatically starts a second motor after a short time delay. The purpose of this control circuit is to provide a predetermined amount of time after starting the first motor before the second motor starts. This circuit would be useful especially in cases where the two motors were very large, and starting them together could cause problems on the facility power system. Auxiliary contacts on the motor starter would not be used in this case. A time-delay relay would be used instead. The pin-out information for the time-delay relay is on the left of the diagram.

## Control Circuit 11. Single Motor Starter with Local-Off-Remote Operation with a Single Pushbutton Control Station Remote

Similar to the hand-off-automatic, two-wire motor control circuit, the local-off-remote control circuit in Figure 14-11 will run the motor manually in the local selector switch position, but in the remote selector switch position the pushbutton control station will operate the control circuit. The local-off-remote switch is most often placed close to the motor being controlled when the pushbutton control station is a distance away, so the operator can start or stop the motor without having to travel to the pushbutton control station.

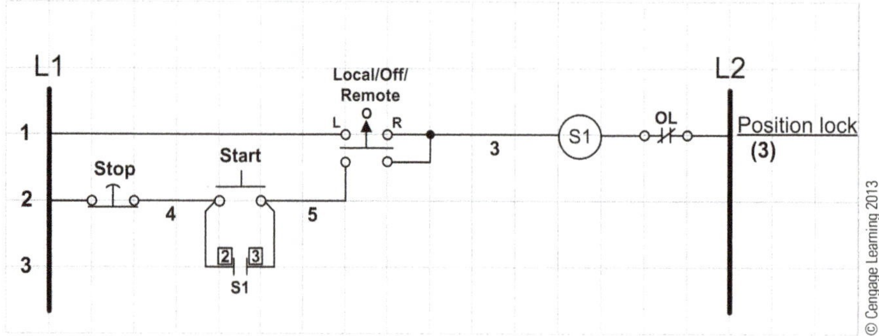


FIGURE 14-11 Circuit 11. Single motor starter with local-off-remote operation with a single pushbutton control station remote

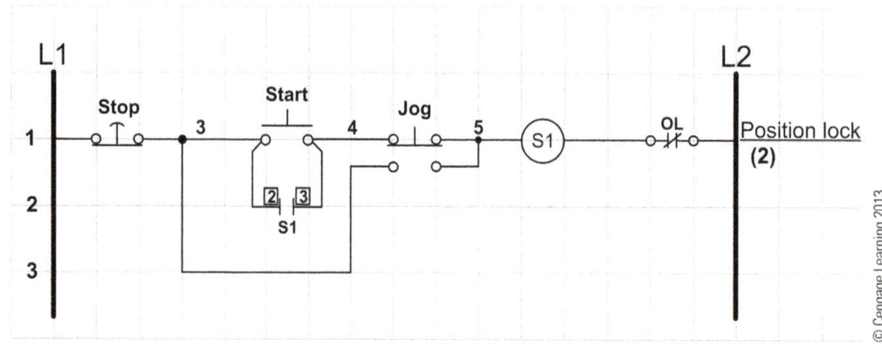


FIGURE 14-12 Circuit 12. Single motor starter with a two-circuit pushbutton jog switch

## Control Circuit 12. Single Motor Starter with a Two-Circuit Pushbutton Jog Switch

The purpose of the jog function, shown in Figure 14-12, is to prevent the holding contacts from holding the contactor control coil energized so that the motor continues to run. With the jog function, the motor only runs as long as the jog pushbutton is pressed, and the motor stops as soon as the pushbutton is released. This function would be used for situations where the operator is trying to position something, and wants the motor to run for very short periods of time.

The jog circuit uses three momentary contact pushbutton switches. When the jog pushbutton is not pressed, the switch bar makes contact between the top two terminals, which allow the start and holding circuits to work as normal to energize control coil S1. When the jog pushbutton is pressed, control coil S1 will become energized, but the holding circuit will not be in the circuit to keep control coil S1 energized when the jog pushbutton is released. When the jog pushbutton is released, control coil S1 will become de-energized.

## Control Circuit 13. Single Motor Starter with a Two-Position, Maintained-Contacts Selector Switch Jog

This jog circuit in Figure 14-13 uses two momentary-contact pushbutton switches, and one two-position selector switch. The purpose of this control circuit is to have a stop-start pushbutton control station to operate a motor, but then add a run-jog selector switch to determine if the start pushbutton will have a start or jog function. An aspect of this circuit that is sometimes confusing is that the second circuit of the two-circuit pushbutton is not utilized. Instead, the run-jog selector switch simply completes the holding circuit when the selector switch is in the run position with the switch bar across the top two contacts, and breaks the holding circuit when the selector

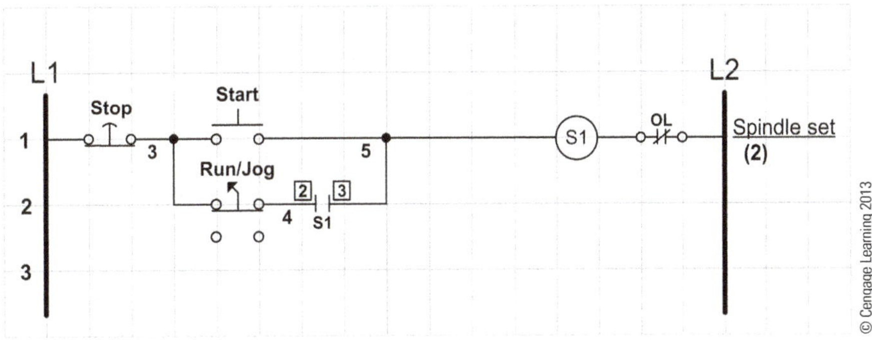


FIGURE 14-13 Circuit 13. Single motor starter with a two-position, maintained-contacts selector switch jog

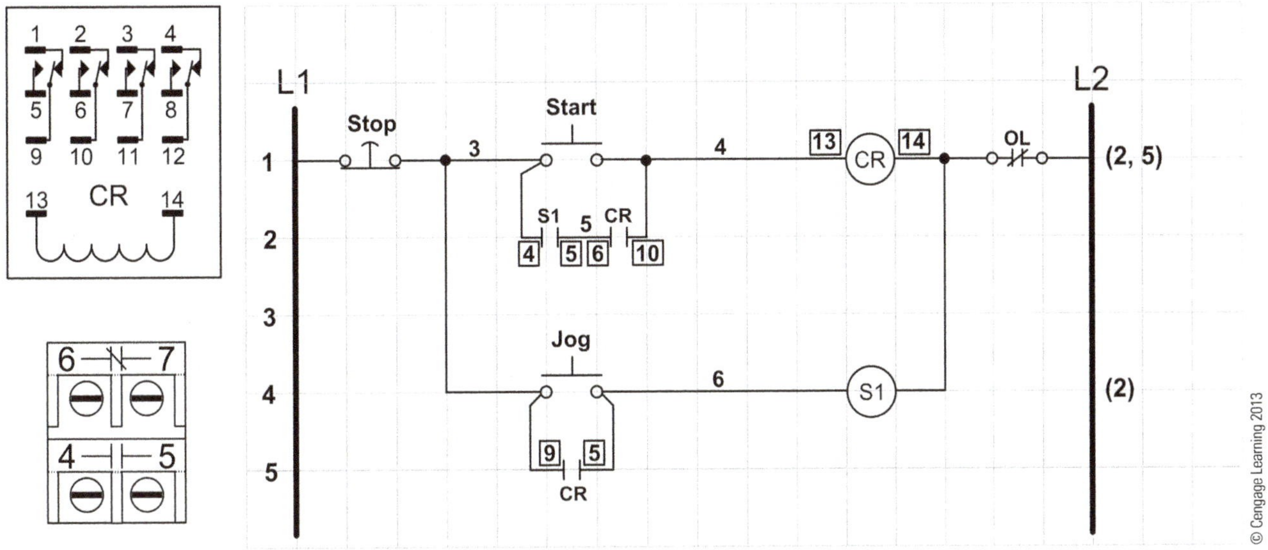


FIGURE 14-14 Circuit 14. Single motor starter with a control relay jog

switch is in the jog position with the switch bar across the bottom two contacts.

## Control Circuit 14. Single Motor Starter with a Control Relay Jog

The jog circuit in Figure 14-14 uses three momentary contact pushbutton switches, and is considered to be safer than other jog circuits that use only momentary contact switches. The jog circuit in Circuit 12 leaves the possibility that the jog pushbutton could be released fast enough to remake the holding circuit before the contactor armature has time to drop out. If the armature does not drop out and open holding contacts 2 and 3 before the jog pushbutton remakes the circuit across the top two contacts of the switch, it could cause the motor to continue to run rather than jog. This is a very rare problem, but it does remain a possibility with this circuit. By adding the relay to the circuit there is no danger of creating a holding circuit when the jog pushbutton is pressed, so the timing concern of other three pushbutton jog circuits is overcome.

For the jog function, notice that the contact connected across the jog pushbutton is not S1; it is a CR contact. Energizing control coil S1 with the jog pushbutton will close the normally open S1 contact in rung 5 but will not energize the CR control coil, so it will not act as a holding contact for the jog function. For the run function, when the start pushbutton is pressed the CR control coil will become energized, which will close the normally open CR contacts in rungs 2 and 5. When the normally open CR contact in rung 5 closes it will energize contactor control coil S1, which will close the normally open S1 contact in rung 2. In rung 2, both normally open contacts S1 and CR now will be closed and provide a holding circuit to keep control coil CR energized.

Use the CR relay pin-out information shown on the left for CR, and the motor starter contactor auxiliary contacts also shown on the left for S1.

## Control Circuit 15. Single Motor Starter with a Control Relay Jog 2

This second control relay jog circuit shown in Figure 14-15, also is a common method of providing a relay jog function. The only difference between the two circuits is the normally open S1 contact in the second rung of the first circuit. Some people believe it is important to have a contact actuated by the motor starter contactor armature used in the holding circuit, to report the starter's actual status. In the second circuit the motor starter could malfunction, such as by having a burned open control coil, and the control circuit would not operate any differently. If reporting the actual status of the motor starter contactor is not an issue, the second circuit will perform the relay jog function adequately, without the additional contact on the motor starter.

## Control Circuit 16. Electrically Interlocking Two Motors to Prevent Simultaneous Operation

The control circuit in Figure 14-16 shows a control circuit for Interlocking two motors so that either can run, but cannot be run together. Remember that the rung number of the normally closed inhibit contacts, referenced to the right of L2, is underlined to document the normally closed contact design. The normally closed inhibit contact for contactor control coil S1 is in the contactor control coil circuit in rung 4. When control coil S1 is energized, the normally closed inhibit contact in rung 4 will open, and inhibit contactor control coil S2 from being energized at the same time. The same inhibit operation is true for contactor control coil S2 if it is started first. Use the auxiliary contact pin-out information shown on the left for both S1 and S2 motor starters.

## Control Circuit 17. Selector Switch Interlocking Two Motors to Prevent Simultaneous Operation

The control circuit in Figure 14-17 shows a selector switch control circuit for Interlocking two motors so that either can run, but they cannot be run together. The two-position, two-circuit selector switch allows only one control circuit rung to be energized at a time, so simultaneous operation is not possible.

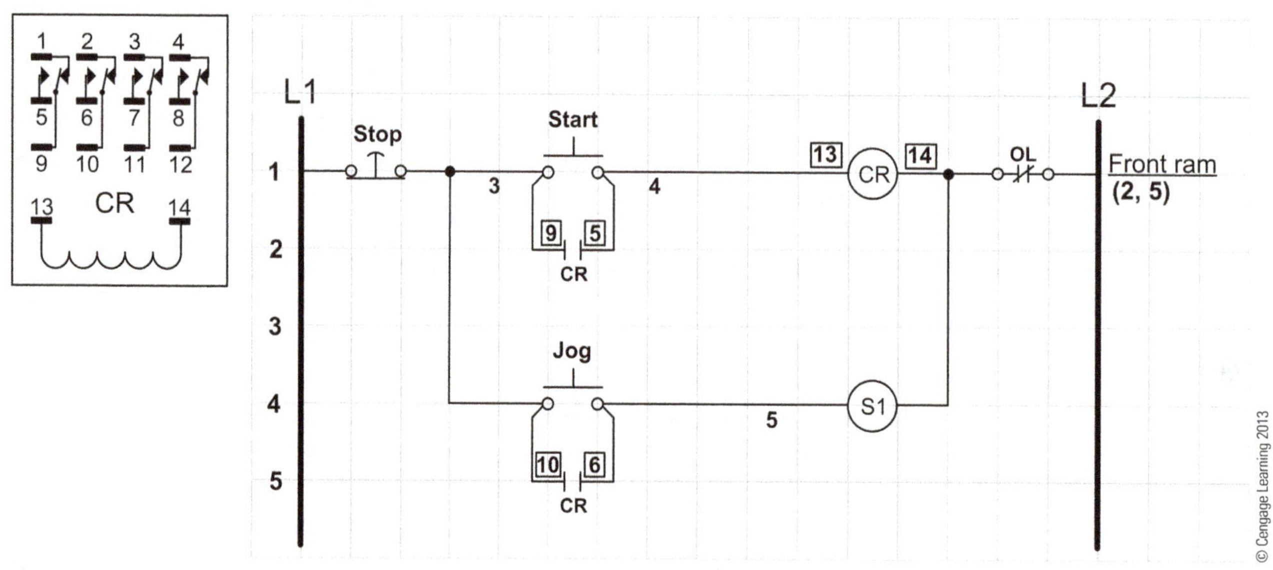



FIGURE 14-15 Circuit 15. Single motor starter with a control relay jog #2

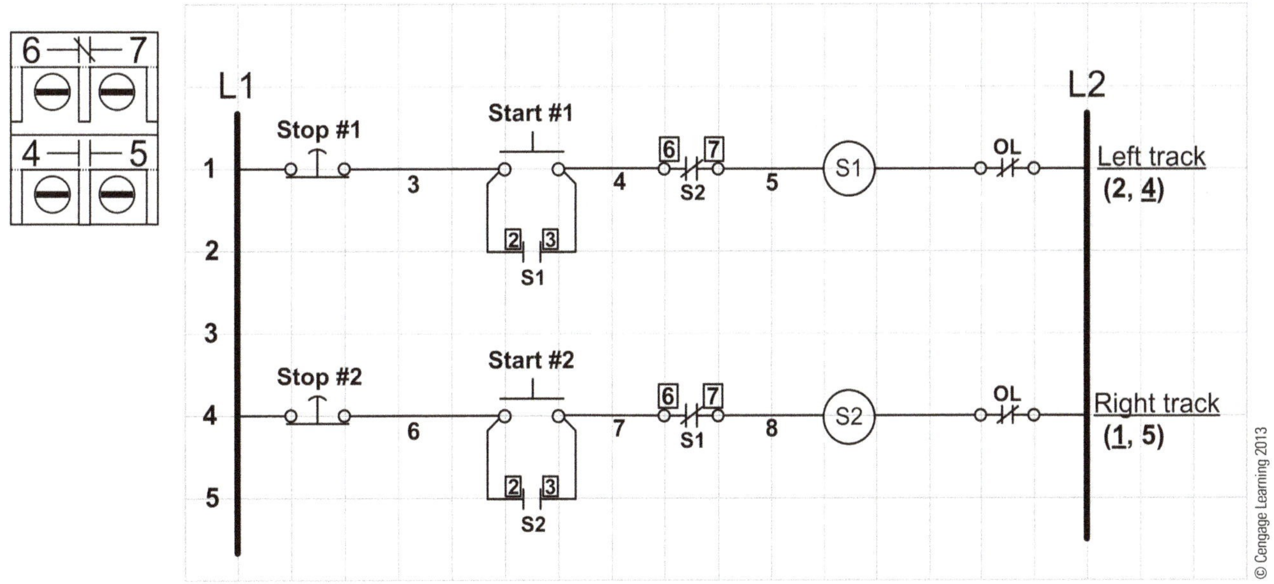


FIGURE 14-16 Circuit 16. Electrically interlocking two motors to prevent simultaneous operation

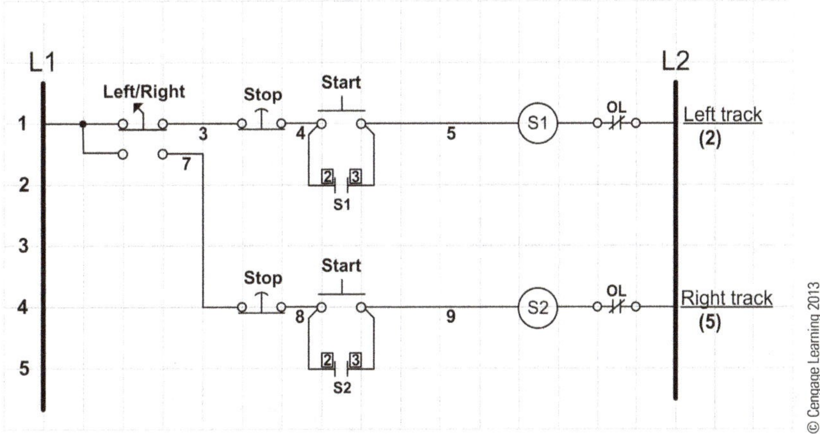


FIGURE 14-17 Circuit 17. Selector switch interlocking two motors to prevent simultaneous operation

## Control Circuit 18. Electrical Interlock Reversing Motor Starter

The control circuit in Figure 14-18 shows a control circuit to electrically interlock a forward and reverse motor starter. This control circuit utilizes the same normally closed inhibit contacts as the last circuit, except in this case there is only one motor being controlled. A reversing motor starter has two contactors and two control coils, one for forward and one for reverse. The two contactors are interlocked electrically with the inhibit contacts because serious damage would occur if both contactors energized at the same time. One difference between this control circuit and the previous control circuit is that the stop button is master for both contactors. (Note: reversing motor starters normally are interlocked mechanically also.)

## Control Circuit 19. Time Delay, Electrical Interlock, Reversing Motor Starter

The time-delay electrical interlock method for a reversing motor starter, shown in Figure 14-19, adds a substantial cost to the control circuit because of the cost of the timing relays, but in some situations the delay may be necessary. The purpose of the timing relays is to prevent the motor from becoming energized to turn in the opposite direction until a time

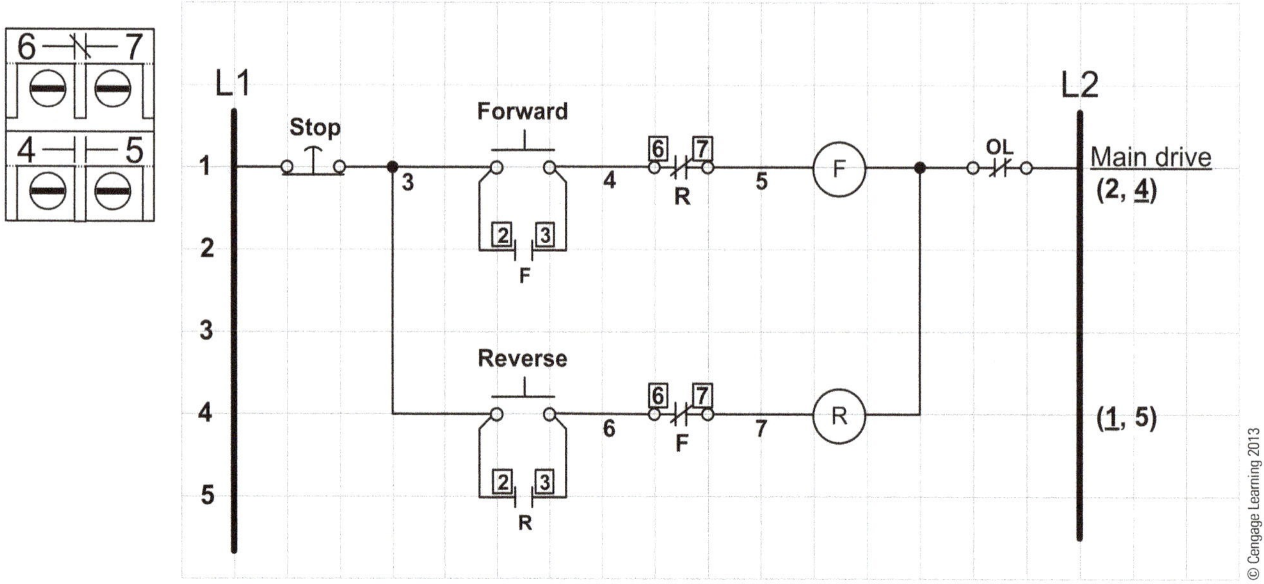


FIGURE 14-18 Circuit 18. Electrical interlock, reversing motor starter

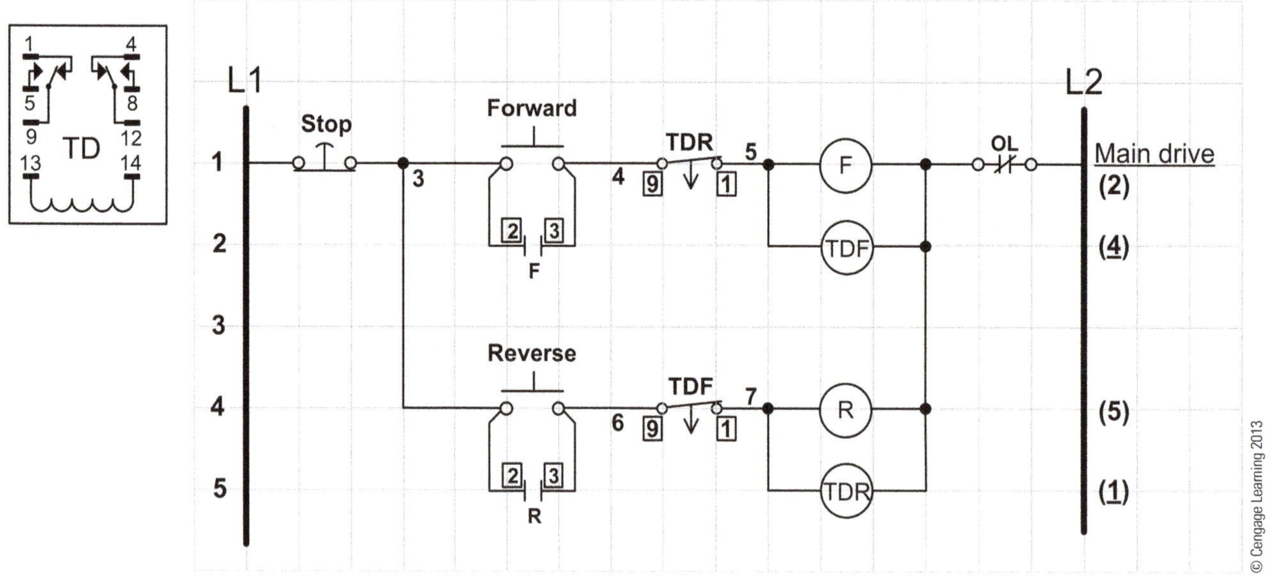


FIGURE 14-19 Circuit 19. Time-delay, electrical interlock, reversing motor starter

delay has lapsed for the motor to stop. For example, in the case of a very high inertia load that takes 15 seconds for the load to stop once it has been de-energized, it would be very hard on the motor and connected mechanical load to be reversed immediately after de-energizing. In this case a time delay electrical interlock would prevent the motor from becoming energized to turn in the opposite direction of travel until a set amount of time has passed to allow the motor to stop turning.

## Control Circuit 20. Selector-Switch Interlock, Reversing Motor Starter

The control circuit in Figure 14-20 shows a selector-switch control circuit, electrically interlocked, forward and reverse motor starter. The difference between this reversing circuit and the last reversing circuit is that the same start pushbutton is used for both forward and reverse, and the decision to run the motor in the forward or reverse direction of rotation is made with a selector switch. Otherwise there is no difference in the operation of the two reversing control circuits.

## Control Circuit 21. Electrical Interlock, Reversing Motor Starter, with Left and Right Travel-Limit Switches

The control circuit in Figure 14-21 shows a control circuit for an electrically interlocked, reversing motor starter, with left and right travel-limit switches to prevent overtravel. If the application were something like a small overhead crane, some type of safety provision would be required to prevent the crane from traveling too far in either direction. In this control circuit, either the left or right direction contactor control coil may be energized when the circuit conditions are normal. Notice, though, if the crane overtraveled in the right direction and opened the right limit switch, the right contactor control coil could not become energized even if the left contactor control coil was not energized at the time. Notice that the normally closed inhibit contacts are each in the rung of the other travel direction contactor control coil, but the limit switches are in their respective rungs. When either one of the overtravel limit switches is actuated open, only the opposite travel direction

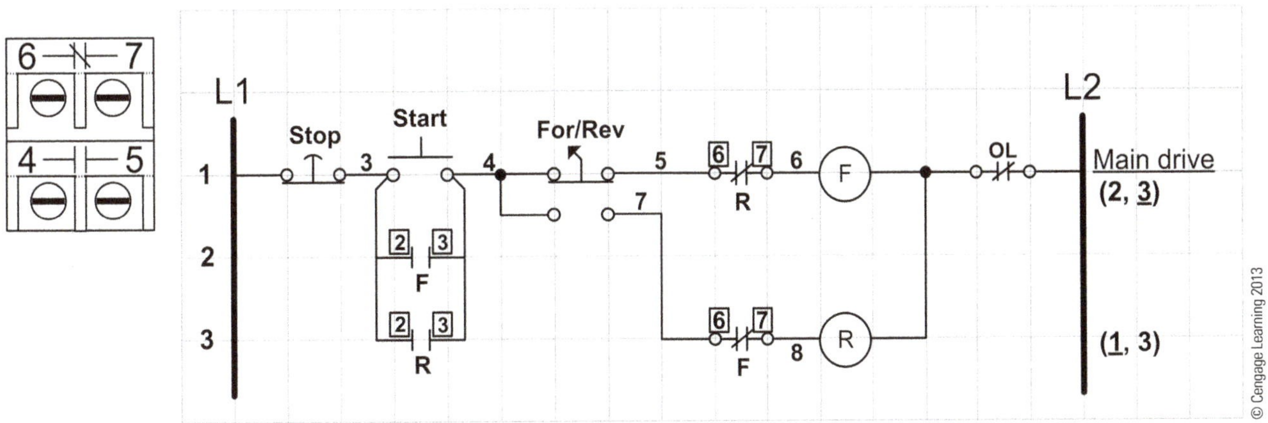


**FIGURE 14-20** Circuit 20. Selector-switch interlock, reversing motor starter

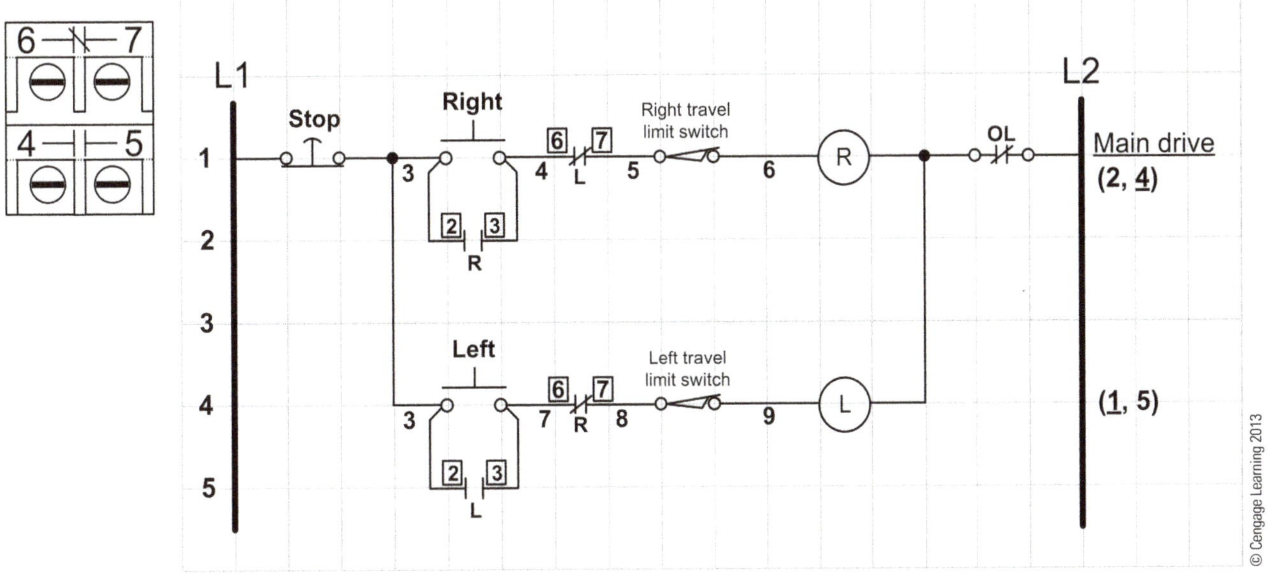

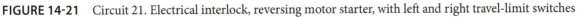

**FIGURE 14-21** Circuit 21. Electrical interlock, reversing motor starter, with left and right travel-limit switches

contactor control coil can be energized until the crane is within its normal operating range again.

### Control Circuit 22. Motor 2 Cannot Start Until Motor 1 Already Is Running

The control circuit in Figure 14-22 shows a control circuit where motor 2 cannot start until motor 1 already is running. This control circuit is useful in conveyor situations where the feeder conveyor cannot start until the main conveyor it feeds onto already is running.

### Control Circuit 23. Motor 2 Cannot Start Until Motor 1 Is Already Running, Version 2

The control circuit in Figure 14-23 shows another way to wire a control circuit where motor 2 cannot start until motor 1 already is running. Other than using a normally open S1 auxiliary contact to control the power to the rung for motor 2, the operation of this circuit is the same as the previous control circuit.

### Control Circuit 24. Motor 1 Can Run Independently of Motor 2, but Motor 1 Must Start Automatically When Motor 2 Is Started

The control circuit shown in Figure 14-24 is also useful around conveyor systems. Motor 1 is the main belt that must run independently of motor 2, which is a feeder belt. Motor 1, the main belt, runs without any effect on motor 2. However, when motor 2 starts, the S2 contact in rung 3 will close to energize control coil S1, but the normally closed S2

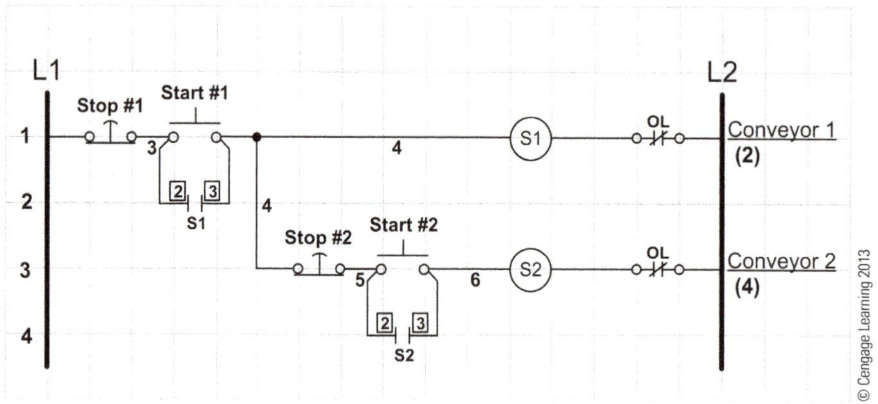


FIGURE 14-22 Circuit 22. Motor 2 cannot start until motor 1 already is running

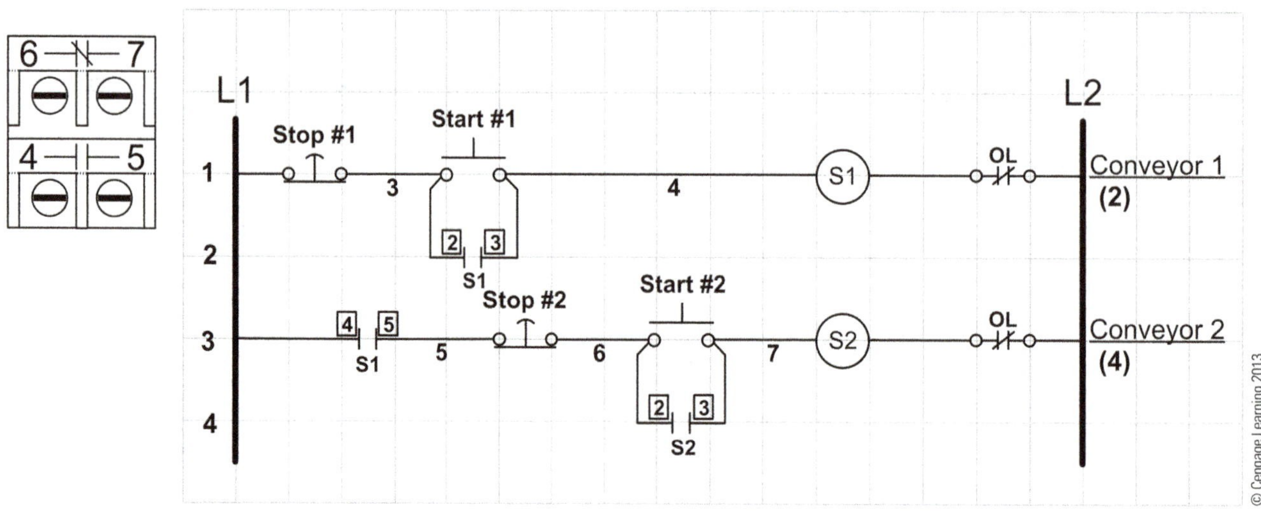


FIGURE 14-23 Circuit 23. Motor 2 cannot start until motor 1 already is running, version 2

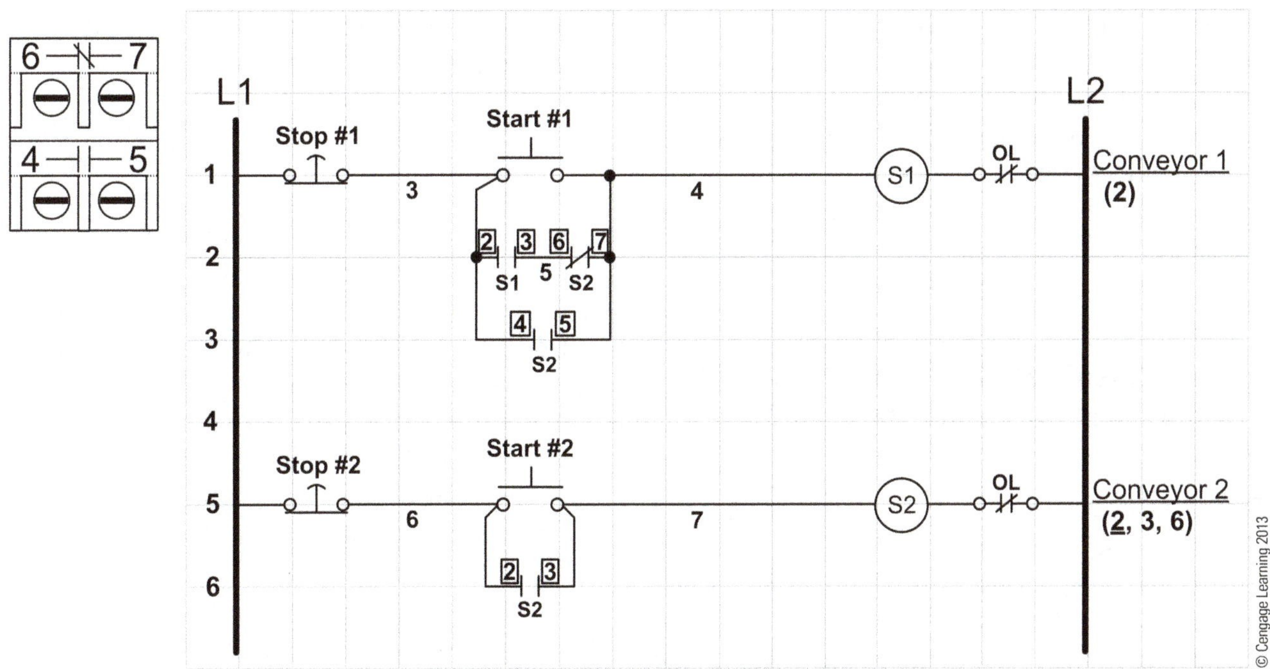


FIGURE 14-24 Circuit 24. Motor 1 can run independently of motor 2, but motor 1 must start automatically when motor 2 is started

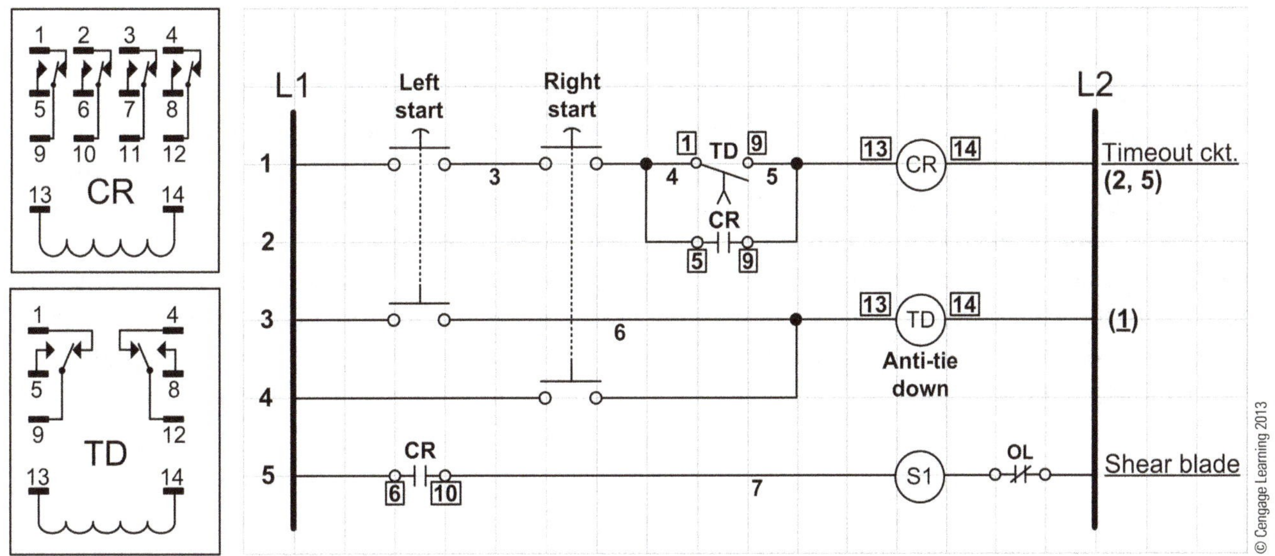


FIGURE 14-25 Circuit 25. Anti-tie down

inhibit contact in rung 2 prevents the normally open S1 contact in run 2 from creating a holding circuit.

## Control Circuit 25. Anti-Tie Down

The control circuit in Figure 14-25 is commonly used to protect process machine operators where they would be injured seriously if the process started when they have their hands in the wrong place. The idea is that the two start pushbuttons would be located in a safe area, and that both of the operator's hands would be needed to start the process. "Anti-tie down" means that the operator could not tape over one of the pushbuttons to hold it down, and then use the other pushbutton to make it a one-hand operation. Pressing just one of the start pushbuttons will not energize control coil CR, but it will start the time delay timing. If the second start pushbutton is not pressed before the normally closed TD time delay contact in rung 1 opens, control coil CR cannot be energized. Anti-tie down circuits do not use holding contacts, so the motor only runs as long as both hands are holding the pushbuttons down.

## Control Circuit 26. Sump Pump Control, Using a Hand-Off-Automatic Selector Switch

The control circuit in Figure 14-26 shows a control circuit for a sump pump control, using a hand-off-automatic (HOA) selector switch. The purpose of this control circuit is to turn the sump pump motor on manually in the hand position, and allow the floats to pump out the sump in the automatic position. In the automatic position, the floats work very much like a three wire stop-start station. When the water level in the sump rises above the low float switch, it acts like the stop pushbutton of a three-wire control, passing control circuit power on to the rest of the rung. When the water reaches the high float switch, control coil M1 will become energized and the M1 holding contacts in rung 3 will close. With the holding contacts closed, control coil M1 will remain energized after the high float switch opens; it will remain energized until the water level dips below the low level float switch, which will de-energize the rung similarly to the stop pushbutton of a three-wire circuit.

## Control Circuit 27. Ladder Rung Control Zone Relay

A ladder rung control zone relay is a means of disconnecting control power to only a portion of the control circuit, normally only a rung or two. The control zone relay circuit shown in Figure 14-27 operates by placing a normally open contact of a relay in the rung that feeds a portion or a zone in the control circuit, which allows the rest of the ladder diagram around that rung to operate as usual. All of the control components beyond the control zone relay contact will be de-energized until the control zone relay control coil start pushbutton is pressed. The indicator lamp in rung 4 will light as soon as the master control relay contact in rung 4 closes, to inform the operator that the controlled portion of the control circuit is energized and ready to run.

## Control Circuit 28. Ladder Rail Master Control Relay

The master control relay (MCR) contact in Figure 14-28 is wired in the L1 power rail of the ladder diagram. The purpose of a master control relay is to provide a method of disconnecting the control circuit power to part, or all, of the rungs of a large, ladder-diagram control circuit. The reasons for using MCRs are usually safety-related, such as an emergency stop to disconnect control power from all rungs with the action of a single emergency stop pushbutton. Whereas the control zone

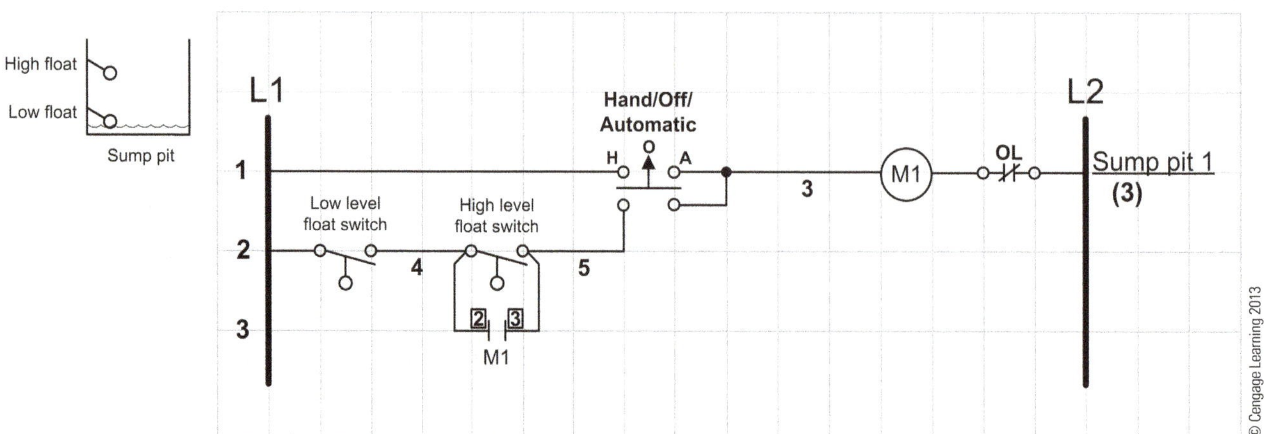



FIGURE 14-26 Circuit 26. Sump pump control using a hand-off-automatic selector switch

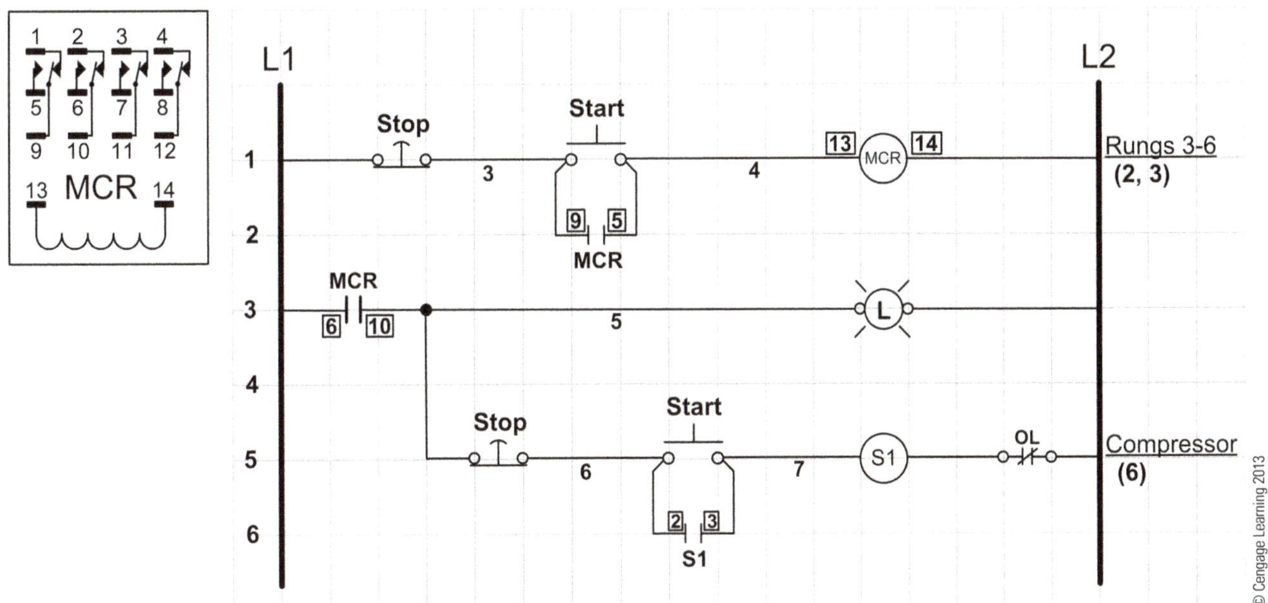


FIGURE 14-27 Circuit 27. Ladder rung control zone relay

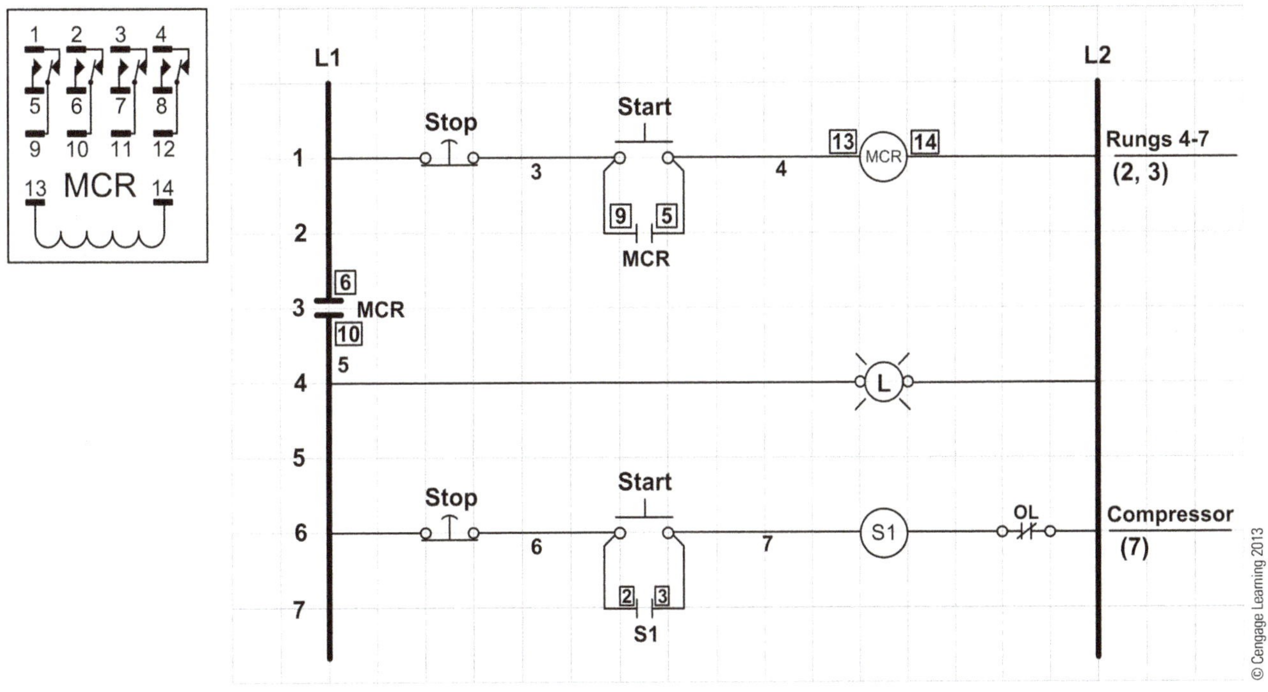


FIGURE 14-28 Circuit 28. Ladder rail master control relay

relay is meant to control the power to only a portion of the rungs in a larger ladder diagram of many rungs, the master control relay is meant to control all the rungs in the ladder diagram beyond the master control relay contact. The indicator lamp in rung 4 will light as soon as the master control relay contact in rung 3 closes, to inform the operator that the control circuit is energized and ready to run.

## Control Circuit 29. Reversing Motor Starter Controlled from Two Locations

The operation of this control circuit shown in Figure 14-29 is not much different than the single pushbutton reversing motor starter in circuit 18, with a second pushbutton station added. Similar to the single motor starter with two pushbutton

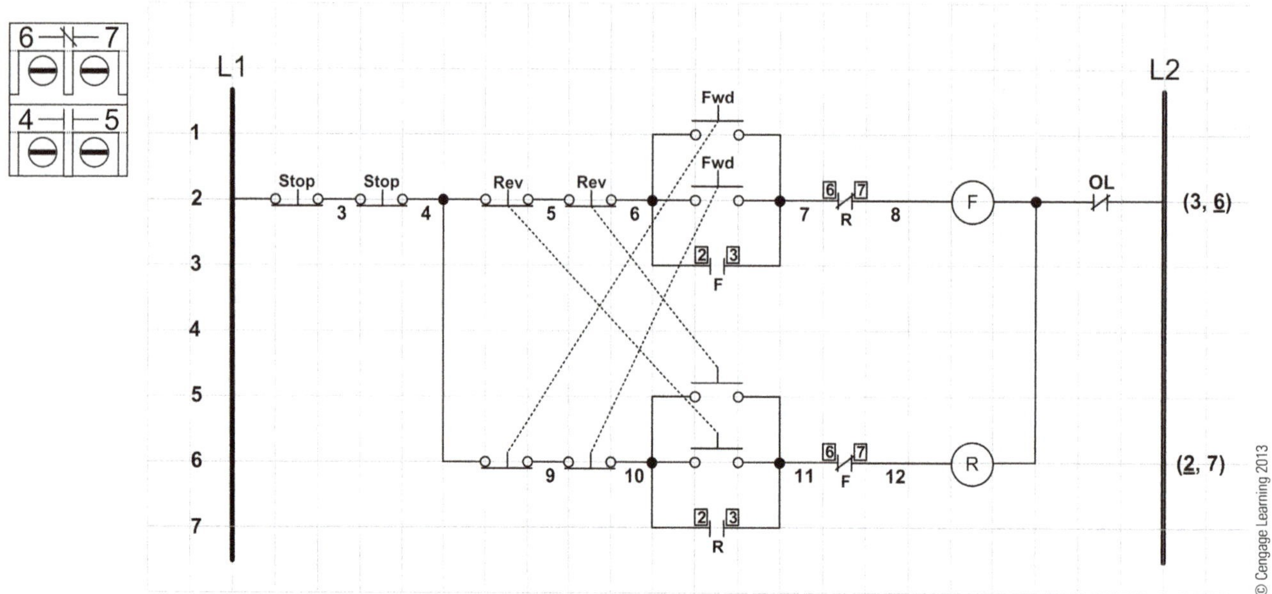


FIGURE 14-29 Circuit 29. Reversing motor starter controlled from two locations

control stations in circuit 4, all of the stop pushbutton contacts from all of the control stations must be wired in series at the start of the rung; and all of the start pushbuttons and the holding contact on the contactor unit are wired in parallel, for both the forward and reverse functions. One difference between this circuit and circuit 18, is that this circuit utilizes two, two-circuit pushbuttons to add an extra level of interlock protection to prevent the forward and reverse functions from occurring at the same time. Each pushbutton has a normally open contact to enable, or start, the function of the rung it is wired in, but each pushbutton also has a normally closed contact wired in the rung of the opposite function to inhibit or prevent it from happening at the same time. This control circuit still requires the normally closed inhibit contacts actuated by the contactor armature to prevent both the forward and reverse control coils from being energized at the same time once the start pushbutton is released, but the two-circuit pushbutton provides the interlock protection until the contactor actuates and opens the respective inhibit contact.

## Control Circuit 30. Jogging, Reversing Motor Starter Controlled from Two Locations

The operation of this control circuit shown in Figure 14-30 is similar to the control relay jog in circuit 14, except that this circuit provides the jog function for both the forward and reverse rotations of a reversing motor starter. When the forward start (Fwd start) pushbutton is pressed, the forward control coil of the starter will become energized. When the forward control coil energized, the forward holding contacts between rungs 2 and 3 will close, which will energize the control relay (CR) control coil, and close the CR contacts in rung 3 to enable the forward holding contacts circuit to keep the forward control coil energized when the forward start pushbutton is released. Notice that the jog pushbuttons for both forward and reverse are two-circuit types, with one normally open and one normally closed contact each. When the normally open contact of the forward jog (Fwd jog) in rung 1 is closed the forward control coil will become

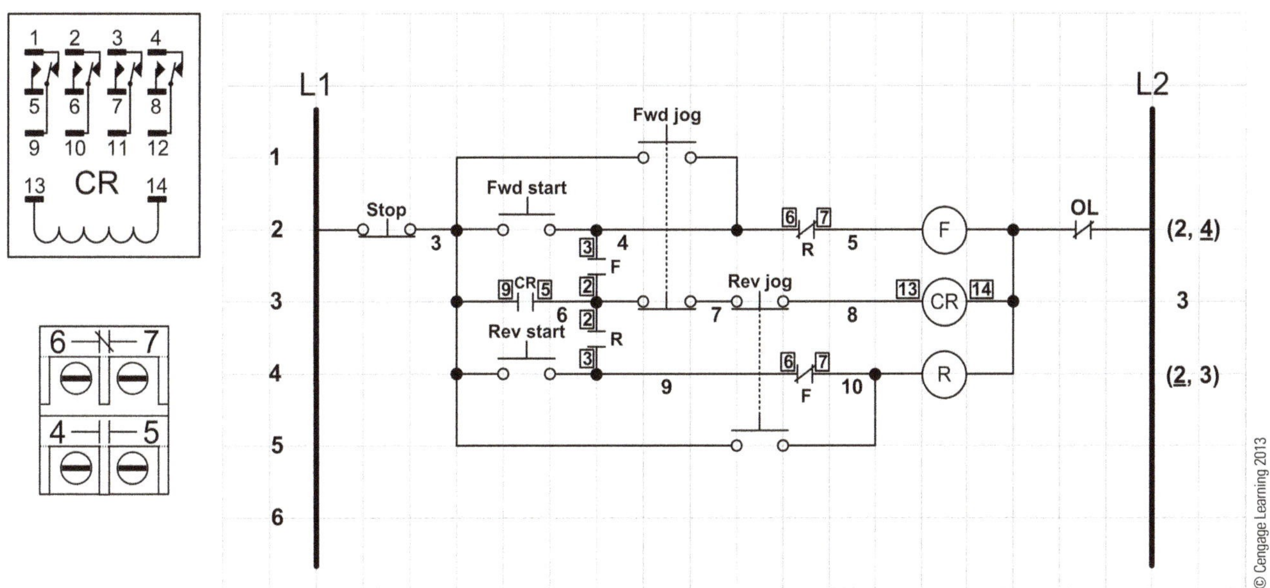


**FIGURE 14-30** Circuit 30. Jogging, reversing motor starter controlled from two locations

energized, but at the same time the normally closed contact of the forward jog pushbutton in rung 3 will open and prevent the control relay control coil from becoming energized. Without the normally open CR contact in rung 3 closing, the holding circuit for the forward control coil is inhibited and the forward control coil will become de-energized when the forward jog pushbutton is released.

The operating principle for both the forward and reverse function of the control circuit is the same. The control relay (CR) enables the holding circuits for both the forward and reverse control coils; without the control relay being energized the contactor control coils will become de-energized when the pushbutton is released. Pressing either of the start pushbuttons will energize the control relay and its respective contactor control coil, and pressing either of the jog pushbuttons will only energize its respective contactor control coil.

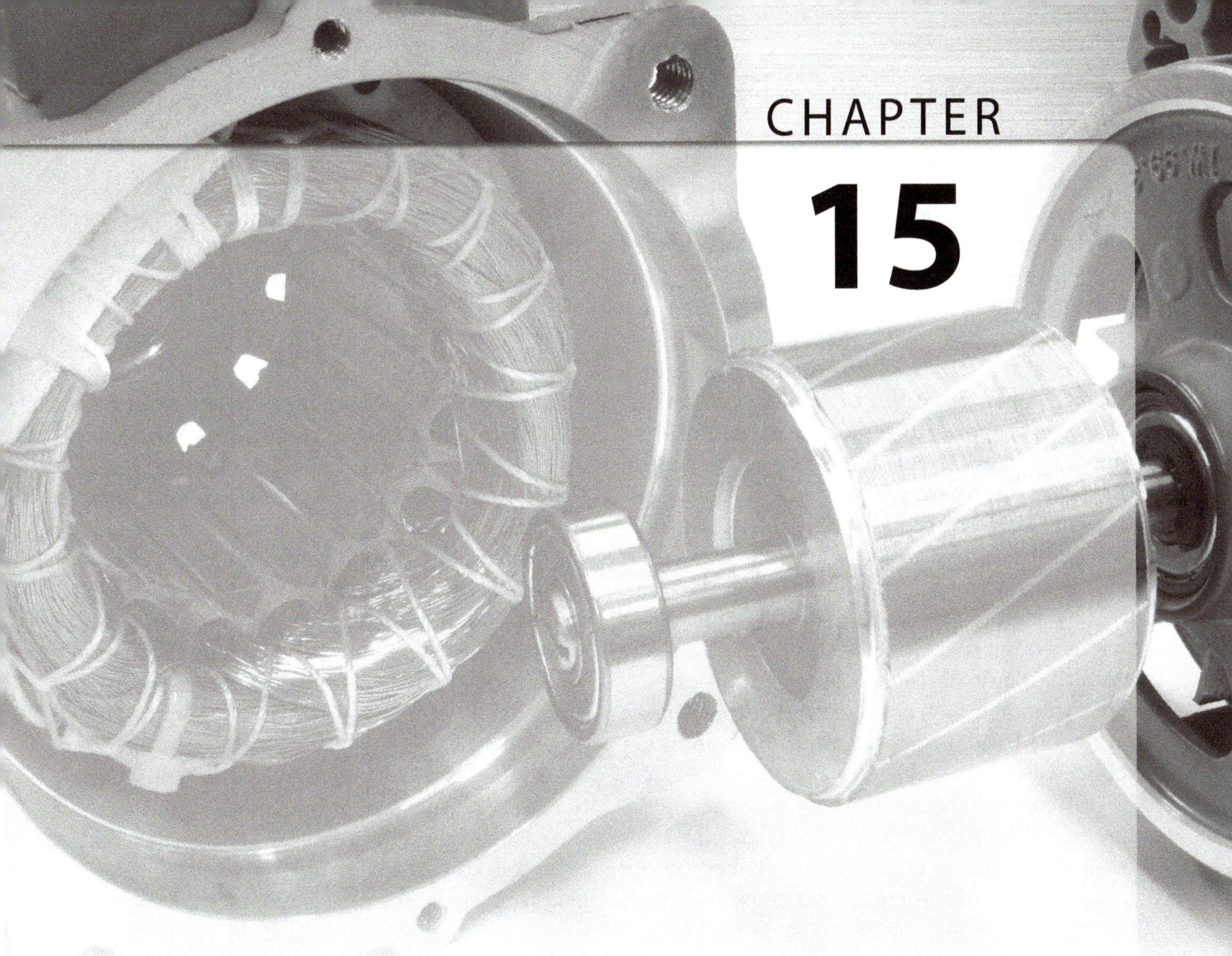

CHAPTER 15

# Variable-Frequency Drive Principles

## PURPOSE

Explain what a variable-frequency drive (VFD) is, describe some of the unique operating parameters of VFDs that differentiate them from magnetic motor starters, and explain the advantages of using them for electric induction motor control.

## OBJECTIVES

After studying this chapter on control circuits the learner will be able to:

- Describe the function and use of a VFD
- Describe soft start, as opposed to across-the-line, full-voltage starting
- Discuss additional motor protection features of VFDs
- Explain the types of electric motors suitable for use with VFDs
- Explain pulse-width modulation (PWM)
- Explain VFD modes of operation (constant torque, constant horsepower, and variable torque)
- Explain the quadrants of motor operation
- Explain regeneration
- Discuss the energy savings of VFD variable torque applications

A formal definition of a variable-frequency drive is a motor controller that controls the primary supply frequency of the electrical power delivered to a three-phase electric motor to control the rotational speed. Less formally, VFDs are electronic motor controllers that, in most applications, can replace traditional magnetic motor starters. VFDs have the capability of changing all electrical power parameters delivered to the motor voltage, current, and frequency to control the motor speed, torque, and acceleration. Controlling the speed of an electric motor not only provides the optimum speed for a particular process, it also allows the motor to "soft" start and stop by ramping the motor speed up and down over a slower specified period of time. In some cases the ramp-start-and-stop operation can extend equipment life, because mechanical stresses on the load and inrush currents of the motor are reduced.

When using the magnetic contactor across the line full-voltage motor starting method, the electrical supply power to the motor is either full on, or completely off. There are no provisions with the magnetic motor starter to change the voltage, current, or frequency delivered to the motor. The across-the-line, full-voltage starting inrush current of a NEMA design B induction motor is about six times higher than the motor's full load current. The starting torque can be 150% of the motor's full-load torque, and the only stop methods available are plugging, and allowing the motor to coast to a stop.

## ADVANCED MOTOR PROTECTION FEATURES

Unlike magnetic motor starters, VFDs have sophisticated electronic circuits that are capable of measuring and monitoring many different electrical motor operating parameters, and protecting motors from many more-damaging power supply conditions. In addition to providing motor overload current protection, many VFDs also monitor and protect for power factor, phase voltage imbalance, motor current imbalance, phase loss, phase reversal, short circuit, ground fault, low voltage, over voltage, power quality, over torque, motor over temperature, and single-phase conditions.

### Three-Phase Motors Only

Most of us think of motor speed control in the context of our motor experience in our homes, all of which are single-phase motors. It is estimated that the average home has about 25 electric motors, and in some cases the motor speed is controlled to fit an application, such as a kitchen appliance or a fan. These usually are specialized, fractional-horsepower motors (universal, shaded-pole, etc.), not the standard induction motors demanded by industry for ruggedness, dependability, simplicity, cost, and application power reasons.

Three-phase motors have stator coils positioned at 120 degrees to one another, and as the three electrical supply phases sequentially energize each coil around the stator, a rotating magnetic field is created. Single-phase motors do not form this same rotating magnetic field. They require additional components to establish a rotating magnetic field: centrifugal switch, capacitor for phase shifting, and an auxiliary start winding. Because of these differences in construction and operating characteristics of single-phase and three-phase motors, only three-phase motors are suitable for use with VFDs.

### VFD Component Parts

VFDs first take the supply power, either single-phase or three-phase, and rectify it into a DC voltage. This DC power becomes the input for the inverter, which turns it into three-phase power with a controlled frequency. This process is accomplished through the four stages of the VFD listed below. Some VFDs have additional components, such as soft charging circuits, filtering inductors, and brake circuits, but this study will concern itself with only the following four components or stages.

**Rectifier (sometimes called a converter).** This is the stage of the VFD that rectifies the AC input power to DC power. Some VFDs include additional filtering circuitry to negate any harmonic conditions that could cause power supply problems.

**DC Bus (sometimes called the DC link).** The DC bus stage of the VFD prepares the DC power from the rectifier stage for use by the inverter stage.

**Inverter.** The input to the inverter is the DC power on the DC bus. The inverter, under control of the control and regulation stage of the VFD, changes the DC bus power to a controlled frequency AC on the output.

**Control and Regulation.** In this section complex mathematical algorithms are used to separate the torque-producing current from the magnetizing current. The control and regulation section of the VFD senses the counter-electromotive force of the motor to make adjustments and regulate the inverter section output.

## VFD Output

Different technology methods of AC motor speed control drive inverters are available today, but the focus of this study will be on only one: the pulse-width modulation (PWM) type AC drive. For general AC motor speed control, the PWM drive is considered the best combination of simplicity, performance, and economy.

## Pulse-Width Modulation (PWM)

The inverter section does not actually produce a sine wave output, because it has only two states it can output: zero volts, and the DC link voltage. A common inverter PWM carrier frequency for a VFD is 12 kHz. What this means is that each second of inverter output is divided into 12,000 segments of roughly 80 microseconds each that can be either on or off. For 60 Hz you would determine that each ninety degrees of the waveform would be broken into approximately fifty-two time segments that could be either on or off. As the output frequency is increased, each cycle waveform will be comprised of fewer oscillations, and as the output frequency is decreased, each cycle waveform will be comprised of more oscillations.

At the beginning of the cycle, to produce a lower effective output voltage, fewer consecutive segments would be turned on. As longer output times are needed to increase the output voltage, more consecutive segments would be turned on. The waveform in Figure 15-1 shows how the inverter section turns the DC link voltage on for shorter and longer periods of time to change the effective output voltage.

The output of the inverter is not a sine wave, but the sine waves in the drawing shown in Figure 15-2 are superimposed on the inverter output to show how the shorter and longer on-times of the inverter would build the semblance of a sine wave output.

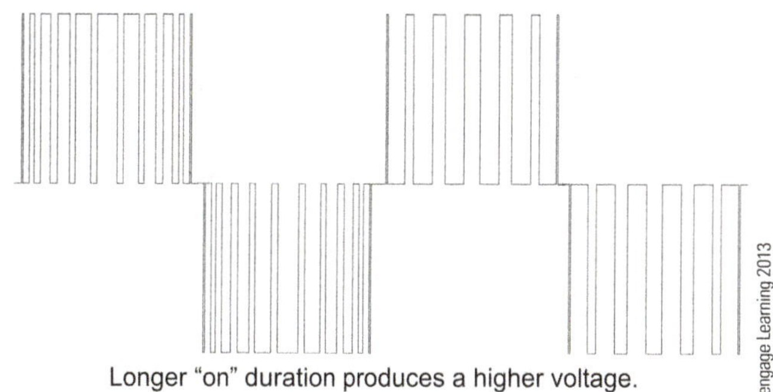

Longer "on" duration produces a higher voltage.
Shorter "on" duration produces a lower voltage.
© Cengage Learning 2013


**FIGURE 15-1** PWM output

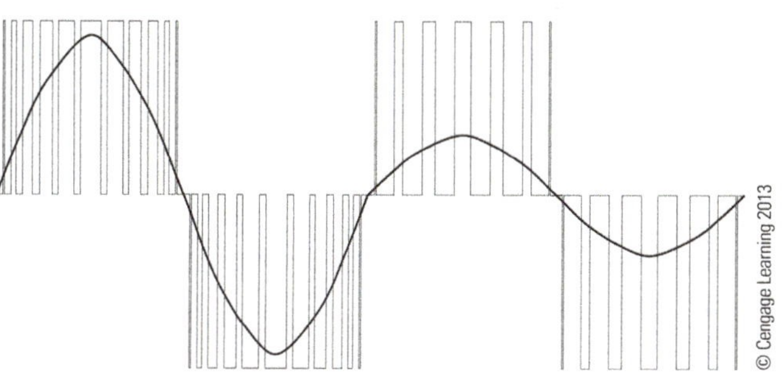


**FIGURE 15-2** PWM output with sine waves

## NEMA Design B Motors

VFDs can operate any three-phase induction motor, but they usually are designed for the operating characteristics of NEMA design B motors. However, some older design B motors may not be suitable for operation with a VFD. There are really two problems with operating an older, non-inverter-rated motor with a VFD. First of all, the inverter output is not a smooth sine wave waveform. It is a series of very fast-rising voltage spikes caused by the fast switching speeds of the inverter electronics, which can break down the insulation on some motors and cause them to fail prematurely. Most new motors have a PWM Inverter rating on the nameplate to indicate that the motor insulation is suitable for this type of voltage waveform.

Second, the cooling ventilation of motors manufactured before the advent of VFDs is designed around the airflow of the integral fan at full RPMs. Where the application requires 100% rated torque at speeds below 50% of synchronous speed for extended periods of time, a separately powered auxiliary ventilation blower should be used to supplement air circulation for cooling purposes. Most new motors purchased now are inverter rated, which means they are designed to provide adequate cooling even at lower speeds.

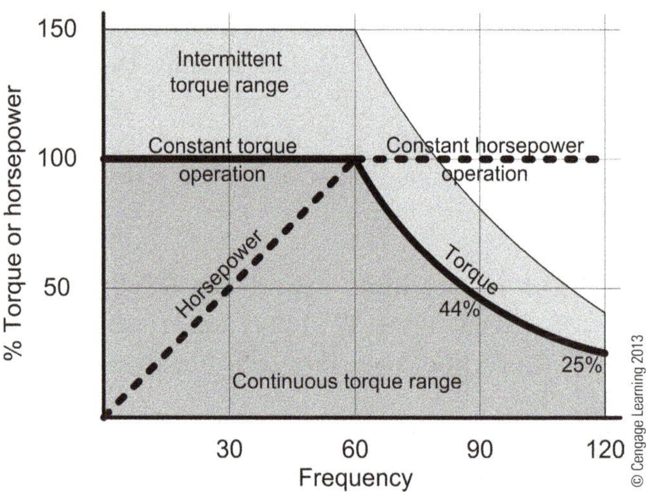


FIGURE 15-3 Frequency-torque chart

## Modes of Operation

In VFD technology horsepower and torque both become controllable parameters with infinite possibilities. One benefit of VFD technology is that the motor can be operated at different speeds above and below its rated synchronous speed. As shown in the chart of Figure 15-3, when a motor is operated at or below its rated frequency it is in the constant torque range; above its rated frequency it is in the constant horsepower range.

## Constant Torque (volts-per-Hertz, or constant flux)

Excluding pump and fan applications, the majority of all general industrial machines are constant torque applications. Constant torque applications require the same amount of driving torque throughout the entire operating speed range. As the speed changes the load torque remains the same. These loads include: conveyors, hoists, drill presses, extruders, positive-displacement pumps.

Refer to the chart in Figure 15-3 and note that up to 60 Hz, the torque is constant: the constant torque range. To maintain constant torque at all speeds, the motor must maintain constant magnetic flux in the rotor-stator air gap, as motor torque is directly proportional to that magnetic flux density. To accomplish constant flux, and thereby constant torque, the VFD utilizes the volts-per-Hertz principle.

The volts-per-Hertz principle is basically a ratio of the voltage applied to the motor, divided by the frequency applied to the motor. For example, a 230 volt motor rated for 60 Hz has a volts-per-Hertz ratio of 3.8 (230 volts/60 Hz), and a 460-volt motor rated for 60 Hz has a volts-per-Hertz ratio of 7.6. As long as the VFD can keep the voltage and frequency ratio delivered to the motor the same through the operating range of the motor, the flux density in the rotor-stator gap will remain constant. The formula that determines magnetic flux density is more complicated, but may be simplified to only the voltage and frequency variables:

$$\text{Flux} = \text{voltage/frequency.}$$

Refer to the constant torque range of the graphic in Figure 15-3, which is any frequency at or below the rated 60-Hz frequency of the motor. The VFD changes the output frequency to the motor to change the speed. If the VFD also changes the output voltage in proportion to the frequency change, the volts-per-Hertz ratio will remain constant. When the frequency is ramped down below 60 Hz the voltage can be reduced to maintain the volts-per-Hertz ratio, but when the frequency is increased above 60 Hz, the volts-per-Hertz ratio will decrease. As the VFD output frequency moves above 60 Hz, there is no corresponding increase in voltage beyond the rated supply voltage available to support the ratio.

## Constant Horsepower Operation (field weakening)

The mode of operation for motors operated above their rated frequency is called constant horsepower, or field weakening. Constant horsepower mechanical loads are applications where the torque loading is an inverse function of the speed: low torque at high speeds, and higher torque at lower speeds. These loads include machines such as grinders and lathes.

To explain constant horsepower operation, the example of a lathe will be used. Refer to the chart and examine the constant horsepower range. Note that as frequency increases in this range, torque decreases. With a lathe, the material speed at the cutting tool must remain at an optimum speed. As the material radius is reduced, the motor speed would need to increase to keep the speed at the cutting tool constant. Reducing the radius by half will cause a corresponding reduction in needed torque. A doubling of speed, and a reduction of torque by half, causes horsepower to remain constant. It may help to look at the horsepower formula to understand this concept:

Horsepower = torque*speed/5250.

## Variable Torque

Variable torque loads is one application where VFDs really shine. Variable torque mechanical loads are machine applications where the torque loading is a function of the speed: low torque at low speeds, and higher torque at higher speeds. The torque increase with speed, however, is not linear; to operate a mechanical variable torque load at a higher speed, a disproportional increase in power is needed to increase torque. These loads include fans, blowers, propellers, and centrifugal pumps.

## Energy Savings

In variable torque applications, varying the speed of an electric motor is usually more about saving energy than controlling the exact speed of a process. Much of the electrical energy consumed by AC motors in industry is for variable torque loads, such as fans and pumps. In these motor applications where exact motor speed regulation is not necessary it is possible to save energy on variable torque loads by slowing the motor speed when the application demand decreases.

In the case of a cooling tower for example, many times it is possible to change the speed of the cooling fan with the temperature of the water in the tower and still achieve the desired result. There will be times that the fan must be run faster to cool very warm water in the tower, but there will be other times when the fan could be run slower to cool water that is less warm. The act of reducing the motor speed when the application demand decreases can significantly reduce the operating costs of the motor.

The principles that demonstrate the effect of motor speed on pump and fan flow, pressure, and horsepower are defined by the affinity laws. The affinity laws show that for variable torque loads, if the motor can be slowed when the load decreases, energy can be saved. The affinity laws state:

1. Flow (air, liquid, etc.) produced by the motor is proportional to the motor speed.
2. Pressure produced by the motor is proportional to the motor speed squared.
3. Horsepower (power) required by the motor is proportional to the motor speed cubed.

It might be assumed that reducing the motor speed to 50% would cause 50% flow and consume 50% of the full speed power, but that would be wrong. The third affinity law tells us that horsepower (which can be expressed in electrical power) is proportional to speed cubed, meaning 50% of full speed would not equal 50% of full speed HP (50% savings), or even 25% of full speed HP (75% savings). No, the cubed factor between speed and horsepower means that 50% of full speed would equal 12.5% of the full speed horsepower, an 87.5% power savings.

There is no other energy savings method that even comes close to saving this much electrical energy. Even using the highest efficiency motors will only produce energy savings of 2 to 6% over standard electric motors. When the motor speed is controlled on variable torque applications, it is common to have overall energy savings of 30% for the application. With some estimates that up to 75% of all electricity consumed in industrial applications is motor related, utilizing motor speed control has the potential to save a lot of energy.

## Quadrant Control

Driving, reversing, and braking motors is explained by quadrants of control, as shown in Figure 15-4. Quadrants 1 and 3 are referred to as motoring quadrants, because both torque and rotation are in the same direction; and quadrants 2 and 4 are referred to as braking quadrants, because torque and rotation are in opposite directions. When the drive is operating in quadrants 1 or 3 the motor functions as a driving force: quadrant 1, forward rotation, and quadrant 3, reverse rotation. When the drive is operating in quadrants 2 and 4 the motor torque opposes the direction of motor rotation, which provides a controlled braking or retarding force called regeneration, or regenerative braking.

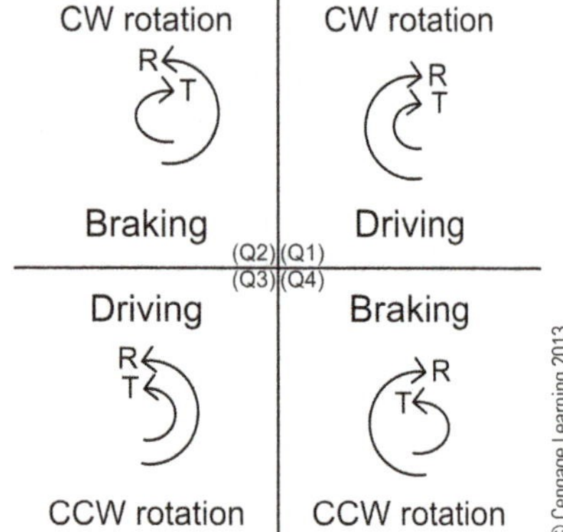


**FIGURE 15-4** Quadrant diagram

## Regeneration

Regeneration is what happens when the induction motor stops functioning like a motor and starts functioning like a generator instead. Regeneration only can happen in an induction motor when two conditions are present: the motor is energized with an AC power supply, and the motor is driven faster mechanically than the synchronous speed of the AC power supply. If a VFD is driving a motor at 60 Hz and then abruptly changes the output frequency to 40 Hz, the motor will be turning faster than the synchronous speed of the 40-Hz signal and the counter-torque of regeneration quickly will slow the motor below the synchronous speed of the 40-Hz supply.

When the controlled ramp to stop is used with the VFD, it may be used to stop the load either faster or slower than the load normally would take to coast to a stop. If the ramp to stop time is set longer than the load would take to coast to a stop, the VFD actually would remain in quadrant one operation, motoring the mechanical load to a stop. If the ramp-to-stop time is set shorter than the load would take to coast to a stop, the VFD would change to quadrant two operation, braking as soon as the load tried to turn the motor rotor mechanically faster than the synchronous speed of the rotating magnetic field.

A magnetic motor starter is considered a single-quadrant control device, because it can operate the motor in quadrant 1, but it cannot operate in any of the other three quadrants because it has no reversing or braking capabilities. A magnetic reversing starter is a two-quadrant control device, because it can operate the motor in either direction of rotation, but it has no braking capabilities. VFDs have the ability to operate induction motors in all four of the control quadrants.

## DC Injection Braking

The DC injection braking mode applies a DC current to the stationary stator windings, replacing the rotating stator field with a stationary DC field; in effect, a 0-Hz power supply. Since the motor's stator field is stationary, the rotor is mechanically driven to turn faster than the stationary magnetic field, and regeneration as described above will cause the motor to stop very quickly. This method of stopping a motor load most likely would be used for safety reasons to stop a low-friction, high-inertial mechanical load that would otherwise continue to coast for a long period of time.

## CHAPTER SUMMARY

- VFDs control the primary supply frequency of the electrical power delivered to a three-phase electric motor to control the rotational speed.
- VFDs are electronic motor controllers that replace traditional magnetic motor starters.
- VFDs have the capability of changing all electrical power parameters delivered to the motor—voltage, current, and frequency—to control the motor speed, torque, and acceleration.
- VFDs have sophisticated electronic circuits that are capable of measuring and monitoring many different electrical motor-operating parameters, and protecting motors from many more-damaging power supply conditions.
- VFDs are only suitable for operating three-phase motors.
- "Soft start" is the name given to the motor starting method when the motor is ramped up to speed at a slower specified period of time.
- VFDs have four stages: rectifier, DC bus, inverter, and control and regulation.
- The pulse width modulation (PWM) output drive is considered the best combination of simplicity, performance, and economy.
- VFDs can operate any three-phase induction motor, but they usually are designed for the operating characteristics of NEMA design B motors.
- Constant torque motor applications require the same amount of driving torque throughout the entire operating speed range (examples: conveyors and hoists).
- To maintain constant torque at all speeds, the motor must maintain constant magnetic flux in the rotor-stator air gap, as motor torque is directly proportional to that magnetic flux density.
- Magnetic flux is determined by the volts-per-Hertz principle, which is basically a ratio of the voltage applied to the motor, divided by the frequency applied to the motor.
- Operation of a motor above its rated speed, where constant magnetic flux cannot be maintained, is called constant horsepower operation, or field weakening.
- Constant horsepower loads are applications where the torque loading is an inverse function of the speed: low torque at high speeds, and higher torque at lower speeds (examples: grinders and lathes).
- Variable torque mechanical loads are machine applications where the torque loading is a function of the speed: low torque at low speeds, and higher torque at higher speeds (examples: fans and centrifugal pumps).
- In motor applications where exact motor speed regulation is not necessary, it is possible to save energy on variable torque loads by slowing the motor speed when the application demand decreases.
- The affinity laws show that the power required by a motor is proportional to the motor speed cubed; a reduction in motor speed to 50% would mean a power reduction to 12.5% of the full speed power.
- Four quadrants for motor rotation control are defined by a motor's direction of rotation and direction of torque. Quadrants 1 and 3 are referred to as motoring quadrants, because both torque and rotation are in the same direction; and quadrants 2 and 4 are referred to as braking quadrants, because torque and rotation are in opposite directions.
- Regeneration is what happens when the induction motor stops functioning like a motor and starts functioning like a generator instead.
- Regeneration only can happen in an induction motor when the motor is energized with an AC power supply, and the motor is driven faster mechanically than the synchronous speed of the AC power supply.
- DC injection braking mode applies a DC current to the stationary stator windings, replacing the rotating stator field with a stationary DC field; in effect, a 0-Hz power supply.

## REVIEW QUESTIONS

1. What does the acronym VFD stand for?
2. What magnetic technology component does a VFD replace?
3. Which of the electrical power parameters delivered to the motor voltage, current, and frequency can a VFD control?
4. What does the term "soft" start mean?
5. What are some additional motor protection features of VFDs that magnetic motor starters do not have?
6. Why are three-phase induction motors demanded for industrial applications?
7. Are single-phase induction motors compatible with VFD speed control?
8. What are the four stages of a VFD?
9. Which type of VFD output is considered the best combination of simplicity, performance, and economy?
10. With PWM, what are the only two output voltage states that the inverter switches between?
11. In order to produce a higher effective output voltage with PWM, does the inverter have to leave more or fewer consecutive time segments turned on?
12. What are two problems with operating older, non-inverter-rated motors with a VFD?
13. What are the three modes of VFD operation?
14. What are examples of mechanical constant torque loads listed in the text?
15. To maintain constant torque at all speeds, the motor must maintain what in the rotor-stator air gap?
16. To maintain constant flux in the rotor-stator gap of a motor, the ratio between what two electrical parameters must be kept constant?
17. What would the volts-per-Hertz ratio be of a 208-volt motor rated for 50 Hz?
18. What is the maximum frequency a motor can be operated at for constant torque operation?
19. Why can't constant torque operation be maintained above the rated frequency of the motor?
20. When a VFD operates a motor faster than its rated speed, what name is given to this mode of operation?
21. Why is the constant horsepower mode sometimes called the field-weakening mode?
22. What is the text's example of a mechanical constant horsepower load?
23. Pumps and fans are examples of mechanical loads that operate in what mode?
24. If a variable torque load can be slowed to 50% of its rated speed when the application demand decreases, how much savings will there be with the power consumption of the motor?
25. The energy savings of high-efficiency motors over normal-efficiency motors is estimated to be approximately how much?
26. How many quadrants of control can a VFD operate a motor in?
27. Quadrants 1 and 3 are referred to as __________ quadrants, and quadrants 2 and 4 are referred to as __________ quadrants.
28. In the braking quadrants, what is it called when the motor is used to generate electrical power to slow the load?
29. Regeneration can happen in an induction motor only when what two conditions are present?

    1. ______________________________

    2. ______________________________
30. How many quadrants of operation is a motor starter with no reversing capability and no braking capability?

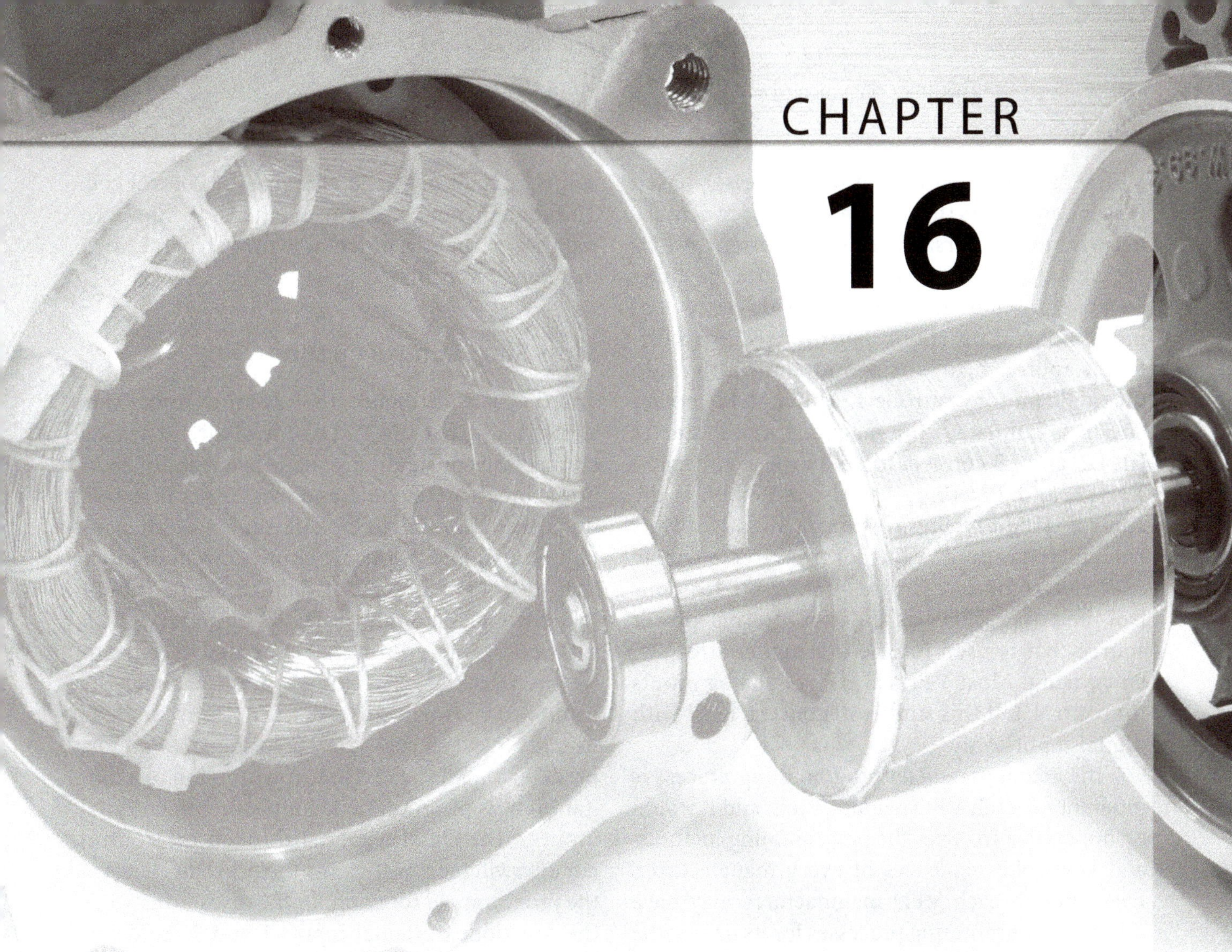


# Commissioning a VFD

## PURPOSE

To understand, generically, how the power and control circuits wiring of VFDs are connected, and the basic commissioning parameters commonly encountered with VFD commissioning.

## OBJECTIVES

After studying this chapter on control circuits the learner will be able to:

- Understand basic power and control circuit wiring for VFDs
- Understand basic operating parameters of VFDs
- Describe the use of default operating parameters
- Explain user-defined input and output
- Describe the digital input and relay output sections of a VFD
- Draw a control circuit on the terminal block from the user manual documentation
- Describe what commissioning a VFD entails

VFDs are electronic microprocessor-based motor controllers that may take the place of magnetic motor starters for most applications. VFDs are considerably more flexible because they may be programmed to perform a wide variety of motor control functions, but they require a different approach to wiring and control than do magnetic motor controllers. Although the power circuit is similar to the magnetic motor starter L1, L2, and L3 connect to the power source, and T1, T2, and T3 connect to the motor leads the control circuit wiring requires a slightly different understanding. VFD control circuit wiring uses electronic circuitry digital inputs for control, and the VFD has its own internal power source that is used for the external control circuit wiring. Most VFDs also have analog inputs for control, but this study will concern itself with digital inputs only.

The actual parameter programming process is different for each VFD manufacturer, and it would be impossible to cover the programming methods and keystroke sequences of every manufacturer. Even though each VFD manufacturer may have different programming processes for its drives, the actual functions between the different drives are very similar. The concepts covered in this chapter will be based generally on the Automation Direct GS2 VFD for example purposes. Any information about the GS2 model found in this chapter, however, cannot be considered reliable for actual commissioning purposes, because all VFD manufacturers continually update, revise, and modify their products to remain current.

## Digital Inputs

What is a digital input? Digital input simply means that the VFD uses electronic circuitry on the inputs that senses only two states: either the control voltage is present on the input, or it is not. Analog inputs are capable of sensing a gradient of different levels of voltage or current, but digital inputs are not. Even though it was not specifically discussed in this way, magnetic motors starters are also a form of digital control; the control circuit components are switches that either do or do not pass the control circuit power to energize the control coil. The digital input method of the VFD, however, requires a slightly different wiring scheme than magnetic motor starters.

## Stop-Start Control Circuit

The diagram in Figure 16-1 is an example that shows both that the VFD control circuit wiring connections normally are made by means of a terminal block on the unit, and what the connection documentation might look like in the user's manual for a simple stop-start control circuit. Notice that the VFD has six user-defined digital inputs, DI1 through DI6, and a DCM terminal, which is the digital common. User-defined input means that the user can program each of the inputs to perform different functions, such as jog, reverse, external reset, hold speed, or go to a predetermined speed. Once the function of a digital input is programmed, connecting that digital input to the DCM terminal through the control circuit switching will cause the VFD to perform the programmed function. In the case of this VFD, the functions of digital inputs 1 and 2 are determined by type of control circuit the user programs the unit for: two-wire control, three-wire control, latching, etc.

## Explanation of Circuit

The external connection of the stop-start pushbutton control circuit for the VFD looks similar to the control circuit of a magnetic motor starter, but the operation is slightly different. When the VFD is powered, the DCM voltage is passed by the normally closed stop pushbutton to digital input 3. In the case of this VFD's configuration, digital input 3 is an enable function; without the DCM voltage on digital input 3 the VFD will not respond to any control commands. Once the VFD is enabled, pressing the start pushbutton will cause the unit to start and operate in response to the programmed operating parameters. Pressing the stop pushbutton will again disable the VFD, but bring the motor to a stop by the programmed stop parameters.

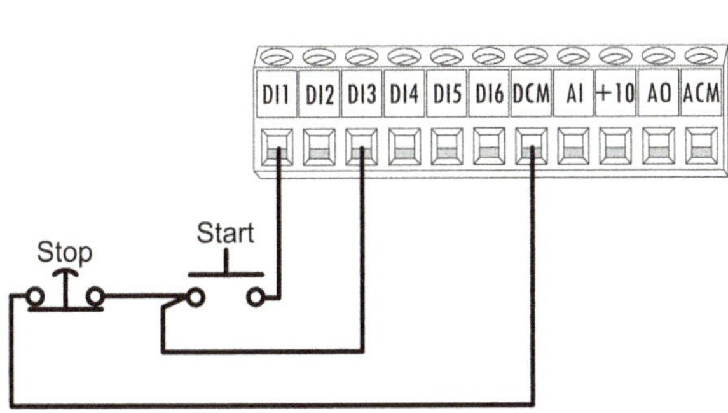


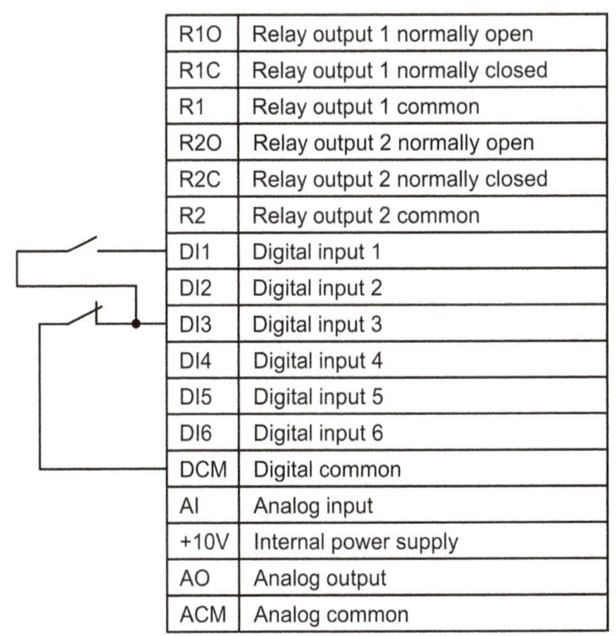

| | |
|---|---|
| R1O | Relay output 1 normally open |
| R1C | Relay output 1 normally closed |
| R1 | Relay output 1 common |
| R2O | Relay output 2 normally open |
| R2C | Relay output 2 normally closed |
| R2 | Relay output 2 common |
| DI1 | Digital input 1 |
| DI2 | Digital input 2 |
| DI3 | Digital input 3 |
| DI4 | Digital input 4 |
| DI5 | Digital input 5 |
| DI6 | Digital input 6 |
| DCM | Digital common |
| AI | Analog input |
| +10V | Internal power supply |
| AO | Analog output |
| ACM | Analog common |





**FIGURE 16-1** VFD stop-start pushbutton

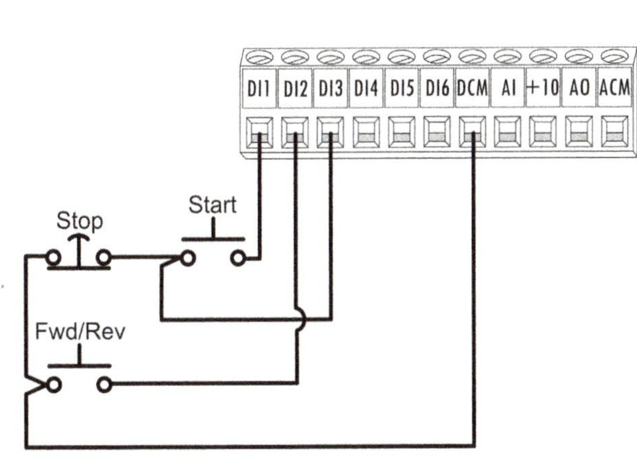


| | |
|---|---|
| R1O | Relay output 1 normally open |
| R1C | Relay output 1 normally closed |
| R1 | Relay output 1 common |
| R2O | Relay output 2 normally open |
| R2C | Relay output 2 normally closed |
| R2 | Relay output 2 common |
| DI1 | Digital input 1 |
| DI2 | Digital input 2 |
| DI3 | Digital input 3 |
| DI4 | Digital input 4 |
| DI5 | Digital input 5 |
| DI6 | Digital input 6 |
| DCM | Digital common |
| AI | Analog input |
| +10V | Internal power supply |
| AO | Analog output |
| ACM | Analog common |



**FIGURE 16-2** VFD stop-start-reversing

## Forward-Reverse Control Circuit

To expand on the definable digital inputs, the stop-start pushbutton control circuit may be expanded to include reversing the motor by simply adding a reversing selector switch, as shown in Figure 16-2. The VFD stop-start pushbutton control shown in Figure 1 also defines digital input 2 as a reversing control. Reversing the motor with this VFD is as simple as inserting a selector switch between the DCM and digital input 2 terminals. When the DCM voltage is not present on digital input 2, the VFD will operate the motor in the forward direction of rotation when started with the start pushbutton. When the DCM voltage is present on digital input 2, the VFD will operate the motor in the reverse direction of rotation.

## Forward-Reverse-Jog Control Circuit

Figure 16-3 is yet another example of how digital input 6 could be programmed to perform the jog function. After programming the VFD to recognize digital input 6 for the jog function, the external control circuit must be wired to connect digital input 6 with the DCM terminal through a jog pushbutton.

## Relay Outputs

The diagram in Figure 16-4 shows that this VFD has two user-defined relay outputs, and how they could be used. In this example the relay outputs are being used to control 120 volt indicator lamps to report the operating status of the VFD. Notice that the relay output contacts for this VFD have transfer contacts; a common terminal with both normally open and normally closed contacts. The relay outputs are similar to the auxiliary contacts on magnetic motor starters, which change states with the movement of the armature. The main difference between the output relays of VFDs and auxiliary contacts of magnetic motor starters is that the user may define which operating parameter the relays are programmed to report, such as drive running, drive fault, at speed, or above desired current.

## Commissioning a VFD

The term "commissioning" simply means to put into service. Commissioning a VFD involves both the wiring connections and the operating parameter programming necessary to make the VFD functional for a particular application. The GS2 VFD has more than a hundred individual operating parameters that may be programmed, but it would not be acceptable to require the user to program all of them for a basic commissioning. Most VFDs will boot with many of the operational parameters preset to default values that the manufacturer believes will make the VFD more user friendly, and facilitate the commissioning process.

Even though they may not be the exact operating parameters necessary for every particular

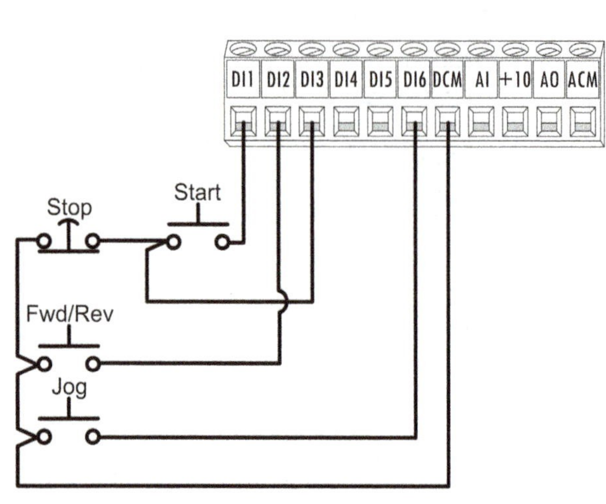


FIGURE 16-3 VFD stop-start-reversing-jog

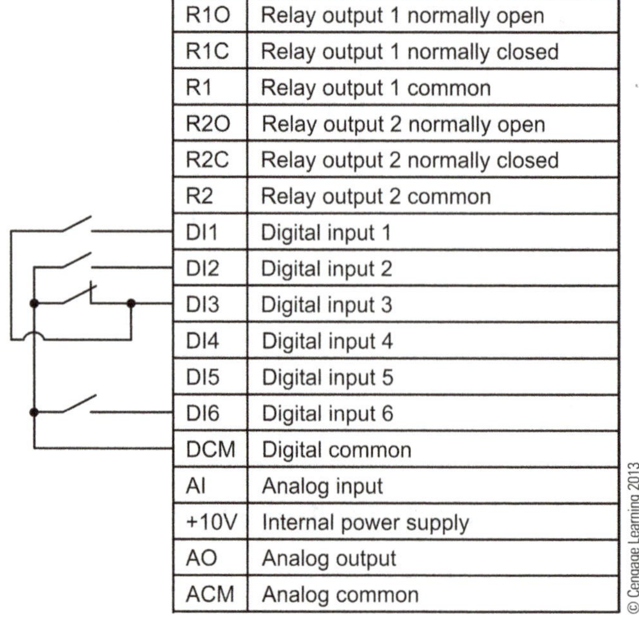

| | |
|---|---|
| R1O | Relay output 1 normally open |
| R1C | Relay output 1 normally closed |
| R1 | Relay output 1 common |
| R2O | Relay output 2 normally open |
| R2C | Relay output 2 normally closed |
| R2 | Relay output 2 common |
| DI1 | Digital input 1 |
| DI2 | Digital input 2 |
| DI3 | Digital input 3 |
| DI4 | Digital input 4 |
| DI5 | Digital input 5 |
| DI6 | Digital input 6 |
| DCM | Digital common |
| AI | Analog input |
| +10V | Internal power supply |
| AO | Analog output |
| ACM | Analog common |

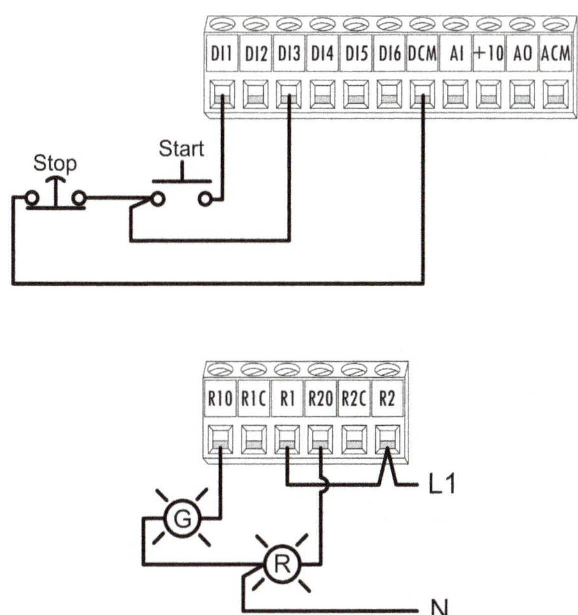


| R1O | Relay output 1 normally open |
|---|---|
| R1C | Relay output 1 normally closed |
| R1 | Relay output 1 common |
| R2O | Relay output 2 normally open |
| R2C | Relay output 2 normally closed |
| R2 | Relay output 2 common |
| DI1 | Digital input 1 |
| DI2 | Digital input 2 |
| DI3 | Digital input 3 |
| DI4 | Digital input 4 |
| DI5 | Digital input 5 |
| DI6 | Digital input 6 |
| DCM | Digital common |
| AI | Analog input |
| +10V | Internal power supply |
| AO | Analog output |
| ACM | Analog common |

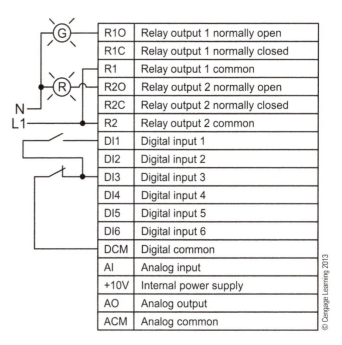

**FIGURE 16-4** Control circuit and relay output blocks

application need, the default values will suffice for many applications. Preset default values for many of these operation parameters allows the VFD to be commissioned with the fewest number of settings, but still be tweaked for optimum application performance by changing individual default operating parameters. The VFD has the following basic operation default parameters programmed:

- VFD display shows the output frequency
- Integral keypad control
- Nameplate voltage of 230 volts
- Nameplate frequency of 60 Hz
- Nameplate RPM of 1750
- Maximum RPM of 1750
- Constant torque operation
- Maximum motor current set to the maximum current rating of the VFD
- Five second ramped start and twenty second ramped stop time

If these operating parameters are acceptable to the user, they do not need to be reprogrammed during the commissioning process. In actual use, however, the user may want different operating parameters, such as external control circuit wiring rather than keypad control, a two-second ramped start and coast to stop, with a reversing selector switch and a jog pushbutton. And in order to protect the motor from thermal overload, the maximum current must be set to the motor's nameplate full-load amperes (FLA). All of these different operating parameters must be programmed into the VFD.

## CHAPTER SUMMARY

- VFD control circuit wiring uses electronic circuitry digital inputs for control, and the VFD has its own internal power source that is used for the external control circuit wiring.
- Digital input means that the VFD uses electronic circuitry on the inputs that senses only two states: either the control voltage is present on the input, or it is not.
- The digital input method of the VFD requires a different wiring scheme than magnetic motor starters.
- User-defined input means that the user can program each of the inputs to perform different functions, such as jog, reverse, external reset, etc.
- The relay outputs are user-defined and programmed to report operating parameters such as drive running, drive fault, at speed, etc.
- The term "commissioning" simply means to put into service. Commissioning a VFD involves both the wiring connections and the operating parameter programming necessary to make the VFD functional for a particular application.
- Most VFDs will boot with many of the operational parameters preset to default values, but they may not be the exact operating parameters necessary for every particular application need.

## REVIEW QUESTIONS

1. What motor control component does the VFD take the place of?
2. Why are VFDs more flexible than magnetic motor starters?
3. What power source is used for the external control circuit wiring with VFDs?
4. What is a VFD digital input?
5. VFD control circuit wiring connections normally are made by means of what?
6. What is meant by the term "user-defined"?
7. What are the two separate control sections found on VFDs?
8. How are the output relays of VFDs different from auxiliary contacts of magnetic motor starters?
9. What is meant by the term "commissioning" a VFD?
10. What is meant by the term "default value" in relation to programming the operating parameters of a VFD?
11. Why would a user change the default parameter settings of a VFD?

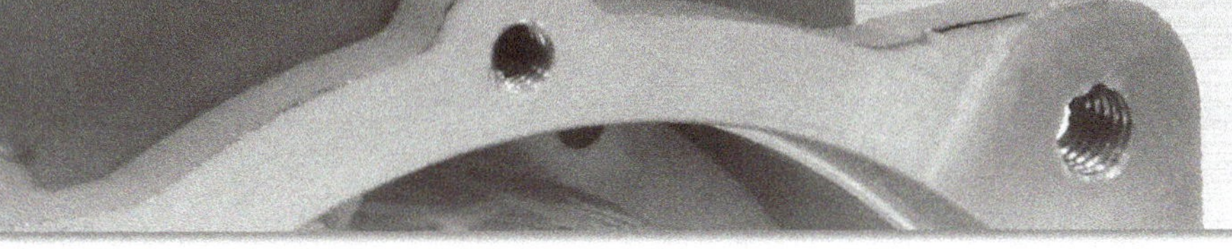

# Index

Page numbers followed by f indicate figures

## I

CPSIA information can be obtained
at www.ICGtesting.com
Printed in the USA
FFHW010231160119
50191760-55155FF

9 780840 024626